THE LAST DRIVER'S LICENSE HOLDER HAS ALREADY BEEN BORN

THE LAST DRIVER'S LICENSE HOLDER HAS ALREADY BEEN BORN

How Rapid Advances in Automotive Technology Will Disrupt Life as We Know It and Why This Is a Good Thing

MARIO HERGER

New York   Chicago   San Francisco   Athens
London   Madrid   Mexico City   Milan
New Delhi   Singapore   Sydney   Toronto

1 2 3 4 5 6 7 8 9   BNG   24 23 22 21 20 19

ISBN   978-1-260-44138-3
MHID   1-260-44138-5

e-ISBN   978-1-260-44139-0
e-MHID   1-260-44139-3

For Gabriel, Darian, and Sebastian.
And for May Kou.

Contents

Introduction   1

Part I
The Last Coachman, or
the First Automobile Revolution

Chapter 1   Electrician, Gunsmith, Physicist   23
            Automobile Pioneers Then and Now

Chapter 2   Love for Cars   29
            Passionate and Fickle

Chapter 3   The iPhone Moment for the Automotive Industry   37

Part II
The Last Novice Driver to the
Second Automobile Revolution

Chapter 4   Data and Facts   53
            About the Automotive Industry

Chapter 5   The Drive Goes Electric   57

Chapter 6   The Future Comes Rolling In   111
            Autonomous and Self-Driving Vehicles

Chapter 7   Artificial Intelligence   137
            America Invents, China Copies, Europe Regulates

Chapter 8   Hey There: Connected Cars in Conversation   215

Chapter 9   Research, Innovation, Disruption   233
             More Money, More Features

Chapter 10   Timescale   247
             What Will Happen to Us and When?

Chapter 11   Wave Effects and Leaps of Faith   279
             Forward, March!

Part III
En Marche! Tools and Methodologies
for Automotive Manufacturers and Suppliers

Chapter 12   Types of Innovation   349

Chapter 13   A Psychologically Safe Environment   361
             Fall Down, Get Up, Go On

Conclusion   "En Marche!" Politics and Society on the Move   375

Afterword   381

Notes   385

Index   425

Introduction

I'm just trying to think about the future and not be sad.
—ELON MUSK

A llow me to introduce you to Max, who just celebrated his first birthday with a lot of cake, colored balloons, and a huge pile of presents. Max is not only a sweet little boy; he is also probably the last person to take a driving test.

You think that's unlikely? Not in your lifetime? Well, I grant you, you may have a point. Actually, I don't know whether it'll be Max, or perhaps Sofie or Julian. It may even be a child from your neighborhood. One thing is certain, however: the last person to take a driving test has already been born. And I collected lots of data and facts, which we will examine more closely in the subsequent chapters of this book. You'll be surprised at how far the development of the electric robotaxi has already progressed.

Max (or Sofie, or Julian) won't even be able to imagine how we could ever have come up with the idea of wanting to own and drive a car, one of those contraptions that are difficult to handle with their pedals and steering wheel, which kept us from working or dedicating ourselves to video games during the drive, and one that in addition claimed many lives not just in the United States but around the world every year. Just how antiquated was our life then? Well, just as antiquated as we now perceive a ride in a horse-drawn coach. The coachman sat outside in front, exposed to the elements, and had to look at the horses' behinds while jolting over bumpy roads.

Even today, the joy of driving is often diminished when once more we are caught up in traffic during the rush hour and we're struggling to cope with tiredness, looming deadlines, and the search for a parking space. In the future, traffic will be even more concentrated in metropolitan areas than today. Sixty percent of the world's population will live in cities by 2030.[1] In the United States, 80 percent of the inhabitants live in cities already, compared with 74 percent in Germany and 66 percent in Austria.[2] The demand for transport services in the cities will grow. The available space and today's infrastructure won't be capable of handling that additional demand. After all, there isn't enough room even now to accommodate more cars in cities, to provide more streets and parking lots.

Silicon Valley alone today has in excess of 600 autonomous cars on public streets, managed by more than 60 manufacturers, and the number in the United States has already reached 1,400 vehicles. More than 1,000 companies are developing technologies for autonomous cars. At the same time, the center of the automotive industry is moving to one of the most expensive locations, where multiple manufacturers are producing or working on electric cars, trucks, and buses—Tesla, Lucid Motors, NIO, and Proterra, to name but a few. You can find more than half a dozen test tracks within a few miles of one another in Silicon Valley. In China, 25 million electric mopeds are produced in one city alone, and three dozen manufacturers are building electric cars. On a global scale, there are six autonomous taxi fleets in trial operation today, already transporting passengers. California started allowing driverless autonomous cars on its streets in 2018, even without any person at all on board.

Since 2016, Tesla has included hardware allowing for a self-driving function in all its vehicles. A software update expected in the next two or three years will allow any car produced by Tesla—so far more than 500,000 vehicles—to travel fully autonomously. At the same time, the first taxi companies are closing because they cannot compete with Uber and Lyft anymore. And the bounty for engineers having the much-desired expertise in artificial intelligence, sensor technology, or self-driving algorithms is at $33 million.

The new developments mainly come from two regions: Silicon Valley and Asia. While Silicon Valley seems to proceed according to some kind of natural evolution and will migrate from the American way of life, where everyone owns a car, to a lifestyle with electric robotaxis, some Asian countries skip entire levels. In China, for example, many people have become reasonably wealthy in just one or two generations, moving from simple farmers and workers into a newly formed middle class. And this middle class wants cars, or at least access to individual means of transportation. All the signs indicate that something is about to happen there that resembles the events after the Eastern bloc crumbled. Hungary had a better cell phone system than Germany. While the German Telekom wanted to amortize its investment into dedicated subscriber line (DSL) cable installations, Hungary did not have any such encumbrance, and instead of laying expensive cables in the ground, the country immediately built cell phone towers. And thus an entire technological generation was skipped completely. Detroit, Germany, Europe in general, and many other regions with strong traditional automobile manufacturing are lagging behind in all the areas of the new automobile industry. They do not have any leading role anymore because innovation is something that happens elsewhere. Germany invented the automobile and built the best cars, but apparently the future is being planned without them. Detroit is a shadow of its former glory. Even today, Germany's traditional manufacturers are lagging behind, and the distance is steadily increasing. This has little to do with others using some kind of magic formula or simply being high-fliers. The problem is not the fact that there are companies from out of nowhere attacking the traditional automobile industry but rather automotive engineers joining digital companies and creating a completely new form of mobility experience.

As I will outline in this book, the danger for traditional automobile companies is not so much coming from new technologies as from the attitude and mindset. Even though I am from Europe and have lived in the United States only since 2001, I am acutely aware of mindset differences in cultures and countries, and more so between the pioneering automotive industries and those in regions such as Silicon Valley, Israel, China, and the rest of the world.

A 2015 article that I had written on the differences between German and Silicon Valley innovation approaches for developing cars by comparing Tesla and Porsche led to some resonance that took me by surprise—and led to this book. If you follow similar comments in other media, you will quickly notice just how much the topic of automobiles hits the public nerve. One element that is striking in the midst of the furious exchange of opinions is the merciless criticism expressed on the topic of this proud economic sector. The announcements from manufacturers regarding new electric vehicles and the statements from top managers in the automobile industry calling self-driving vehicles a "hype" are heaped with scorn and ridicule. Perhaps the car manufacturers should use this as food for thought. They are about to lose the trust their own compatriots place in them—once and for all. The diesel emissions scandal, the unbelievable scope of the price fixing, and the "fraud collaboration" among the German manufacturers, the bankruptcy of Detroit's automobile industry in 2008, and several other severe mistakes made in recent years just serve to aggravate the situation.

Accordingly, I found it a logical next step to investigate this topic in more detail, to describe the current state of developments and combine the individual puzzle pieces into one coherent picture. And this despite the fact that I am myself anything but a car fanatic. Personally, I think driving a car is a waste of time; I would much prefer to spend my time reading. I was born and bred in Vienna, a city with an excellent public transport system, and thus at first saw no reason at all why I should have a driver's license. Consequently, I waited until I was 22 to take my driving test. My first car had to wait until I was forced to buy one on moving to California. Even while I lived in Germany for a couple of years, a country that also has a very good system of public local and intercity transportation networks, I found the car to be more of a problem than helpful in my daily life. Of course, occasionally I would have need for one; some things just were easier using a car. By contrast, when I think back to how often my car was damaged in those narrow streets of the historical city center of Heidelberg and how difficult it was to find a place to park, I know that I would have preferred life without a car even then.

I understand that there are still many drivers who enjoy their time behind the wheel, listening to the radio, relaxing, thinking deep thoughts, or listening to an audiobook. However, I can do the same on the bus or a train. But what will it be like in a vehicle that drives for me?

I've been living in Silicon Valley since 2001, an area mostly known as a place of pilgrimage for computer nerds and origin of so many things that we today perceive as perfectly natural, both at work and at home. Computers, smartphones, Facebook, and Google are just some of the new technologies from Silicon Valley, and it would be easy to reduce this relatively small place in California with its barely 3½ million inhabitants to just that.

In recent years, I became aware of the rapidly growing number of activities concerning automobiles. There is not a single day that I do not see one of Google's self-driving vehicles in or around Mountain View, and Google by no means is the only company that has such cars out on the streets of Silicon Valley. During times when a startup called Tesla can cause Apple-style hype with its Model S and buyers are waiting in line at dawn to get on the list for the latest model, the Model 3, which is being built at one of the most expensive locations in the world, Silicon Valley, one should really wake up to the signs. And if you then start reading about Apple's ambitions in the automobile sector, about Chinese manufacturers with billions of dollars available, and about the flood of hundreds of automobile startups, then you cannot possibly try to deny that something is up. The more I looked at this phenomenon, the clearer the picture became. The days of cars as we know them today are coming to an end. We are right in the middle of the second automobile revolution.

The signs are there. All the parts a robotaxi would require are available, that is, the ones that have provided us with the picture of the self-driving electric Uber. It is just a question of time for the combination of sensors, algorithms, artificial intelligence (AI), and apps to really take off. In many countries, discussions about new ways of thinking about automobiles have only recently started—discussions that previously would have been unthinkable. This means that awareness on the part of the public and political institutions is also changing. A technological revolution that is closely linked to an adaptation of behaviors and rules leads to disruption, to the destruction of a market.

Watch out for the telltale signals. Once there are several of them together, the disruption is already under way. A revolution is happening, one that will fundamentally alter our relationship with our "sacred cow," the automobile, and one that will have a similar, if not greater, effect on our economy and our society than the transition from horses to engine-driven vehicles. The first question is not whether but when those changes will affect us. A look at the exponential curve to which technological developments are subject and an evaluation of the facts already in evidence in Silicon Valley tell us that those changes are closer than many may think. And the second question that inevitably follows is, will Detroit or Germany still play a role in and after this second automobile revolution? Why do German manufacturers, which until now built the best cars in the world, and Ford, which revolutionized the way cars are mass produced, suddenly appear so antiquated? And how could they avoid sinking into oblivion?

Harvard professor Clayton Christensen has concerned himself with this phenomenon for several years now. His studies show that 50 to 80 percent of the top companies in a sector are no longer among the top ten in the next generation after a disruptive innovation has taken place. These findings are similar regardless of the industrial sector he analyzed. According to this logic, at least half the corporations that are the home of brands such as General Motors (GM), Ford, Honda, Toyota, Hyundai, Volkswagen, Mercedes, BMW, and Porsche will no longer be independent or no longer even exist at all.

I admit that from a traditional carmaker's point of view, this may seem very unlikely, but this is the same kind of thinking we saw in Detroit, the American automobile center, a decade ago, when large cars and pickup trucks were the key to success. And similarly, nobody at the Nokia headquarters in Espoo near Helsinki, Finland, would have thought it possible: they regarded the iPhone with great skepticism. And people in Rochester, New York, were equally sure that digital cameras would never pose a danger to the film and photographic paper industry. Nevertheless, today Kodak and Nokia are names that are used as excellent examples for lessons in economic studies that illustrate missed chances. Do we really want GM, Ford, Volkswagen, Daimler, and BMW to become synonyms for companies that did not see the big changes coming? Do we want them to lose the glamour of being the companies that invented the automobile, that opened the great wide world for people to explore, and that fired up our appetite for travel adventures?

We can agree that the best cars are built in Germany, the most beautiful sports cars come from Italy, France contributes the most elegant designs, Sweden is the first in safety standards, Japan is totally focused on reliability, and America created a car lifestyle. Unfortunately, the criteria for what makes up a really good car are changing right in front of us. A car's safety soon will no longer be mainly determined by a stable passenger cabin and an airbag but rather by the algorithm that guides the driverless vehicle. An elegant and beautiful design is less important if I am sitting in a taxi. Reliability will have more importance for fleet managers than for passengers. And the public is likely to base its decision on what constitutes a good car more on the integrated entertainment system, an area that, in the past, car manufacturers have completely ceded to others. In the future, we will perceive the car not so much as a single object but as an element in a larger transportation service provider offering.

Just as the best photographic paper lost its importance when people stopped printing their digital photos and the most superb telephone keyboard was replaced by touch screens and voice commands, the automotive sector will undergo an extreme change whose effects are not limited to that one economic sector alone. Our understanding and handling of mobility will change drastically, and cities, regions, and other players in this game will have to adapt to new conditions. A number of industries will be obsolete, and new ones are going to develop.

In the following chapters, we will take a closer look at all this: how it all started, how the automobile modified our daily life and our towns, the new requirements that have to be met, the technologies behind them, the legal framework that is affected, the behavior it will influence, and the effects all of this will have on our society, the job market, and business sites and the economy. We in Detroit, Asia, and Europe have the same technologies and processes available as everyone else, so the reason why we are lagging behind is our behavior, our mindset. Accordingly, we will take a closer look at this aspect in the last part of this book so that we can see how every one of us can and must contribute to the development of an innovative, entrepreneurial mindset for the benefit of our society and humanity as such.

About Davids and Goliaths

> If we have data, let's look at data. If all we have are opinions, let's go with mine.
>
> —JIM BARKSDALE, CEO OF NETSCAPE

Goliath, that invincible giant, had no reason to believe that David, this little shepherd, could pose any threat at all. David was not even a soldier, and he stood there without any heavy armor, totally unlike a real fighter. To Goliath, the simple fact that his opponents did not send an experienced soldier to the battle was ridiculous and reeked of despair. Still, Goliath lost and never even saw it coming. The fight was over before it had really begun.

This story of the outsider winning against an overwhelming enemy sounds good, but as a matter of fact, Goliath never really had a chance. Malcolm Gladwell, author of *David and Goliath: Underdogs, Misfits, and the Art of Battling Giants,* describes the starting point on the basis of the original text. And now it is perfectly clear that Goliath was a very sick man. The things he said and the way his contemporaries described him make it likely that this man of 8 feet height suffered from gigantism, an illness with considerable side effects. Goliath therefore was nearsighted, and could only see his opponent clearly when he was close up. His joints were aching, and he needed aides to bring his shield to the fighting arena. His impressive height earned the respect of his opponents, and his long arms ensured sufficient distance in a normal sword fight, so he could strike his enemy without being hit.

David, however, was an underdog who was underestimated in two ways. First, because he was a simple shepherd, he selected a sling as his weapon, unworthy of a real soldier. Second, David was of normal size, much smaller than Goliath, but more nimble. His weapon allowed him to attack his opponent from a distance, and very effectively at that. A trained sling shooter may get a stone flying approximately the velocity of a bullet fired from a pistol. Even if David could not place a "golden shot" at his first try, he was quick enough on his feet and far away enough to retry just as often as he had stones available.

So if you look at the starting scenario with this information, Goliath was destined to lose from the get-go: he brought a knife to a gun fight. Precisely because David went to the battle using an unorthodox weapon, he had advantages—not despite this choice. David did not follow the usual rules of a duel between sword fighters, which said that the opponents had to come into close physical contact. And David did not care a bit that it was "unsoldier-like" behavior to use a slingshot. After all, he was no soldier—he was a shepherd.

The outsider who apparently has no chance, who then employs unorthodox means, who does not comply with the rules, and who does not care what experts think about him and as a consequence surprises all those around him: this is the central element of many such scenarios depicting a David and a Goliath. We tend to root for the underdog but should actually also feel sorry for Goliath—at least sometimes! However, we often have to deal with giants who think too much of themselves because of their past successes. This book is also about such Goliaths and Davids; we will learn why you should never underestimate Davids, why giants are more vulnerable than they appear, and why it may be too late for some of them already. But Davids will play a role, too: why some Davids should not let their victory carry them away or rest on their laurels. Davids can quickly turn into Goliaths themselves and then fall victim to the next David. The crucial factor for the underdog's victory is the fact that he changed the rules, because those who let their opponent choose the weapon have only a 30 percent chance of winning. Those playing their own game have a 65 percent chance.[3]

Just how much the David and Goliath relationships have been reversed in the automobile industry is illustrated by an example of a delegation from a German premium car manufacturer visiting Silicon Valley. The cars from this manufacturer are among the most desired in the world and feature unrivaled quality, a fact that regularly helps to polish the annual reports of the parent corporation. Prototype vehicles covered in camouflage film, which the manufacturer uses on the streets for testing, are coveted by car fanatics and car magazine photographers and discussed in detail. And then you see the delegation members of this manufacturer suddenly turning their heads to follow every Tesla Model S and X, observe them running excitedly to the garage where Google keeps its fleet of self-driving vehicles, and watch them cling to the bars blocking the entrance like little children at the door to a sweets shop—just to get a photo of ugly little cars that look like balls of metal and plastic. Critical change is in the air, and traditional car manufacturers can no longer deny it, although they are doing their best to appear as if everything is under control and that they are on top of it all.

A Volkswagen (VW) employee traveling in a borrowed Tesla from southern Germany to a meeting at the company's Wolfsburg headquarters in the north really roused interest among his colleagues. They all stood around the car, wanted to test-drive it, experience the acceleration, see how much room there was in the vehicle, and play with the large touch screen. Normally, this is a scene you see when other car manufacturers examine German premium brands—or possibly when German manufacturers are impressed by a Ferrari. The difference is that a sports car is a luxury for most people, but when you see a Tesla, you immediately understand that this is the future coming. And the future is already here, much sooner than expected.

It's not Tesla hitting the traditional automakers; it's the future that's hitting them.

The example of an industry related to one of humanity's longest-held dreams shows just how quickly something like this can happen: flight. Orville and Wilbur Wright were two bicycle mechanics from Dayton, Ohio. Even as boys they had watched birds, observing how the animals moved their wings in order to stay up in the air. The two brothers were so deeply involved in their observations that they were able to imitate the various movements with their arms. Needless to say, their first attempts with wing-like contraptions attached to their arms with belts only gained them some scraped knees. Their neighbors regarded them as totally crazy. They had to be—only madmen had nothing better to do than stand outside for hours on end and look at birds.

And yet slowly but steadily, the pair tinkered with objects that would fly and eagerly absorbed any publications by other pioneers of aviation such as Otto Lilienthal and Octave Chanute. They even built the first wind tunnel to study aerodynamics. After countless attempts at gliding, they accomplished the first successful motor-driven flight ever in Kitty Hawk, North Carolina, on December 17, 1903, with a duration of 59 seconds. The distance traveled was 852 feet. The local population only knew about this several days later: the message was either presumed to be fake or thought to be irrelevant. The reaction in Paris, France, however, was in total contrast to the indifference at home. The Paris Aéroclub had already been aware of the Wright brothers' work through their correspondence with Chanute, and the club invited the brothers to France for a demonstration. Their home country only became interested in them after this journey.

While the Wright brothers followed their inventive intuition largely undisturbed by the public, another American flight pioneer was at the center of attention. Samuel Pierpont Langley was a renowned scientist, director of the Smithsonian Astronomical Observatory, and member of the American Academy of Arts and Sciences, as well as the Royal Society.[4] He was desperate to achieve a feat similar to that of his friend and colleague Alexander Graham Bell, who had invented the telephone. Langley regarded manned flight as the next threshold to be overcome and saw it as his chance to immortalize his name. Given his good network of relationships and good reputation, he had received $50,000 from the War Department and an additional $20,000 from the Smithsonian Institution to build an "aeroplane." The *New York Times* followed his work closely and regularly reported on his progress. However, all his attempts did not have the desired result, and quietly and determinedly, the Wright brothers won the race. When Langley heard about the Wright brothers' success, he immediately stopped all his aeronautical activities. While he had been focused on his personal gain and reputation, the Wright brothers had concentrated their efforts on allowing humans to fly.[5]

But why am I writing about the pioneers of flight when this book is about automobiles? The simple answer is that this example serves to show a pattern that is typical for many disruptive innovations.

First, disruption is often introduced into an industry by outsiders, not by the experts in that field—outsiders who at first are thought to be naive, detached from reality, or simply completely crazy. Nevertheless, these outsiders are exactly the right people to perceive matters clearly and without prejudice and to provide unconventional approaches. And because they are not involved in the history of their respective disciplines and have no obligations to anybody in the hierarchical structure, they are able to approach matters without showing any respect or deference to anyone. They do not have to worry about the unwritten rules or be afraid to alienate someone to whom they owe something. They can look at the problem to be solved from a more general perspective.

This approach is called *first principles* or *thinking in basic terms*. The basic term cannot be traced back to any other term and relates back to the original problem that was posed. The question that is closer to the basic term and the problem to be solved is not, "How can I improve a carriage?," but rather,

"What do we need carriages for?" And with this approach, you soon discover that the quantum leaps necessary for solving the problem do not involve a step-by-step improvement of existing technology but the discovery of completely new starting points.

This way of thinking, however, requires more mental energy. An innovative leap almost always surprises the experts in the industry, who never tire of pointing out the difficulties or the impossibility of any venture but continue to think only within their own limited framework.

Second, disruptors usually are not so much interested in personal fame but rather want to bring forward the actual cause. They want to change the universe or make the world a better place and help people. This is precisely what Tesla CEO Elon Musk meant when he talked in an interview with the German *Handelsblatt* business newspaper about his reasons for stopping his collaboration with Daimler and Toyota.

> The problem that we found with programs we did with Toyota and with Daimler was that they ended up being too small. They basically just calculated the amount they needed to keep the regulators happy and made the program as small as possible. We don't want to do programs like that. We want to do programs that are going to change the world.[6]

Musk wants to make the world a better place and wants to help people improve their life circumstances. In the German-speaking world, *Weltverbesserer* (literally "improvers of the world") are regarded as naive, starry-eyed dreamers of pipe dreams who will probably end their lives in an asylum. Being called a *Weltverbesserer*, therefore, is not really a compliment in Germany. But what would you then call people not aiming for that? World deteriorators/impairers? That would certainly have some internal logic.

It is not without reason that management is understood to be something in sharp contrast to traditional entrepreneurs. In his book *The Future: Six Drivers of Global Change* Al Gore quotes a survey in which CEOs and CFOs were asked whether they would consider a good investment now if it meant that they would not meet their targets for the next quarter. It won't surprise you to hear that 80 percent of the people interviewed said no.[7]

In his work, behavioral economist and 2017 Nobel Prize winner Richard Thaler points out the internal conflict in companies surrounding the macro and micro assessment of risky projects. During a meeting of 23 managers with the company CEO, the managers were asked whether they would start a project if the chances for success were 50 percent. If the project turned out to be really successful, there would be a gain of $2 million per project; in case of failure, the potential loss would be $1 million. In total, there would be 23 independent projects. The results: of the 23 managers, only 3 opted for the risk; the other 20 rejected the idea.

When the CEO was asked how many of the projects he would authorize, he immediately replied, "All of them!" And from his point of view, this absolutely makes sense. Half of the 23 projects probably would fail and lead to a total loss of $11.5 million. The successful other half of the projects, however, together would make the company $23 million. This means that at the end of the day, the results would be a plus of $11.5 million. Managers were asked to give their reasons for not undertaking such projects, and they said that in the cases where the projects were successful, all they had to gain was a pat on the back and a small bonus, but if they failed, they not only would lose their reputation in the company but would have to face being fired in a worst-case scenario. For them, the risk was much too high for the potential gain.[8]

Even if the company CEO is aware of the fact that from a macro point of view, the company should approve all 23 projects, the company's focus and reward system are aimed at the micro view (i.e., the individual project). When you think this through, you see that a failed project that nevertheless was tackled involves a large commitment and should at least be rewarded in the same way that a successful project is rewarded, because, after all, someone was willing to risk something here. From the company's macro point of view, this would make a lot more sense. And this leaves us with the astonishing conclusion that we should rather "punish" the mediocre projects with average success, which are those aimed at keeping within safe limits, and which are in any case pushed back by extremely successful projects. We are surrounded by mediocrity because a lot of people don't have the courage or do not receive sufficient incentives to embark on something extraordinary. The micro point of view that is deeply embedded in our working life today punishes all those who take risks and regards failure as a personal and not as a learning experience.

Coming back to all the news about the diesel emissions scandal, illegal price fixing, and government subsidies, we might come to the conclusion that a wrong approach and the usual reward systems simply induce people to "muddle through." People strive for short-term gains and rewards, and such aims are not necessarily in line with corporate missions. Furthermore, it appears that highly profitable car manufacturers have also become experts in getting access to public subsidies. Porsche received more than $6.8 million from the German government, Daimler received more than $68 million from economic incentive programs, and BMW received subsidies of $50 million from 2010 to 2012.[9] This list could be continued ad infinitum.[10] It is a fact, however, that those amounts were only used for minimum innovation—if you take the results so far as a measure—and more effort seems to have gone into finding reasons to explain why a new concept could not possibly work. The main motivation seems to be less a desire to create a better world and more a striving for power and glory, or at least for a good feathering of one's own nest. This works until the moment when external players show you how it's done if you just have the necessary willpower and endurance.

Third, directly after a new technology is accepted, many new players appear on board in an extremely short time. Only one year had passed from the moment Wilbur and Orville Wright had completed their first public flight presentations in the United States and France with great attention from the public when there were already 22 pilots signing up for the first great flying competition in Reims, each one with his own flying machine.[11] Ever since Google's self-driving vehicles appeared in the news headlines and electric Teslas started to fascinate their new owners, dozens of new players have joined the game. In 2019, there were more than 1,000 companies working on technologies for self-driving vehicles. We will have a closer look at those companies a little later, but there is one thing we can already disclose: the vast majority of traditional automobile companies are not among the leading companies, although they would like to make us think that they are.

Cars have become an integral part of our society. The kind of transportation we choose is not just a statement on our personal status but also a comment on our society. We spend more time in traffic than we do enjoying holidays, or eating with our families, or having sex. Traffic has become a kind of lifestyle. Such a lifestyle forces innovation on us. Entire industries, from Starbucks' drive-through and food-to-go establishments to audiobooks, only developed or introduced adjustments when the mobile society required such services.

But before we get into all this too deeply, let us take a step back and return to the romantically transformed "good old times." Just when the Wrights were busy with their pioneering work on flying devices, another age-old kind of transport underwent a drastic reconfiguration engineered by outsiders. And it all revolved around the horse.

From the Horse Manure Crisis to Climate Change

Anyone who has visited Vienna, Austria, surely knows them: the special hackney carriages there, the *fiakers*. I am talking about those horse-drawn carriages you can hire in the town center to enjoy a trip through the history of the town slowly and comfortably, powered by two horses. Some attentive tourists are bound to have noticed a chute-like leather bag underneath the horse's tail. This contraption is called a *horse diaper* and was introduced to avoid horse droppings all over the streets of the city.

This idea, which today invariably makes us smile, actually addressed a huge issue that burdened city administrators more than 100 years ago. As the cities grew, the number of carriages on the streets increased, and thus so did the number of horses on the streets. Around 1900, there were 11,000 carriages in London serving as taxis, plus several thousand horse-drawn streetcars, horse-drawn buses, and countless transport carts for all kinds of goods and products. At least 100,000 horses were required to keep the citizens of London and New York City moving. And those horses left traces. A horse's solid digestion products amount to 15 to 30 pounds every day, plus half a gallon of urine. Just try to imagine how a metropolis at the time must have smelled. Try to envision the danger of epidemics and the slalom courses pedestrians had to take to avoid tracking too much of all that refuse into their homes. The dried horse manure was kicked up into the air on hot summer days and turned into a gluey mass when it rained. It was also a preferred breeding ground for house flies. Whatever its state, horse manure was definitely unpleasant—and surely you now also realize that *mudguard* is a bit of a euphemism.

Nevertheless, what was stinking horse manure for one observer was a valuable resource and fertilizer for another. Entire professional groups made a living collecting, recycling, and selling horse manure. And now think of all the specialists working in the horse industry, such as farriers, bridle makers, carriage builders, horse breeders, stable managers, feed producers, veterinarians, and horse trainers, all working to keep the horse business running. The use of horses as means for transport and work meant that the average lifespan of a horse was just two to three years. Animals that collapsed or died on the streets often could not be removed immediately. Instead, they were left where they fell for several days until their cadavers had dried out somewhat and were easier to transport. It is almost impossible for us today to imagine the stink and the unhygienic conditions, especially during the hot summer months— fortunately impossible, you might want to add.

It comes as no surprise that the London *Times* in 1894 predicted a 9-foot-thick layer of horse manure on every single street of the city just 50 years ahead. The great horse manure crisis of 1894 was the initial impetus for the first international urban planning conference in New York in 1898. The conference was dedicated to outlining solutions for this imminent danger.[12]

Mind you, this was before the number of horses had reached its peak. In the United States, "peak horse" was reached in 1915: that was the year with the greatest number of four-legged transport devices (>21 million horses).[13] There was one horse to every three Americans. And exactly 100 years later, car manufacturers sell more vehicles than ever before. In the United States, we count 260 million vehicles; there are 43 million in Germany and 2 billion all over the world—the highest number of automobiles that ever moved on our globe. We have reached "peak car" or are just about to do so. At the same time, though, the automobile industry is facing the biggest upheavals in its entire history. Is it possible that the car manufacturers are about to come to a similar end as the horse industry, now, at the height of their economic power?

Management consultant James C. Collins, author of *How the Mighty Fall*, analyzed well-known cases of companies that faded into irrelevance within a fairly short period of time, although they had just before been the most successful corporations in their field.[14] Using global players such as the Bank of America, Motorola, Merck, Hewlett-Packard, and Circuit City, he identified five levels a corporation could pass through (although some can be skipped). All those corporations had a history of success that tricked them into thinking they were invulnerable. This is followed by arrogance and a rampant desire for more and the denial of danger and risk: all this together caused the companies to make more and more mistakes and to rely on old success models for too long, until it was too late (see Figure I.1).

The automobile industry is currently at level 3. The number of cars sold is higher than ever before; there are always more powerful, more economic, larger models available; and strongly growing promising markets in Asia hide the fact that the traditional sales markets in Europe and North America are stagnating. The latter point, however, is not at all due to the fact that people no longer need any means of transportation—quite the contrary. But change is in the air. We are witnessing a change in how we feel about accessing or owning a means of transport, as well as its availability; the importance of the drive and the experience of a drive differ from day to day. In order to understand the underlying dynamics, we have to take a closer look at who manages the companies and the time when they were founded. One aspect that is

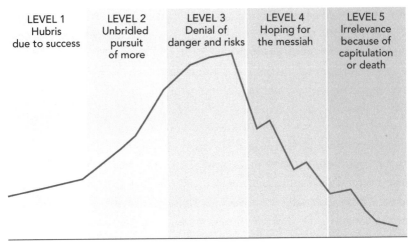

FIGURE I.1   The five steps of decline

apparent immediately: the CEOs of the companies that pose the most serious competition for traditional car manufacturers today are also the founders of their companies. In contrast, traditional automobile companies are managed mostly by, well, managers, which really is all you need to know to explain the difference. Managers are not entrepreneurs.

Throughout its history, the United States has always managed to encourage and enable entrepreneurship. A list of the most famous American entrepreneurs includes:

- Alexander Graham Bell (Bell Labs)
- Thomas Edison (General Electric)
- Henry Ford (Ford)
- Andrew Carnegie (Carnegie Steel Company)
- Walt Disney (Disney)
- Thomas Watson (IBM)
- Bill Hewlett (Hewlett-Packard)
- David Packard (Hewlett-Packard)
- Gordon Moore (Intel)
- Bill Gates (Microsoft)
- Michael Dell (Dell)
- Jeffrey Bezos (Amazon)
- Steve Jobs (Apple)
- Larry Page (Google)

- Sergey Brin (Google)
- Mark Zuckerberg (Facebook)
- Elon Musk (Paypal, Tesla, SpaceX)

Just one glance at this list shows that there are a number of names on it that we all have known for decades. And now let us think for a moment about German entrepreneurs. Do any names come to mind? Here is a (most certainly incomplete) list of the most important names:

- Carl Benz (Daimler-Benz)
- Karl Rapp (BMW)
- Ferdinand Porsche (Porsche/Volkswagen)
- Rudolf Diesel
- August Horch (Audi)
- Claude Dornier (Dornier)
- Werner von Siemens (Siemens)
- Karl Albrecht (Aldi)
- Adi Dassler (Adidas)
- Konrad Zuse (Zuse KG)
- Heinz Nixdorf (Nixdorf)
- Hasso Plattner (SAP)

So do you notice anything in particular when you compare the lists? First, only very few large German companies were founded in the recent decades; the most famous entrepreneurs worked in the late nineteenth and early twentieth centuries. Second, the American list is dominated by technology companies.

There were five technology corporations in the list of the six "most valuable" companies issued in April 2017: Apple, the Google parent corporation Alphabet, Microsoft, Amazon, and Facebook. Those corporate giants alone accounted for a market capitalization of a breathtaking $4.2 trillion in October 2018. In comparison, the market capitalization of all companies quoted on the DAX-30 list was $1.6 trillion—a third of the former amount. Those five U.S. companies are digital corporations, three of them were founded less than 25 years ago, and the other two were founded just about 40 years ago. Two of them have their origins in Seattle; the other three in Silicon Valley. The dominating German corporations, such as Bosch, Siemens, Mercedes, BMW, and Volkswagen, all have 100 years of history or more behind them, with the exception of SAP, which celebrated its forty-fifth anniversary in 2017. Twenty-four companies on the DAX-30 list are older than 100 years; only three are younger than 45 years. SAP is the only German digital company—and the most valuable at that.

So there is absolutely no reason to develop any kind of hubris—that is something that is slowly but surely destroying the companies. The most spectacular example is the Volkswagen Corporation. The drive for power of the owner families and management resulted in ethically and legally questionable decisions leading to the emissions scandal and had the company tottering on the brink of a collapse of trust. This is obvious from the massive downturn in sales (17 percent less!) as well as the employers' ranking among German graduates, who voted Volkswagen down to eighth position.[15] If you are aware of the ownership structure of Volkswagen, of which the founders' families still own more than 50 percent of the shares and 20 percent are owned by the federal state of Lower Saxony, if you know how massive the influence of the Works Council is (no decision is ever made without its approval), and if you then take a good look at productivity and profitability, you will find that Volkswagen employs 610,000 people to produce the same number of cars as Toyota, but the latter needs only 340,000 employees. Keeping this in mind, the question is not really whether Volkswagen will be divided and disappear from the market, but only when that will happen.

But we were looking at another collapse, actually, the predicted complete coverage of all streets by horse manure. That event never came to pass—because another revolution prevented it.

Part I

The Last Coachman, or the First Automobile Revolution

Vienna around 1900. Two gentlemen are standing on a street corner. One of those newfangled automobiles goes past. Both gentlemen watch it pass. Then one turns to the other and says dismissively, "Well, I suppose that will be done with soon enough."

The prophesied collapse of the cities under horse manure was averted by an invention that was as disruptive as it was unforeseeable—the automobile. The importance of horses as a means of transport declined when Carl Friedrich Benz built the first truly usable car and started producing it. His wealthy wife, Bertha Benz, played an important role in this endeavor because it was she who had the courage to undertake the first overland tour from Mannheim to Pforzheim and thus provide excellent advertising for this new means of transport. Soon after, the streets that had just a short while ago been covered with horse manure, lost horseshoes, and dead horses were dominated by automobiles.

What is the end result if you combine a hairpin, a garter, and the unheard-of amount of half a gallon of gasoline from a pharmacy with a woman's courage, ingenuity, and chutzpah? The first city trip in an automobile.

This is a German success story that would not have happened if not for a woman who invested her dowry in her husband's startup, thus becoming a venture capitalist, and who consciously transgressed the rules and regulations of society to get the company off the ground.

In 1888, the locals between Mannheim and Pforzheim rubbed their eyes in disbelief: a carriage without any horses was jogging along the dusty country road. At the wheel was a woman accompanied by two teenagers. Bertha Benz had let her sons Eugen and Richard accompany her on her trip to see her mother in Pforzheim in her husband's automobile but had neglected to inform her husband, Carl Benz, or to obtain the necessary permits from the local authorities. Admittedly, the distance between Mannheim and Pforzheim is only 66 miles, and nowadays that is just a stone's throw away, but in 1888, that was a very impressive achievement. The mother and her sons traveled over unpaved roads for a total of 12 hours and more than once had to push the car uphill. Moreover, every few miles the radiator had to be refilled with water. It was Bertha Benz's ingenuity that made the journey a success—an ingenuity that was definitely equal to her husband's: she used a hairpin to free a blocked fuel line and her garter to insulate a worn ignition. The news traveled fast, and the family's trip became a cross-regional topic of interest, with extensive accounts in the newspapers. This advertising campaign was a stroke of genius on Bertha's part. Over the next decade, her husband's company would manufacture hundreds of vehicles per year.

Whether Carl Benz did indeed invent the automobile—or whether others had had the idea before him—is ultimately of no importance. The idea of placing a steam engine or a gasoline-powered motor on top of a carriage was one other people had had as well, which is also a common phenomenon when you look at the history of inventions. The so-called adjacent possible refers to the fact that something is in the air, so to speak, just waiting for someone to hit on it. All the individual components are already available, and sooner or later somebody will have the idea of putting them together. The telephone, the battery, the ship's propeller, and, yes, the motor vehicle each was invented simultaneously by several people who often were not aware of each other or lived in different countries or in fact on different continents.

This fact was analyzed by two researchers from Columbia University as early as 1922. They found more than 140 examples of independent innovations and discoveries, most of which occurred within the same decade.[1] It is crucial, however, that an invention be made available to the general public. The most wonderful motor vehicle is of no use to anyone if it is just kept in the inventor's garage. It has to be built and sold to users; only then will it turn into an innovation.

Bill Aulet, lecturer of entrepreneurship at the Massachusetts Institute of Technology, defines an innovation as follows:

$$\text{Innovation} = \text{invention} \times \text{commercialization}[2]$$

Before you get to that point, some questions need to be asked, and those questions must be followed by actions leading to a discovery or an actual innovation. Thus

$$\text{Innovation} = \text{questions} + \text{action}[3]$$

The process is similar for all innovations. First, someone comes across something he or she considers worthy of improvement and asks, "Why is this done in this way?" Then the person tries to figure out how it could be done in a better way. The question is modified to, "What would happen if we did it like this?" Once the person has looked at all the alternatives and variations, he or she has a new question, "How do we put this into practice?"

This is the essential achievement of Carl and Bertha Benz (and every other innovator). Thanks to his wife's insistent urging and her daring exploits, ignoring prohibitions and regulations in the pursuit of her disruptive ideas—just as inventors ought to do—the orders came in by the dozen, and Carl embarked on the serial production of the first automobiles. Nonetheless, a woman is responsible for the initial push into the automobile era. This is a fact often relegated to the sidelines, but it is actually a very important achievement in the history of the automobile.

Another little known fact is that it took several decades for it to become clear which kind of propulsion would become the dominant power source. In addition to the gasoline-powered engine, both steam engines and electric motors were available for vehicle propulsion. As a matter of fact, in the year 1900 in the United States, the share of steam engines in vehicles was 40 percent, whereas 38 percent had electric motors and only 22 percent had gasoline-powered engines.[4] It seemed at the time that the propulsion varied according to the task at hand. Whereas steam-powered vehicles were used for heavy loads, electric vehicles were used mainly for public transportation within the city limits. After all, cities were still just about "passable" on foot at the time. By contrast, gasoline engines were ideal for longer drives to the suburbs and the countryside.

Among the American producers of electric vehicles were Anthony Electric, Baker, Columbia, Anderson, Detroit Electric, Edison, Studebaker, and Riker. In the German-speaking world, the Flocken electric vehicle, the Lohner Porsche, and cars made by more than two dozen other companies were manufactured. Detroit Electric produced more than 12,000 electric vehicles over the course of 30 years, advertised as having a range of 80 miles and a maximum speed of almost 19 miles per hour.

If you consider the widespread presence of electric vehicles around 1900, you cannot help but wonder what led to their disappearance. Well, there were several reasons. On the one hand, the continual development of combustion engines ensured more power, a larger range, and higher reliability. While a combustion engine still had to be started with a hand crank—a feat that was both difficult and dangerous because the hand crank frequently started turning by itself when the engine caught, breaking the hand or arm of the poor operator—the invention of the electric starter solved this problem. Gasoline-powered cars became more manageable. On the other hand, the increasing use of cars allowed the cities to grow in size, which, in turn, required a greater range for their automobiles. Vehicles with combustion engines were more suitable in that situation.

After the end of World War I, the triumph of vehicles powered by combustion engines was complete. Electric cars and steam-powered vehicles disappeared from city streets. In 1939, Detroit Electric, once the most successful manufacturer of electric vehicles, finally closed down.

CHAPTER 1

Electrician, Gunsmith, Physicist

Automobile Pioneers Then and Now

Be a realist and demand the impossible!

—ANARCHISTS' SLOGAN

Who were the pioneers who swept away the horse industry with their automobiles? If we take a look at the life and education of Carl Benz or Ferdinand Porsche, we immediately realize that they did not start out in the transport industry. Benz was a mechanical engineer, Porsche originally trained as a plumber and electrician, and Nicolaus Otto, father of the Otto motor, was a merchant who taught himself all the relevant knowledge he needed to tinker with his inventions. Gottfried Daimler completed his training as a gunsmith before studying mechanical engineering. August Sporkhorst was the owner of a weaving mill, and Robert Allmers was a publisher when they founded Hansa-Automobil.[1] Johann Puch was a locksmith, and Wilhelm von Opel was an engineer.

Ludwig Lohner, by contrast, was one of the few who came from a family of carriage builders. Heinrich Lohner had fled from Napoleon's troops in 1821 and moved from Alsace to Vienna, where he founded his company. Jacob Lohner & Co. manufactured horse-drawn carriages and luxury coaches and was even nominated purveyor to the Imperial Court in Vienna. In 1897, the company produced the first electric vehicle in collaboration with Ferdinand

Porsche.[2] Only a short time later, Lohner focused on the construction of airplanes and tramways and later on building motor scooters. In the United States, we find Studebaker, who started out as a coachbuilder and then produced cars until the 1960s (see Table 1.1). In fact, it was Fred Fish, the son-in-law of one of the five Studebaker brothers, who took that path after he became chairman of the company.

TABLE 1.1   A Selection of Automobile Pioneers and Their Training and Education

NAME	LIFE	TRAINING
Robert Allmers	1872–1951	Publisher
Herbert Austin	1866–1941	Technician
Carl Friedrich Benz	1844–1929	Mechanical engineer
Bertha Benz	1849–1944	Venture capitalist, cofounder, engineer, nonconformist, test pilot
Ettore Bugatti	1881–1947	Engineer
Gottlieb Daimler	1834–1900	Engineer, industrialist
Albert de Dian	1856–1946	Mechanic, Germanist
Henry Ford	1863–1947	Mechanic
Frederick William Lanchester	1968–1946	Engineer
Hans List	1896–1996	Mechanical engineer
Ludwig Lohner	1858–1925	Coachbuilder
Wilhelm Maybach	1846–1929	Design engineer
Nicolaus Otto	1832–1891	Merchant
Ferdinand Porsche	1875–1951	Plumber, electrician
Johann Puch	1862–1914	Locksmith
Louis Renault	1877–1944	Mechanic
Charles Rolls	1877–1910	Engineer
Frederick Henry Royce	1863–1933	Engineer
August Sporkhorst	1870–1940	Owner of a weaving mill
Wilhelm von Opel	1871–1948	Engineer

Regardless of whether we take a look at automobile pioneers from the German-speaking world or from France, England, and the United States, hardly any of them had worked in the coach and carriage or horse industries before. How is it possible that outsiders pushed out established companies and that those companies were unable to manage the transition into a new era?

Harvard professor Clayton Christensen examined this phenomenon a number of years ago. When he conducted a study of storage media across several generations and looked at the manufacturers of magnetic tape, floppy disks, and memory sticks, he noticed that among the manufacturers of the new generation of media, 50 to 80 percent were newcomers. The previously dominant manufacturers only rarely managed to continue into their respective next technological era and defend their dominant position.[3] This applies to all industries in which there is a disruptive innovation that shakes the foundations of said industry, independent of the context. Kodak and Polaroid completely missed the moment when the world turned to digital cameras. The video rental chain Blockbuster obstinately remained concentrated on renting videos from stores until it was too late to challenge Netflix. As late as 1975, Eumig was the biggest producer of film projectors in the world, but all that became obsolete when video recorders were introduced into the market. In 1982, the company filed for bankruptcy, holding 100 percent of the market share in a market that had decreased to zero. In 2007, Nokia was the uncontested market leader in cell phones, dominating a third of the market, but only one year later, the company's inventory turnover dropped sharply. Apple's iPhone had started its triumphant advance. From 1956 to 1981, 24 companies were removed from the Forbes 500 list every year. Between 1982 and 2006, the number rose to 40 in every given year.[4] Every two weeks, a company disappeared from the Standard and Poor's (S&P) 500 Index, which meant that 75 percent of all listed companies had changed within a period of 16 years.[5] Any company missing the moment or lagging behind lost. Perhaps you might want to explain it with Google's unofficial motto: "If you are not fast, you are f**ked."

Studebaker and Lohner were the exceptions who managed the changeover from coachbuilder to automobile manufacturer. Usually, the upheaval is not initiated in the company's own industry but is triggered or at least accelerated by other industries around it. In 1859, the first oil well was drilled in the United States. At the 1876 World's Fair in Philadelphia, mechanical, agricultural, and scientific achievements were presented—these in addition to the first typewriter, the bicycle, and Heinz tomato ketchup. The bicycle allowed an entire generation of mechanics to find work, many of whom would later apply their insights to the development of the automobile and the airplane. By 1900, almost a third of all patent applications submitted in the United States offered improvements to the bicycle. Without typewriters, modern companies with their mass production and their constant need for documents would be almost inconceivable.

This pattern becomes obvious again in the second automobile revolution. The indicators are there in the background; we see them in the modern pioneers and in the progress made in essential technologies. Tesla's CEO Elon Musk is a physicist. Google's Larry Page and Sergey Brin are computer scientists, as is Shai Agassi, who founded the battery-swapping company Better Place. The founders of Uber and Lyft, as if to make a point, did not originally work in the transport or taxi industries either. Six of the eight founders of Drive.ai hold postgraduate degrees in AI.[6] Sebastian Thrun, winner of the DARPA Grand Challenge and cofounder of Google's self-driving division, used to be a professor of AI at Stanford University. Kyle Vogt, cofounder of GM Cruise, is a roboticist, and Anthony Levandowski, whose startup 510 Systems was incorporated into the Google self-driving division, and who had a leading role in its development together with Thrun, is an industrial engineer.[7] The breakthroughs in AI occurred almost simultaneously. The price of sensors fell, and their performance increased. A greater amount of memory and faster processor speed allow for the rapid processing of incoming data.

But let's go back to David and Goliath again for a moment. Just how exactly did those alleged outsiders manage to shake solid industries, change them permanently, and oust the top dogs from the business? What was the weapon they used to overthrow their opponents? The answer may well be a new approach to (specialist) knowledge and the right mindset. Please do not get me wrong: sound expertise is important for innovation and creativity; it is one of its cornerstones. However, it can become dangerous when you no longer see the forest for the trees, when you delve in too deeply and fail to recognize solutions when they are outside your own specialized field.

For the coachbuilders, horse breeding and carriage building were the center of attention, whereas with the construction of automobiles, the focus shifted to the engine and everything connected to it. At the same time, expert knowledge of horse (transportation) became superfluous.

Although the first drivers still had to tolerate many disadvantages compared with the tried and tested transport system with four legs and two to four wheels, the experience as a whole became different. Disruptive innovation should not be equated with purely technological innovation. Although the latter was available, with all respective consequences, other factors gained at least the same level of importance. Vehicles could cover longer distances faster. The physical limitations of horses no longer played a role. Expenses for stables, feed, veterinarians, and stable hands could be cut, and the stench of the stables and the accompanying hygiene problems could be avoided as well. Admittedly, the first motors still gave off sooty smoke and were terribly loud, and there was no well-developed system of gas or service stations where pioneer drivers could get technical help in case of breakdowns. However, those negative aspects would be remedied over the years, and mobility by automobile proved to be much more advantageous than by horse carriage.

Counterarguments were made at every progressive step; this is as true today as it was then. Experts discourage the use of electric vehicles or self-driving cars and warn of their dangers. The network of charging stations is insufficient! Whose fault is it if a self-driving car is involved in an accident or kills someone? How dangerous can a battery be when it goes up in flames? How do you stop a self-driving car in which someone has planted a bomb?

The new automobile pioneers of the digital era regard problems first and foremost as purely software-related problems, however, and apply methods and principles from the software industry with which automobile experts are not familiar. Instead of waiting for perfection, they just release a beta version. Reid Hoffmann, cofounder of the social network LinkedIn and the internet payment service PayPal, comments: "If you are not embarrassed by the first version of your product, you've launched too late."[8] The added value is not so much derived from "bending metal" but from programming new software codes. All of this allows you to question the fundamental use of a car. What exactly is it good for?

Consider the slogans the car manufacturers have been feeding us for decades, such as "Progress through technology" and the simple "Das Auto," which were only abandoned when the emissions scandal dominated public discussion. They sound out of touch at a time marked by traffic congestion and environmental pollution. BMW and its "Joy of driving" totally overlook the fact that far fewer people have a passion for driving than the company believes. This is probably due to an inherent selection process for new employees. After all, who is likely to apply for a position with GM/Ford/BMW/Daimler/VW/Audi? People who like to drive, naturally. So what the car manufacturers forget completely is their underlying mission, which is not to make joyful driving possible. Nor is it to find solutions to the problems of transportation or mobility. A car is meant to create connections among people, places, and things in the physical world. The car is a "connector." I do not drive into town because I love driving but because I want to meet friends. I do not drive to work just because driving is so wonderful but because I need the interaction with my customers and colleagues in order to achieve something together.

Mobile devices can take over this task in many cases today. An iPhone is a virtual connector between people. If I have to drive the car myself, I cannot connect well with other people at that time because I have to keep my eyes on the road. It is obvious just how strong this desire for contact is if we consider how many people use their smartphones while driving, fully aware of the danger this poses for themselves and others.

Traditional automobile manufacturers are shooting themselves in the foot without even noticing. The importance of this particular sector for the local economy makes politicians eager to make concessions to this industry time and again and to create advantages for them. Lobbyists know perfectly well how to play on such readiness. Eased emission regulations, little or no punishment in cases of violations, subsidies favoring domestic manufacturers—all this under the pretense of maintaining jobs and promoting business locations. All the while people believe they are safe but are completely overlooking the tsunami that is about to arrive.[9] Just like "helicopter parents" wanting to protect their young from disappointments and ending up making them dependent on their parents, local governments try preventing new companies from entering into the competition and endangering the existing industry.

Politicians worry that long-term measures might lead to short-term "punishment." A diesel ban might lead to withdrawal of affection on the part of car manufacturers as well as enraged diesel car owners if they are kept from entering certain areas of towns, and who will therefore refuse to reelect those politicians. The truth is that this fear is rather unfounded. According to a study conducted at Columbia University, voters get used to an unpopular measure after six to nine months and forget their initial resistance.[10] Instead, politicians often succumb to "vetocracy," meaning that it is easier to prevent something than to make something happen.[11] The promise of a well-paid position with one of the lobbying companies after the end of a political career helps pass a number of laws that do not have the general public's interest at heart, when all we really wish for is people who look further than the next election or quarterly report. The Iroquois Confederacy's *Great Law of Peace* considers all the effects of a measure under discussion until the seventh generation.[12] In our era of quarterly figures and shareholder value, seven generations seem as far away as the dinosaurs—just in the other direction.

CHAPTER 2

Love for Cars

Passionate and Fickle

Don't trust statistics you didn't fake yourself.

—POPULAR SAYING

We are about to reach the seventh generation since the invention of the automobile, and we have to admit that our forebears have done a rather bad job. Although we no longer suffocate in horse manure, we do suffocate in car exhaust fumes, and we have never before been as dependent on our vehicles as we are now. Cars caused us to plan cities for vehicles, not for humans.

We are constantly reminded that a car is a symbol of freedom and that we are a nation of car enthusiasts. Cars are a standard conversation starter, a topic on which everyone can share an opinion. Most of us cannot imagine living without our own car; yet this impacts many areas of life. For many years, priority for cars appeared to be the ultimate credo for traffic management. A hundred years ago, all you saw on the streets were horse-drawn carts and pedestrians. If you ask people today who the roads are for, the answer will almost invariably be, "They're for cars."

Yet if you look closely, you will find that this new means of transportation was not as undisputed as we nowadays think it was. As early as 1923, 42,000 citizens of Cincinnati, Ohio, signed a petition that demanded that cars be limited mechanically to a maximum speed of 25 miles per hour.[1] Obligatory license plates and driving permits were only introduced under pressure from mothers who lost their children to car accidents. The Swiss canton of Graubünden banned cars from 1900 to 1925, with 80 percent of the population stubbornly refusing to accommodate such transportation: too loud, too smelly, too much damage to roads and generally threatening jobs in tourism, especially among coach drivers and the costly and barely finished railway.[2]

Accidents happened, in which pedestrians were the victims in three-quarters of all cases; they were even blamed for such accidents. Fatal traffic accidents were considered *force majeure*, no longer cases of negligent homicide. A change in public opinion was triggered in 1923 by an initiative pushed by the National Automobile Chamber of Commerce (NACC), which had a massive influence on the press. Until that time, newspapers assigned the guilt for accidents to the car drivers, thus causing an image problem for the rising automobile industry. The NACC therefore began to send accident forms to the newspapers to fill out and send back to the NACC, allegedly to obtain a better overview of traffic accidents. What actually happened was that the NACC interpreted the results with a bias—favoring the car—and the tide began to turn within the same year. Basically, accidents were the victims' fault.[3] The "fake news" published by the NACC—that does ring a bell, doesn't it?—was used to manipulate the public and the people in charge in order to pass the desired legal changes. Because pedestrians crossing the road limited the "joy of driving," the automotive industry pushed to have "jaywalking" inserted into the catalog of traffic violations.[4] After all, streets were first and foremost intended for cars, weren't they? Consequently, it was prohibited to cross streets unless there were clearly marked points and crossings, as it is still prohibited today.

According to the U.S. Department of Transportation, 94 percent of all accidents are due to human error. In the United States alone, about 40,000 people die in traffic accidents every year. In addition, the statistics show more than 2.31 million injuries. Annual damages are estimated to be as high as $1 trillion.

Our way of traveling today seems to generate more problems than it solves. Automobiles are of a high quality in the early twenty-first century, but there are too many of them. Every vehicle that is produced and sold causes problems both during production and during use: from the ecological footprint to the space required for roads and parking lots, from damage due to accidents to maintenance efforts, human behavior or error, and finally the exploitation of necessary resources. Accordingly, automotive manufacturers and transportation service providers have been under pressure from Silicon Valley companies such as Tesla Motors, Google, Apple, and the now-liquidated Better Place with their disruptive technologies. Uber, Lyft, and other ridesharing platforms are changing the way we experience transportation services. The fact that things must change is additionally confirmed by the decreasing number of licensed drivers in Western countries. Some time ago, the German magazine *Spiegel Online* published portraits of young Germans who introduced themselves as objectors to cars and explained why they would not consider obtaining a driver's license.[6]

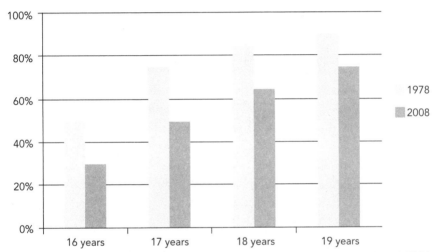

FIGURE 2.1   Percentage of licensed drivers per age group in 1978 and 2008
Source: U.S. Department of Transportation, http://www.fhwa.dot.gov

In Switzerland, only 59 percent of the individuals aged 18 to 24 had a driver's license in 2010 compared with 71 percent in the same age group who held a license in 1994.[7] The share of individuals holding a driver's license has decreased in the United States as well. Whereas only 8 percent of the 19-year-olds did not have a driver's license in the 1970s, this share had risen to 23 percent in 2008. In addition, it must be noted that young people who have a driver's license tend to drive less. The share of 20-year-old drivers in the total amount of miles driven decreased from 20.8 to 13.7 percent.[8] Children travel more miles by car than ever before, but nevertheless, fewer of them take a driving test when they are old enough to do so than people of the same age 40 years ago.

With the decreasing number of licensed drivers, the number of young car owners falls as well. In Stuttgart in 2000, 12,600 individuals aged 18 to 25 years had their own car, whereas in 2015, this number had decreased to 5,000 despite the fact that the general number of people in this age group had risen by 10 percent.[9] A survey by Goldman Sachs commissioned in 2013 confirms these results. Barely 15 percent of the people interviewed definitely wanted to own a car. The others either simply did not or stated that they would only buy one if they really had no other choice.[10] Accordingly, the average age of new car buyers is rising as well. In 1995, the average age of a new car buyer was "only" 46.1 years; in 2015, the age had risen to 53 years.[11]

At the same time, we are witnessing a deteriorating emotional relationship with cars in general. Today, it is the cell phone that we personalize and decorate.[12] Whereas one's own car used to be a symbol of unlimited freedom and uniqueness, today it poses a liability more than anything else. The feeling of the "wind in your hair" dissipates quickly when you are moving forward at a crawl in a traffic jam or despair while hunting for a place to park.

At the same time, the use of public transportation is steadily increasing. In the United States, the share of public transportation is now at its highest level since the 1960s.[13] This may not be very meaningful for the United States because its public transportation network is much less developed than in Europe and far less reliable; nonetheless, there is support from an unexpected direction to push this increase: I am talking about the smartphone. A route-planning app for local public transportation is comparable to a navigation system for drivers. Urban means of transport from different operators that published schedules that passengers had to laboriously interpret and combine have now become accessible and clear. This change can result in massive economic advantages: 85 percent of the expenses for a car are usually lost to the local economy.[14] If you manage to make some savings there, people are likely to spend more for their housing. There is nothing more local than that.

The significance of this social transformation is often overlooked, and the general reaction to it is very slow to come. So it is no surprise that it is a software company that taunts traditional automobile manufacturers with its self-driving cars, whereas none of the latter has a practical solution or even an idea of how one could be found. Meanwhile, Google has recognized the trend—not only because many people who object to cars work there and commute to work by company bus but also because the company consciously tackles difficult problems that may have a dramatic impact on society. In other words, as long as automotive manufacturers focus only on more elegant designs or a lowered chassis and more engine power because those are the things they know about, others have to step in to manage the challenges and needs of our time in an innovative way.

In the meantime, automakers focus their lobbying efforts on pushing for generally high emission thresholds because their gas or diesel engines could only meet more challenging threshold values if they really put some oomph into the effort (or could not meet them at all, as we all have come to find out). Then along comes Tesla and builds an electric vehicle that not only is more environmentally friendly but also makes combustion vehicles look bad in a number of other key ways. *Consumer Reports* magazine in October 2015 was forced to modify its assessment scale when a Tesla Model S scored 103 points in its system that only allowed for a maximum of 100 points. BMW enjoys homages to being the "German Apple" in the *Manager Magazin* instead. And all the while one has the nagging feeling that BMW's electric "i series" was developed by a business unit that would have much preferred to tinker with traditional cars some more.[15] It is also worth noting the words chosen for the presentation of the car both within the company and for its external partners because this allows you to draw conclusions about how important—or, rather, unimportant—the issue at hand is considered. Something presented as "image building" sounds more like a cosmetic kind of embellishment than a serious kind of project. Yet this is exactly what the vehicles look like and exactly what their performance data show. BMW simply ticked a category on its list and had done its duty.

Many of the things I just mentioned are matters that we are already aware of, either consciously or subconsciously. This unnerves us, but it also shows us new options. Unfortunately, we lack the tools we need to draw something sensible from this. So let us delve deeper together into the topic to prepare for the next few chapters to follow the different threads and draw the right conclusions.

Signals, Trends, and Foresight Mindset

Watch out for the telltale signs! Initially, these are small or local innovations with a potential to quickly scale in terms of size, effect, and geographic distribution. Signal clusters can tell us a better story about the future. The best means to describe something and make it comprehensible in a way that cannot be directly predicted is a narrative. We cannot predict with certainty because there are simply too many possible scenarios for the future. If we imagine a whole range of possible results, we can see that only very few major changes really have the capacity to come true tomorrow. If, by contrast, you consider a period of 10 or 100 years, technologies and societal changes make many things appear possible. The further you look ahead, the greater is the potential but also the less reliable is the prediction.

The iPhone was launched a bit over 10 years ago! Without this technology, the rise of companies such as Uber, Twitter, and Facebook, as well as application software such as Google Maps, Pokémon Go, and Tinder, would have been utterly unthinkable. The iPhone not only paved the way for new technologies but also changed our societies and their rules. Nowadays, we read about how fewer young people are getting their driver's licenses, and yet many people are still injured or killed in traffic accidents, and hundreds of companies are spending billions to develop self-driving systems. We are looking at a cluster of signals. Signals show various action strategies that at first appear possible, then seem plausible, and eventually are ever more probable until they become reality.

If we do not want to be steamrollered by this development but instead wish to play an active part, we have to question our own assumptions about the efficacy of today's strategies in a future world and create a long-term perspective as well as identify previously undetected strategic options and avoid mid-term threats and uncertainties. These are the steps to take if we want to influence the future. There are many possible ways into the future; and as soon as you take one road, more paths open up in front of you as you leave behind others that are not leading to your ultimate goal. The image to understand the "adjacent possible" is passing from one room into the next. It is hardly possible to simply skip a room. It is possible, however, to opt for one of several doors in order to step from one room into the same number of other rooms. As soon as you pass through one door, the other ones are lost from sight. One innovation unveils new paths for further innovation, while other things become a little less probable. The invention of the battery starter in the early twentieth century paved the way for the combustion engine while at the same time closing the door to electric vehicles for a long time.

The Institute for the Future in Palo Alto, California, specializes in recognizing the signals and the descriptions of trends, or the so-called big trends. And Singularity University in Mountain View, California, is on the lookout for exponential trends. Both are trends with the potential to impact the lives of at least 1 billion people in just a few years. Among them are AI, three-dimensional (3D) printing, augmented and virtual reality, and nanotechnology.

Technology does not develop in a vacuum. It is influenced and driven by realities and external factors. It is also the connector between the dimensions of daily life: the distribution of wealth, education, government, politics, health, the economy, the environment, journalism, the media, and society. Only then can it fully exploit its potential.[16] Scenarios that describe the future using signals must take all those dimensions into account. Narratives around people h 'p us to better understand trends, imagine the future, and make it happen. The Google company Waymo presented a video, for example, that showed how a blind man uses a self-driving car and thus not only becomes more able to adapt but is also able to enjoy his professional and social life much more.

Signals sometimes come with a bang. The iPhone has become a synonym for the moment that heralded the end of existing phone makers and announced disruption. One such "iPhone moment" occurred in the spring of 2016.

CHAPTER 3

The iPhone Moment for the Automotive Industry

Coffee lets me think faster, tea lets me think deeper.

—APHORISM

It is a Thursday, a rainy one in some areas, but still there are long lines of people who've been waiting there from the early morning hours, all of them waiting for their opportunity to order a car they have not even seen yet. March 31, 2016, became *the* iPhone moment for the automotive industry, its historical watershed. Just as Apple's presentation of the iPhone in 2007 was the trigger for the decline of the alpha companies Nokia and RIM within a period of just two years, Tesla's Model 3 may well turn out to be the "final call" for established automobile manufacturers.

Reactions on the part of the experts were dry and restrained. "Traditional manufacturers have made greater progress than many may think now. Finished solutions are ready and just waiting—like other collaborations in cell manufacture—for the moment when the market is actually ready for electric vehicles."[1] You will hear similar statements from "car experts" from every corner of the industry as well as politicians. The only effect such empty phrases seem to have, however, is to reassure the speakers and tell them that everything is going to be all right instead of letting the facts speak for themselves.

They are conveniently overlooking the fact that it is not sufficient to wait until a market is ready. The easy acceptance of innovations has reached a point that surprises the uninitiated. From the invention of the telephone in 1878, it took 75 years until the number of users reached 100 million. For the cell phone 100 years later, it took a mere 16 years to reach the same number of users. The internet reached the threshold of 100 million users in a period of 7 years, Facebook reached them in 4½ years, and Candy Crush Saga managed it in just 15 months.[2] In 2016, the augmented reality game Pokémon Go beat them all and reached 100 million users in less than 2 weeks.

People are good at predicting constant linear growth. Unfortunately, this kind of growth does not happen very often in the history of humankind. Exponential growth, however, is something we find difficult to assess; it exceeds our imaginative powers. This is where the singularity comes in, a phenomenon we will discuss later in more detail.

Let us go back 150 years and take a good look at the "Victorian internet"—namely, the telegraph. Samuel Morse had just installed a 37-mile test line between Washington and Baltimore. Four years later, almost 12,000 miles of cables had been installed, and in 1858, the first transatlantic submarine cable between the United States and Europe was in operation. The celebrations for the opening became so animated that New York City Hall burned down as a result of sparks from fireworks and the guests barely escaped the flames.[3] Some more examples: the first postcard was sent in 1871. Two years later, there were already 72 million of them.[4] In 1896, the citizens of New York first admired moving pictures, and by 1910, the fledgling film industry produced 200 short films per week.[5] This latter feat succeeded despite Thomas Edison's unceasing attempts to protect his patent by filing complaints through his lawyers.

Future researcher Lars Thomsen calls the speed of a trend the "popcorn effect." Referring to a quote by Steve Jobs, who said that "it's too late once you've identified a trend," Thomsen compares the process with corn put into a pan with a bit of oil to make popcorn. It will take a while to reach the temperature of 180°F and the corn starts popping, but once you get there, it happens all at once. At first, you do not see a trend, because nothing is happening. Experts and the competition are waiting. Then the first "popcorn" pops, and people say that it is an anomaly. Then another one pops, and suddenly they all go off at once. This is the moment when experts and the competition fall over each other in a futile attempt at action, but it is already too late—the new market has opened, and the old one has been destroyed.

The traditional life cycle for the introduction of innovations—basically, who starts using it and how quickly—is made up of five groups of people. *Innovators* are the ones who introduce us to the innovation. They are the smallest group and make up about 2.5 percent of the population. The *early adopters* are the next group, about 13.5 percent of the population. They are ready to accept half-baked innovation because they either desperately need those solutions, are honestly interested, and/or want to impress their friends with their experimentation and coolness. In the *early* and *late majorities* we find 34 percent of the total population each, and the *laggards* amount to 16 percent—those who only make the change because they have to.[6]

The traditional model starts to crumble with the digital revolution and with the support of the technologies available to us. The time for an innovation to be comprehensively embraced is reduced, and the population falls into two categories only: *test users* and *the majority*.

Confirmation of this can be found in various industries. The way we watch movies at home has changed radically. While we first found small video rental stores when playback devices such as VCRs were invented, they were then displaced by chains such as Blockbuster. This took about 20 years. Then along came Netflix, a provider from Silicon Valley that did not care for shops but sent DVDs by mail. The process as well as the business and profit model underwent a modification. A monthly fee allowed film fans to keep their rental expenses under control. Blockbuster, for example, made $500 million a year at its peak in 2004 in late fees alone. Netflix, founded in 1998, abolished late fees altogether. A few years later in 2010, Blockbuster went bankrupt. Netflix, however, did not want to rest on its laurels and could not, either, because three financially sound and technologically advanced power centers, Apple TV, Amazon Instant Video, and Google TV, were already waiting in the wings. Accordingly, Netflix introduced a streaming solution in 2007 that allowed users to enjoy their films directly over the internet.

All those companies did not wait for the market to be "ready" but simply went and established their own market. They then invested inconceivably high sums of money, at least from a European point of view, to monopolize those markets. If they had made the mistake of waiting, we probably would never even have heard of them. They simply would not exist.

Peter Thiel, a well-known investor, called the creation of one's own market as "going from 'zero to one'"[7]—zero, because those startups establish something completely new, and one, because frequently they not only dominate the new market with their innovative technology but also actually monopolize it. Among the kinds of startups he referred to, for example, are Facebook, Google, Twitter, and LinkedIn, as well as Alibaba, Uber, and Airbnb. Even if they were not the first, they managed to almost completely cover their segment thanks to network effects. This is also the way to create the most value. All the others who want to get a slice of the pie later on are "going from 'one to N.'" They are primarily fighting in niche markets with smaller margins.

In the early 1990s, American author Robert L. Stine had the idea for a series of horror books for children aged 7 to 12. This was a novelty, and the publishers regarded the idea with caution at first. Won't the children be overly frightened? Not really—the unique realization of the plot, providing goose-bumps moments but never actually placing the protagonists in real danger, plus lots of humor (without any finger-wagging), created a market that had not existed before in precisely this way. In the first year, a million copies of the books were sold every month. Since then, Stine has sold 350 million books worldwide. He went from zero to one. Other publishing houses and authors tried to imitate him, but they never reached the same figures. They followed from one to N.

We are witnessing once more the first indications of just how quickly markets can change in the field of electric vehicles. While the electric car was first belittled as an uncomfortable alternative for nerds, more than 400,000 customers ordered Tesla's Model 3 in the course of only a few days. At the same time, there is news from several countries that the local authorities not only are thinking about prohibiting the use of cars with combustion engines but are already implementing measures to this end. By law, only electric vehicles will be driven on Norwegian streets by the year 2030. The Netherlands plans to prohibit the sale of combustion engines, a proposal that goes even further by aiming to implement a total ban by 2025. Similar head-lines make the news in many other countries as well.[8]

With regard to the efforts and recognizable trends in the field of electric mobility, it is hardly surprising that even skeptics admit that combustion engines will be a thing of the past before long.[9] Mike Fox, chairman of the Gasoline and Automotive Services Dealers of America, succinctly stated in the summer of 2016: "If Tesla can deliver on its current promises with the Model 3, gas vehicles are history—it's horse and buggy days."[10] While other manufacturers have announced similar vehicles of their own make for the end of the decade, more than half a million Teslas are already fully functioning and on the streets. And that is not all. Tesla aggressively advances its infra-structure of charging stations with the aim of doubling it within the year. At the same time, the company prepares for the expected demand for batteries, constructing its gigafactory in Reno, Nevada. The factory started production in 2016. Given the cost reductions achieved by the size of the production output, Tesla hopes to obtain a 50 percent reduction in the price of its batteries. From whom will the other automobile manufacturers get their batteries? As long as they do not have their own production sites, obviously they will have to buy from CATL, Panasonic, and LG. And battery manufacturers may pressure them to abandon their own battery factory plans, as Volkswagen had to learn in early 2019. This shows that Tesla is moving into position on all levels, ready to monopolize a market or at least become a player nobody can afford to ignore. With this in mind, the losses the company incurred so far, approximately $3.4 billion, are put into perspective. The income that may be expected as a result of its market dominance will more than equal the investments made. Traditional automakers may come to deeply regret their half-hearted attempts. We should keep in mind that the automotive industry is a key industry for Germany, with 800,000 jobs and $420 billion in annual revenues. Austria is facing a similar situation, with 700 production companies in this sector and a total of 450,000 employees if we also consider the industrial sectors upstream and downstream.[11]

Here is another element illustrating the dilemma of car manufacturers: if we add Google's expenses to Tesla's losses, which amount to an annual $30 million to $600 million since Google introduced its own self-driving technology program, we end up with expenses of approximately $7 billion spent on autonomous and electric vehicles by just those two technological leaders. Volkswagen in comparison had to put aside more than $25 billion in penalty payments for the first settlements with American authorities. Keep in mind that this does not settle all the lawsuits in the United States and most certainly does not include the compensation payments expected in Europe and Asia. Basically, Volkswagen now has to spend a sum for penalty payments that is many times greater than the sum Google and Tesla used to develop new technologies—and all that just because someone took a fraudulent shortcut to keep from working too hard.

Other manufacturers are not faring any better either. General Motors (GM) has spent more than $16 billion since 2012 to buy back its own shares.[12] Yes, read that again, $16 billion spent not to invest in a battery production site or to put more effort into the development of new technologies but just to buy back its own shares. The company prefers taking shares off the market instead of pushing the share price by innovations and investments in the future of the company. This amount spent is no small sum either; it represents no less than 30 percent of GM's market capitalization. Economists regard share buyback programs as a means to destroy money. The money is not used to create any new value. Shareholders interested in long-term success and innovation experts understand this as a sign of a lack of innovative ideas on the part of management. So what happens is that such an action ensures a short-term rise in the share price, the dividend, and management bonuses by artificial scarcity.

On April 10, 2017, Tesla's share price rose to unprecedented heights; it overtook Ford and GM in the market rating with $51.44 billion. Ford, a car manufacturer with more than 6.65 million vehicles sold in 2016, a turnover of $151.8 billion, and pretax earnings of $10.4 billion according to its annual report, reached only $44.7 billion.[13] GM, with a turnover of $166.3 and 10 million cars sold, made $50.15 billion. However, Tesla actually showed a loss of $773 million and sold fewer than 80,000 cars in 2016. Daimler had a market capitalization of $80 billion, BMW $61 billion, and Volkswagen $77 billion. One June 8, 2017, Tesla was worth more than BMW for the first time, and on December 5, 2018, Tesla's market capitalization surpassed even Daimler's.

Investor James Montier called shares such as Tesla stock "story stocks," bonds that tell a story and give us a vision of the future. The important elements are not the past or current results, but the potential of future gain. Uber, Waymo, Amazon, and Airbnb received similar assessments that bear no resemblance to their current income or profit.[14] The tragic dilemma is that our educational system focuses on the execution, on finding solutions for known problems, and that this is what brings rewards. By contrast, this leads to a lack of training in visionary terms—thinking across boundaries, asking new questions, and finding the answers.

So what is the story told by a company like Tesla? No, it is not that it is building electric cars—that would be a tremendous misunderstanding. The bigger story behind Elon Musk's work with Tesla is to make humanity independent of fossil fuels. Starting with an electric car was just that—a start. Another step was building a factory to equip the cars with economical batteries. The next was to set up charging stations for the cars. And so on. This explains why the company acquired solar cell producer SolarCity and began production of accumulator storage systems for households. This makes the story coherent and allows for additional options that far exceed the field of activity of a simple manufacturer of electric vehicles.

The automotive industry's approach differs drastically from Tesla's. In Germany, the government is the driving force, and manufacturers are doing only the bare minimum required to just about comply with the demands of the authorities. Not surprisingly, nothing moves, and even the already unambitious objectives are not met. Although the German government wants to have 1 million electric cars on the roads by 2020, the reality is far from this goal. In 2015, this reality meant only 25,500 real electric vehicles and 130,000 hybrids on the road—out of a total of 45 million automobiles.

University professor Hans-Peter Lenz, former chairman of the Institute for Combustion Engines and Automobile Construction at the Technical University in Vienna, estimates the number of jobs that depend on engine production in the automotive sector to be up to a third of all jobs globally. This would mean more than 300,000 jobs that could become redundant among the German manufacturers. The subcontractor company Bosch estimates that 100,000 of its own jobs are in danger. Retraining measures have been started and pose a big challenge for the human resources division.

Such a prediction about mass redundancies may seem strange today if you look at the record sales the German manufacturers can boast of. But remember, GM was once the biggest U.S. automobile manufacturer and had to declare insolvency in 2009. Kodak still dominated the market for film and roll film in 1996, holding 80 percent of the market and making almost $16 billion in revenues. Then it went downhill and declared insolvency in 2012. Nokia was the leader in the cell phone market in 2007 with a market share of over 30 percent. That was also the year that the iPhone started its victory tour and showed Nokia its limits.

But how do potential car buyers see Google and Apple? Can the "newcomers" expect the same trust consumers have for companies such as Mercedes and VW? The consulting agency Capgemini interviewed more than 7,000 consumers in seven countries to find out how likely they would be to switch from a current automobile brand to the offers made by a technology company. The greatest readiness to take this step was found in rising industrial nations with a large share of young people, such as India (81 percent), China (74 percent), and Brazil (63 percent)—in some cases, more than 50 percent acceptance levels. Consumers in France (38 percent), Germany (32 percent), the United States (28 percent), and the United Kingdom (26 percent) had the most reservations. The breakdown according to age is interesting to see: 65 percent of 18- to 34-year-olds, 49 percent of 35- to 49-year-olds, and only 26 percent of the age group 50+ could imagine such a change for themselves.[15]

So what will it really be like? It is difficult to predict unless we invent the future ourselves. Everything people hope and expect depends on their personal attitudes and needs. Or on who pays them. Some automobile experts complain bitterly that the discussion about the combustion engine is filled with too much hatred.[16] Others cannot decide and settle for indicating ranges for the share of autonomous and electric vehicles in the future.[17] In his contribution to *New Yorker* magazine in 2011, Adam Gopnik classified technology commentators into three groups: the "never betters" (the future comes and is going to be wonderful), the "better nevers" (best if the whole thing had never happened), and the "ever wasers" (new things come and go).[18]

Never betters firmly believe that we are on the verge of a new utopia, a land of milk and honey where people suddenly all become good and treat each other well, all thanks to new technology and sudden inspiration. The better nevers mourn for past greatness and simplicity, for a time that supposedly was so much safer and offered more stability than our period today. Ever wasers dryly observe that change was always part of human development, just like the predictable reactions to it. They believe that people accept new things faster than we think but at the same time complain about them constantly.

Complaints such as not being able to hear the comforting hum of the engine are eerily like the laments of old about the loss of the comforting, well-known sound of horses whinnying and their hooves clacking on the road. The fear of data breaches in networked cars seems irrational in a world of cell phones and the internet, in which many citizens already freely grant access to any kind of personal data to third parties.

So what is the influence that Silicon Valley companies have on the entire ecological system, and what is it we can expect of them? Let us take a look at the individual technologies and analyze their impact to answer this question.

Part II

The Last Novice Driver to the Second Automobile Revolution

The future started yesterday and we're already late.

—JOHN LEGEND

A modern car differs from the first motorized carriage the way an iPhone differs from the telegraph. Both of the latter serve for communication, but the way in which it is accomplished in either case is very different, not just in terms of looks and operation but also in terms of the underlying infrastructure. Instead of traveling by wire lines strung over wooden posts, our data travel by satellite and fiber-optic cables.

Sufficiently advanced technology seems a lot like magic.[1] Just imagine the impression an iPhone would make on a person living in the nineteenth century. At the time Bertha Benz undertook her first excursion by car in 1888, she drew people's attention not just because of the new technology but also because many of the onlookers could only imagine that a carriage moving without horses must be a work of the devil.

An invention has the power to expand human consciousness. A perfect example of this is the eyewitness testimony of British actress

Fanny Kemble, who described her first ride on a train. Three weeks before the first railway line between Manchester and Liverpool was put into operation in 1830, 21-year-old Kemble had been invited to join the maiden trip. She described the locomotive, the carriages, and the tunnels in detail and then positively raved about her experience riding at the breathtaking top speed of 35 miles per hour:[2]

> The carriage . . . was set off at its utmost speed, thirty-five miles an hour, swifter than a bird flies (for they tried the experiment with a snipe). You cannot conceive what that sensation of cutting the air was; the motion is as smooth as possible, too. I could either have read or written; and as it was, I stood up, and with my bonnet off and drank the air before me. The wind, which was strong, or perhaps the force of our own thrusting against it, absolutely weighed my eyelids down. (I remember a similar experience to this, the first time I attempted to go behind the sheet of water of Niagara Falls; the wind coming from beneath the waterfall met me with such direct force that it literally bore down my eyelids, and I had to abandon any attempt of penetrating the curtain of spray till another day, when the conditions would be less hostile.) When I closed my eyes this sensation of flying was quite delightful, and strange beyond description; yet, strange as it was, I had a perfect sense of security, and not the slightest fear. . . . This brave little she-dragon of ours flew on . . . when I add that this pretty little creature can run with equal facility either backward or forward, I believe I have given you an account of all her capacities.

Although almost 200 years later we may feel amused by this account, it has many things in common with the reports by journalists who travel in a driverless car for the very first time. Their descriptions are full of emotions, fear, and astonishment faced with this miracle of technology, although they do not fail to assert how quickly they got used to it all. But let us turn back to the situation today again for a moment. We have already come a long way.

Since the invention of the automobile in the late nineteenth century, important improvements have been made. First and foremost, the vehicle itself was affected: from its fuel consumption to passive and active safety measures to comfort and increasing degrees of digitalization. This type of innovation occurred gradually in that the vehicle itself, or the way in which it moved and operated, was not generally questioned. Let us take a look at safety. Seatbelts, airbags, softer materials, antilock brake systems, deformable zones on the chassis, and the removal of sharp or protruding parts inside the vehicle, as well as on the outside, are just some of the incredible number of changes. The bumper, often underestimated and reduced to its exterior shape, underwent a change from a shining but hard, decorative element to a soft first-impact zone with embedded sensors.

The production changed, too, most notably with Ford switching from single production to the revolutionary assembly-line production. This move enhanced efficiency and made automobiles much cheaper and available to the masses. Although the sequence could hardly be called flexible at the time and the first automobiles all looked more or less alike, today you will find different vehicles on the same assembly line—a convertible following a sedan and an SUV, for example. Just-in-time delivery ensures that the parts arrive at the factory just when they are needed, which reduces storage costs. The increasing automation of production involving the introduction of assembly robots and a decreasing number of workers present in the production shop reduces labor cost and leads to a consistent level of quality. The same robot may complete tasks of varying complexity, for example, fitting a seat or inserting a panoramic roof. The way in which automotive brands outsource an increasing number of tasks to their suppliers and the way in which system components are delivered and simply put together in the factory require precision and assurance of quality standards across company boundaries that were unheard of even within the same company just a few decades ago.

The few hundred parts the automobile pioneers required to put together their first cars have evolved into a highly complex value-added chain involving 30,000 single parts or more until a modern combustion engine car is finished. Production has become a convoluted interaction of manufacturers and suppliers and is conducted *just in time* or *just in sequence*. Final assembly basically consists of fitting together components such as engines that are delivered as complete elements in the correct sequence. Doors just have to be mounted.

A procedure such as this requires a very deep *vertical integration*, which means that the manufacturers instruct the suppliers about the software to be used and the degree of access the manufacturer wants to have on the systems. Orders are not just passed on but rather are directly written into the supplier's system and adjusted continually. Thus, even if, legally speaking, there are different companies working together, the automotive manufacturers and their suppliers are a unified organism.

The changes, however, involved more than just the vehicle, the technology, or the way in which a car is produced. Financing models from car loans to leasing schemes were created to promote sales. Insurance companies introduced offers such as full insurance to replace losses in case of damage.

Another quite surprising and unpredicted phenomenon connected with the automobile was the development of a rating system for restaurants. Tire producer Michelin had the idea to suggest attractive destinations for outings by car to their clients, pioneer motorists, and presented French restaurants and hotels in a printed travel guide. As the guide became more popular, restaurant ratings were introduced—a measure that would have a huge impact on the quality and image of French gastronomy. And all that, just because Michelin wanted to sell more tires.

Problems came in the wake of progress. As during the "horse manure crisis," it became obvious that the enemy of what is good is not what is better but rather too much of the good stuff. There are 1.2 billion cars on our globe that basically do nothing for 22 to 23 hours per day except remain parked; that is, they are not used. And the vehicles that are on the road are often "parked" in traffic jams. Americans spend 175 billion hours in their vehicles every year.[3] The loss in productivity resulting from time lost in traffic is expected to double in Germany, Great Britain, France, and the United States by 2030 and to reach $293 billion, $124 billion in the United States alone. The cost of all this waiting is estimated for those countries to amount to a staggering $5 trillion.[4]

During the second automobile revolution, software is at the center of the disruption. Intelligent battery management ensures that electric vehicles are able to reach the next charging station. The high-volume data calculations required to steer a self-driving vehicle safely through traffic and to its destination are another application area. Passengers thus can turn their attention to the electronic entertainment system during the ride, relax, or work, as they prefer. It is not surprising, therefore, that smartphones and communication between vehicles and with objects also require a complex software solution. A taxi app is not just a simple application but instead a very real tool used to plan and coordinate offer and demand. If each approach individually has a disruptive effect on the immediately affected industries, just imagine what their combined effect will be. Could we be faced with electrically operated self-driving vehicles ordered via an app? We will take a closer look at Uber cars in a little while to examine this.

With its almost 800,000 employees, the automotive sector represents the most important industry in Germany.[5] In addition to the 450,000 workers employed directly by the car manufacturers, there are 4,500 suppliers with a total of 300,000 employees. Because the industry requires a lot of research, more than 90,000 researchers and developers account for more than a quarter of the entire research and development staff in the German economy. If you also add drivers, car dealers, and workers in the oil industry and in parking management, the result is that directly or indirectly one out of seven employees in Germany works for the automotive industry. The German association of automotive manufacturers, the Verband der Automobilindustrie (VDA), estimates that the jobs of 5.4 million Germans are indirectly related to the automotive industry, and an economic performance of about $460 billion annually is linked with it. Car exports account for more than half of all German exports.[6]

However, these figures must be put into perspective and interpreted. For example, the number of people working in the automotive industry, as quoted by the VDA, also includes taxi drivers, traffic officers, and car insurance companies.[7] Similarly, the amount spent on R&D offers few insights into the innovative power of a company, especially once you discover that Volkswagen was the company with the highest R&D budget worldwide in 2014 and 2016, outspending Samsung, Amazon, Google (Alphabet Inc.), and Apple until 2017[8] (see Table II.1).

Outsiders cannot help but follow with amazement the events unfolding in the scandal involving Volkswagen and marvel at the kind of discussions we are having. The public appears to be completely aware of the fact that this cheating is a sign that something in Germany and the automotive industry stinks to high heaven, even if people only occasionally pay attention to the automotive industry. Whenever I ask a trade delegation or the executives of an automobile club who of them already test-drove an electric vehicle, more than half of those present raise their hands. And if I then continue to ask who of them believes that electric vehicles are the future for us, more than 80 percent of the hands go up immediately. Everybody knows it, and yet the car "bosses" only unwillingly want to face this new reality.

TABLE II.1   Top 25 Companies with the Highest R&D Expenditures

2018 RANK	COMPANY NAME	R&D EXPENDITURES ($U.S. BILLIONS)				
		2014	2015	2016	2017	2018
1	Amazon.com Inc.	6.6	9.3	12.5	16.1	22.6
2	Alphabet Inc.	7.1	9.8	12.3	13.9	16.2
3	Volkswagen Aktiengesellschaft	12.2	13.9	14.2	13.8	15.8
4	Samsung Electronics Co., Ltd.	13.4	13.9	13.5	14.3	15.3
5	Intel Corporation	10.6	11.5	12.1	12.7	13.1
6	Microsoft Corporation	11.4	12.0	12.0	13.0	12.3
7	Apple Inc.	4.5	6.0	8.1	10.0	11.6
8	Roche Holding AG	9.5	10.2	9.8	11.8	10.8
9	Johnson & Johnson	8.2	8.5	9.0	9.1	10.6
10	Merck & Co., Inc.	7.5	7.2	6.7	10.1	10.2
11	Toyota Motor Corporation	8.6	9.5	9.9	9.8	10.0
12	Novartis AG	9.7	9.7	9.5	9.6	8.5
13	Ford Motor Company	6.4	6.7	6.7	7.3	8.0
14	Facebook, Inc.	1.4	2.7	4.8	5.9	7.8
15	Pfizer, Inc.	6.7	8.4	7.7	7.9	7.7
16	General Motors Company	7.2	7.4	7.5	8.1	7.3
17	Daimler AG	6.4	6.9	7.2	7.8	7.1
18	Honda Motor Co., Ltd.	5.6	5.7	6.2	6.5	7.1
19	Sanofi	5.7	5.6	6.1	6.2	6.6
20	Siemens Aktiengesellschaft	4.8	4.8	5.3	5.8	6.1
21	Oracle Corporation	5.2	5.5	5.8	6.8	6.1
22	Cisco Systems, Inc.	5.9	6.3	6.2	6.3	6.1
23	GlaxoSmithKline plc	5.3	4.7	4.8	4.9	6.0
24	Celgene Corporation	2.2	2.4	3.7	4.5	5.9
25	Bayerische Motoren Werke Aktiengesellschaft	4.9	5.0	5.1	5.2	5.9

Psychologists call it *cognitive dissonance* if one's firm belief in something does not change even when exposed to contradicting facts. It seems to be an unshakable truth that Germans build the best cars—and then a newcomer such as Tesla comes along and challenges them. Engineers perfected the combustion engine over the course of 100 years, and the best experts are found in the traditional head offices of carmakers. Now, suddenly, however, all this expertise is obsolete? Or how can you seriously promote the development of autonomous cars if you have spent the last decades telling your customers and yourself that the "joy of driving" is the ultimate goal in life?

People who have based their entire career on their own expertise and now discover that this knowledge is outdated and useless invariably have to experience some existential doubt. A very understandable human reaction to this is to deny reality and provide "alternative facts" instead: "The iPhone is just a passing hype"; "People always want to use a keyboard"; "Nobody will want to look at photos on a screen; only images printed on paper are the real thing"; "The automobile is going to disappear; I predict a long future for the horse." Well, we all know how that went.

So you see, I am not going to mince my words talking about the traditional automotive manufacturers in general and the German ones in particular. It may seem that I am praising Tesla or Google to the skies and want to make Daimler, BMW, and others look bad, but the time for patting ourselves on the back and talking ourselves into believing that we "are awesome" has passed with the discovery of the diesel emissions and price-fixing scandals. The velvet gloves must come off, even and especially when dealing with showcase industries. The gentle treatment helped to shape their behavior like that of an obnoxious brat whose arrogance damages the child and us. We have to speak up clearly, and because apparently nobody wants to do this in politics or in our society, I am doing it with this book. Mind you, I am not doing it because I like it, but because I would like to do my share to prevent the decline of an important industry in my home country while my country of choices is showing how to do it. There is too much at stake for my family, friends, and acquaintances, and we all become better when we can learn from each other.

Let us take a look at the individual technologies and the implications for the German-speaking economic region as a model for other regions on the globe.

CHAPTER 4

Data and Facts

About the Automotive Industry

A billion here, a billion there, and pretty soon you're talking real money.

—Everett Dirksen

Eighty million new cars sold every year amount to an incredible $1,500 billion. This number is the result of the labor of the 14 largest automotive manufacturers, the so-called original equipment manufacturers (OEMs), a group of more than 50 brands, all of them household names for us, including Volkswagen, Toyota, Daimler, General Motors, Ford, Fiat, Chrysler, Honda, BMW, Nissan, Hyundai, Peugeot, Renault, and Kia.[1]

Despite this, the manufacturers actually do not build the majority of their vehicles themselves anymore. Many components are delivered by so-called tier 1 suppliers. Bosch, Magna, and Continental deliver anything from a comprehensive windshield wiper system to a complete door or trunk on site, ready for installation. The same wipers can be used on a Mercedes, a BMW, or a Toyota. Tier 1 suppliers work for more than one OEM, ensuring lower costs. Mercedes uses a component such as a windshield wiper 400,000 times for its C class cars, and it is much cheaper for the OEMs if the same solution is sold to four other manufacturers so that the wiper manufacturer can produce 2 million wipers instead of only 400,000. Tier 1 suppliers, in turn, do not produce all the components themselves but instead order them from downstream suppliers, which are referred to as tier 2 suppliers because they supply indirectly to the OEMs via third parties. There are approximately 1,500 supplier companies in Germany alone.[2]

The reason why there are only 14 very large OEMs and not hundreds of them is the required capital intensity in this industry. Building a car is an extremely costly endeavor. The machine and tooling costs for an automobile plant alone can reach up to half a billion dollars. This amount includes metal stamping machines and conveyor belts, as well as many other things that make the production such a complex issue. After all, thousands of parts have to be assembled to build a car and then comply with regulations and function with minimal defects for approximately the next 15 years. In addition to the machine costs, there are the workers' wages and expenses for design, marketing, and testing. This quickly adds up to $1 billion to $2 billion in initial costs that have to be spent to produce a vehicle. Meanwhile, the cost factor increasingly shifts from hardware to software, or as former Netscape founder and now venture capitalist Marc Andreessen aptly put it, software is "eating" the world. Electronic controls, entertainment electronics, and applications in connection with the slow switchover to self-driving electric cars have gained importance. It is therefore not surprising that there are only a few successful newcomers in this field listed on the stock exchange. A total of 54 years passed between the flotation of Ford in 1956 and Tesla Motors in June 2010.

The automotive industry is not just one of the most capital-intensive but also one of the most research-intensive industries. Sixteen percent of the global research expenses are incurred by the automotive industry.[3] For every vehicle sold, more than $1,100 is spent on research.[4] OEMs and tier 1 companies put a total of $100 billion per year into R&D, two-thirds of which are borne by the former. Only the software and the pharmaceutical industries spend more money.

Furthermore, you cannot really limit the automotive industry to the production area, which is only at the beginning of the automobile value-added chain. Sales, maintenance, and disposal are downstream services ensuring employment. In North America, almost 1.1 million people are employed by 16,500 car dealerships.[5] In Germany, the last count in 2018 revealed 36,750 body shops, car dealerships, garages, and car services.[6]

Financing models from car loans to leasing ensure that the automotive industry works without a hitch. The manufacturers and dealers either have their own bank licenses or work closely with financial institutions. All this serves to some extent as a lubricant, keeping sales as well as production and subsequent services well oiled and running.

And we already looked at how OEMs advance into new business segments: Tesla operates charging stations, and Daimler and BMW operate a joint car-sharing system called SHARE NOW. There are several manufacturers that signed collaboration agreements with ridesharing companies: Toyota with Uber, GM with Lyft, Apple with Didi Kuai, and Volkswagen with Gett. Mercedes simply went ahead and acquired the entire mytaxi company.

In addition to the vertical integration forming a hierarchical system between manufacturers and the supply industry, there is an increasing need for horizontal integration (in which nobody has a position of highest authority) because other factors play a role: competing manufacturers that have to coop-erate to create standards for charging stations, for example; hundreds of com-puter chips and wiring that must be united to reduce errors and cost; and app ecosystems we desperately need. Autonomous driving needs an operating system, and companies such as Google and Apple simply have more expe-rience with that. After all, that's their core competency. Companies from Silicon Valley have a huge advantage in horizontal integrations and over the years have developed processes that allow them to switch between competition and collaboration. This is a massive culture shock for the traditionally secretive automotive industry.[7]

The character of the industry in which car manufacturers and suppliers operate started changing with every step. You no longer simply build cars; you are forced to ask yourself the fundamental question of why you really do this! One of the answers you get is that the industry has to provide mobility solutions. But what is a *mobility solution*? And this is where the generational conflict kicks in. Cars used to be a symbol of freedom and independence, but today they are more a source of headache and anxiety. Traffic jams, futile searches for parking, fuel costs, and environmental pollution turn the dream of one's own car into a nightmare.

Automotive managers who spent their entire careers in companies that valued great new engines and regarded "bending metals very precisely" as the most important task now have to relate to representatives of a new gener-ation who have no wish to drive themselves or to own a car and who expect solutions to the problems just listed. The demands are for no emissions, no searches for parking, and no desire to sit behind the wheel. Such expectations shake the self-image of an entire industry whose representatives like to pretend that all of this is just a passing fad. As a result, implementation is done rather half-heartedly. Automotive manufacturers have been promising us electric vehicles for years, and all we get at the next presentation is another diesel model and big "muscle" SUV.

All the progress made in building automobiles cannot hide the fact that up to 95 percent of the energy a vehicle needs to run is used to move the mass of the car itself; modern vehicles weigh about 1 to 2 tons. Only 5 to 10 percent of the energy used is needed to move the passengers (assuming an average weight of 165 pounds per passenger plus some luggage). Even the lightweight designs developed in recent years did little to lower the weight of vehicles, because additional safety measures and installations for increased comfort such as electric window switches and electronic controls more than made up for any savings in weight.

The transportation industry made up 27.7 percent of the total energy consumption in the United States in 2015—second place after power plants (with 38 percent). At 79 percent, moving people and goods (and vehicles) is at the same time the most wasteful use of energy.[8]

Among the standard fuels for individual transport today are gasoline and diesel fuels. Over the years, alternative fuels such as flex-fuel were introduced; these alternative fuels use up to 25 percent methanol or ethanol. The mixture can be used if the engine can handle such a mix. The advantage of flex-fuel is that it comes (in part) from sustainable sources. The methanol or ethanol used can be obtained from cane sugar, corn, or organic waste. The share of new vehicles with flex-fuel capacity was particularly high in Brazil and reached up to 90 percent, thanks to several tax incentives. Despite its number of advantages, flex-fuel has not really caught on because automotive manufacturers and oil corporations took action against it. Consequently, most of the cars sold in the United States are technically equipped to drive with flex-fuel, except that this function is often switched off electronically.[9]

CHAPTER 5

The Drive Goes Electric

It is difficult to get a man to understand something when his salary depends upon his not understanding it.

—UPTON SINCLAIR

A cool summer breeze surrounds us in Emeryville, California. Squashed between Oakland in the south and Berkeley in the north, Emeryville is mainly known for the animation film studio Pixar that gave the world *Toy Story* and *Finding Nemo* as well as *Cars*. We are not here to be entertained by a talking cartoon car but to test-drive the i3, BMW's only EV (electric vehicle) on the market. An acquaintance at the BMW Technology Office in Mountain View, whose role is to follow the latest Silicon Valley trends and report them to headquarters, explains the vehicle and its functions. One tap on the accelerator, and the car noiselessly shoots forward. It moves briskly despite carrying the weight of four passengers. The brakes take some getting used to. There is no neutral position as we know it from standard engines. Instead, the car slowly brakes as soon as you take your foot off the accelerator. This allows the car to recover brake energy and charge the battery. This modification allows for new, playful driving styles: how far ahead of a red traffic light should I take my foot of the pedal so that I do not have to use the brake itself?

This short drive is enough to make it very clear to us that this is the future of automobile driving. The days of combustion engines really seem to be coming to an end. And this is very necessary indeed. Countries such as China face a human-made environmental collapse in the near future. The environment was ruthlessly exploited to ensure growing prosperity there, and now the environment is taking its revenge. The country suffers from immense pollution levels caused by factories, power plants, and cars. Those levels exceed the threshold limits several times over; blankets of smog never dissipate. The damage to the environment more than offsets the economic growth.[1]

Other countries have some very good political reasons for their attempts to reduce their dependence on oil. Israel, for example, has been in conflict with its Arab neighbors ever since it was founded and has no wish to finance its enemies by being dependent on their petroleum. It is therefore not surprising that there is a whole range of Israeli startups in the automotive industry.[2] By contrast, Norway's motives for readjusting its direction are completely different. Not long ago, Norway turned itself from being a small nation of fishermen into one of the richest countries in the world, thanks to the oil deposits found there. Norwegians know, however, that this resource will not last forever, that they have to prepare for what is to come once the oil runs out. It is hardly surprising, then, that those countries are the avant-garde of electric mobility.

Shanghai and Tel Aviv are interesting places to visit for adventurers, and the same applies to the technologically inclined. The importance of fast and affordable cars for the population can be estimated by looking at the number of mopeds and motorcycles. If you look closely, you will notice the drive-train technology in particular: many of those vehicles are powered by a square battery block with a handle located under the seat. While the scooter is parked on the street, the driver removes the battery and takes it home to charge it.

About 10 years ago, authorities in the Chinese city of Qingdao decided to promote electric motorcycles and penalize vehicles with combustion engines. Further incentives to a local enterprise to produce electric mopeds and free electric power for local government employees resulted in the production of 25 million electric motorcycles annually in that city.[3] By the way, Qingdao was a German trading post in the nineteenth century. Back then, the German Empire taught the Chinese how to make German beer. Today, we should learn from the Chinese how to make electric mobility happen.

However, the efforts at electrification in China are not limited to motorcycles and mopeds. BYD, Byton, NIO, Qiantu Motors, and Qoros are some of the newcomers in the automotive sector.[4] Their announcements of ambitious goals such as the intended production of hundreds of thousands of EVs in the next few years are usually met with naked disbelief in Western countries.

The contract manufacturer Foxconn, known as the one to produce Apple's iPhones, for example, invested $811 million in the development of an EV.[5] Future Mobility announced an autonomous EV for 2020.[6] A look behind the ownership structures of these corporations reveals interesting facts. LeEco, Lucid Motors, and Faraday Future were initially financed by Chinese entrepreneur and billionaire Jia Yueting.[7] At Lucid Motors, Chinese investors were involved in the first rounds of financing, followed now by Saudi investors.[8] BYD, by contrast, has a joint venture with Daimler and already sells the outcome of this shared development, the EV Denza, on the Chinese market. In 2017, BYD sold more than 108,000 EVs, more than its prominent competitor Tesla.[9] In 2018, though, Tesla sold more than 245,000 vehicles, thanks to the successful ramp-up of its mass-market sedan Model 3. Detroit Electric has been resurrected and is shamelessly trying to clone Tesla's model for success, planning to first build a roadster, then a sedan, and eventually an SUV (see Table 5.1).

China is already the biggest market for EVs in the world. In addition to electric motorcycles, there are more than a million electric cars. The delivery of 400 electric buses for the city of Shenzhen in December 2015 caused a traffic jam of several miles on the highway. Not only did Shenzhen replace all its more than 16,000 diesel buses with electric buses, but a year later it also replaced all its more than 20,000 taxis with electric taxis. Twenty percent of all the public buses in China are already EVs.[10] The CEO of Proterra, an electric bus manufacturer in Silicon Valley, states that electric buses are already cheaper in operation than combustion engine buses. He expects local public transportation providers to buy only electric buses by 2025.[11] In March 2017, several dozen American cities including New York City, San Francisco, and Chicago announced intentions to purchase 114,000 EVs worth $10 billion for police, trash collection, and other municipal institutions.[12]

Even if the former holds true, the infrastructure for EVs in the United States and China is still insufficient. There is a lack of charging stations in China, and power mainly comes from polluting coal-fired power stations. There are two long-term advantages to buying EVs: first, the shift to green energy can exploit the sustainability potential of EVs, and second, manufacturers gain experience in battery technology and electricity needs for e-mobility. This gain provides advantages for both China and the United States in the chase for dominance in this technology.[13] Chinese manufacturers were not able to compete with their internal combustion engine vehicle

TABLE 5.1   A Selection of Manufacturers of EVs

COMPANY	COUNTRY	TYPE
Apple	US	Passenger cars
BAIC	China	Passenger cars
BYD	China	Passenger cars, buses
Byton	China	Passenger cars
Chongqing Sokon Industry Group	China	Passenger cars, vans
Detroit Electric	US	Passenger cars
e.Go	Germany	Passenger cars
Faraday Future	China	Passenger cars
Future Mobility	China	Passenger cars
Karma Automotive	China/US	Passenger cars
LeECO	China	Passenger cars
Lucid Motors	US	Passenger cars
NextEV	China	Passenger cars
Nikola One	US	Trucks
NIO	China/Sweden	Passenger cars
Proterra	US	Buses
Qiantu Motors	China	Passenger cars
Rimac	Croatia	Passenger cars
Sondors	US	Passenger cars
Sono Motors	Germany	Passenger cars
Tesla	US	Passenger cars, trucks
Thunder Power	Taiwan	Passenger cars
WM Motor	China	Passenger cars

offerings on a global level. Their dream of a globally exporting automobile nation did not come true. However, everybody is starting out from zero with EVs. Because the established manufacturers also had to struggle with the new technology, China recognizes its chance to establish a reputation as a leading nation at least in this field.[14]

The country is ideally suited for this: It has the largest deposits of rare earth elements needed for batteries. In addition, it has the production sites, and demand for individual mobility skyrocketed because of the country's economic growth. At the same time, the country is combating massive environmental pollution, and EVs are part of the solution. Although today's infrastructure and quality may not always meet our Western standards yet, the first Japanese knock-off cameras and cars after World War II did not either. Today, Japanese products are leading-edge products all over the world, and German and American enterprises copied Japanese production processes in the automotive industry during the 1990s. We may well expect a similar development in China.

The seriousness with which China approaches e-mobility is apparent in its public subsidies and the speed with which change is being implemented. Taiyuan, a city with 4 million inhabitants, exchanged all its 8,000 combustion engine taxis for electric taxis by BYD in the course of just one year. Generous incentives made the change more palatable to the taxi companies. Almost two-thirds of the list price of $38,600 was sponsored; thus the price per vehicle dropped to $13,900. At the same time, the city built more than 2,000 charging stations, and an additional 3,550 are under way. The latest news is that Beijing has started to exchange all its combustion engine taxis. This means that the 70,000 taxis in the greater Beijing area are to be replaced by electric taxis. The cost is estimated to be about $1.3 billion.[15] Shenzhen finished replacing its entire taxi fleet in December 2018.

Americans and Europeans can only stare open-mouthed at the news about the delivery of hundreds, if not thousands, of electric buses to the public transit systems of Shenzhen and Tianjin. We are used to seeing representatives of our public transportation services in the media, celebrating the fact that they authorized a test run for a few electric buses, or, in a best-case scenario, actually order one (ambition and environmental protection have little room here), whereas China goes ahead and electrifies entire public transport systems in one fell swoop.

In 2015, almost 189,000 EVs were sold in China. In 2016, the production reached 312,000 EVs, almost exclusively produced for the domestic market, thus making China, not the United States, the leading country for EVs.[16] Europe holds a distant third place when it comes to electric mobility. Among the ten biggest manufacturers of EVs are four Chinese companies—not joint ventures with Western producers but independent companies such as BYD, Kandi, and Zotye, companies most Westerners have never heard of.[17] The Chinese government announced that a share of 8 percent of EVs must be met for new registrations from 2018 onward. This really caught German manufacturers off guard. Anybody who has firsthand experience with the pollution in this country will understand why the authorities are pushing the issue so hard. Germany also suffers from high pollution levels. The particulate matter measurements in many German city centers are frighteningly high nowadays. For a long time, nobody knew who was responsible for this increase—until fingers could be pointed at the polluters in the course of the Volkswagen emissions scandal disclosures. We are in urgent need of changes in other countries as well.

Many people are no longer willing to wait until manufacturers remove combustion engines from their product lists. Norway, the Netherlands, Austria, and India have announced a time frame for a sales ban on combustion engine vehicles.[18] German cities suffering under high levels of nitrogen oxide emissions continue their heated discussions about a general ban on diesel vehicles in the city centers.[19] Just a few months ago, this kind of talk would have been unthinkable. These cities, however, are forced to pursue such deliberations because they are facing penalties and infringement proceedings initiated by the European Union (EU) if the threshold values are exceeded.

German manufacturers were not always so lethargic in terms of the development of and research concerning EVs and batteries. There were pilot projects in several regions, where solid subsidies from the government helped conduct tests and set up charging stations.[20] However, these activities ceased as soon as it became apparent that "pretending" also brought in money, namely, subsidy money. Still, many incentives are available, including a bonus from the German federal government for people buying EVs and money for installing a network of charging stations. In addition, the manufacturers received subsidies for their research. They received hundreds of millions of dollars, money that such highly profitable enterprises really should have been able to pay out of their own pockets without public help.[21]

Moreover, the results are less than satisfactory. German manufacturers mainly produce compliance vehicles, that is, vehicles that help to lower the fuel consumption of their fleet but are no real competition for Tesla and the other producers. German manufacturers still believe in the future of the combustion engine, just as Germany's last emperor William was convinced that the future lay with horses.[22] Manufacturers only started taking electric cars into consideration again, albeit reluctantly, when Tesla started to be successful. It hurts when a newcomer from the United States sells more cars than you do, and in a premium segment of an important market to boot.[23]

German manufacturers were too self-absorbed for too long. Mercedes and BMW were competing for the position of market leader, Volkswagen was busy trying to cheat everyone with its emissions scam, and the owning families Piëch and Porsche were each trying to outmaneuver each other to gain control within their company.[24] German manufacturers have yet to prove that they heard the alarm bells ringing, although they all have made great announcements of coming new models. There is even talk about 30 different models from one manufacturer. It would be nice to see just one competitive, purely electric vehicle from each one.

Meanwhile, others are handling the electric mobility business. BYD, Tesla, Nissan, and Renault not only sell more vehicles and become very profitable doing so but also produce better electric cars.[25] Some German enterprises are so desperate at this point that they are willing to pay to have their technology installed. The Silicon Valley startup Lucid Motors received a paid exclusive contract for providing the batteries for the Formula E Racing—the electric equivalent of Formula 1 race cars—despite the fact that Porsche would have been ready to grant huge discounts for doing the same.[26]

The number of German engineers we find in key positions in those young automotive enterprises is no surprise, so why are German corporations lagging behind? After all, the technology, the experience, and the expertise are all here; we were instrumental in inventing and developing them. And yet people here desperately hold onto their combustion engines or, even worse, their beloved diesel motors.[27] The events that came to light in the course of the Volkswagen emissions scandal that involved basically every manufacturer abruptly made people aware of the extent to which they had been deceived for many years. Actual emissions values of those diesel vehicles that were praised as being clean and environmentally friendly dramatically exceeded the acceptable reference values, according to the investigative commission set up by the German government.[28] And that is not all: for passenger vehicles, those values were even higher than for trucks.[29] Everybody now sees that diesel cars could only be pushed onto the market thanks to a pack of lies and still benefit from tax incentives. This obviously leads to a loss of credibility for the industry. And it does not help to pretend that it is all an American conspiracy theory.

And yet the German automobile makers will not let it go. But every minute we waste by not drawing the necessary conclusion—namely, to stop using this technology—is another minute the newcomers from China and Silicon Valley can use to improve their EVs and occupy the market. If the

German automotive industry ends up sinking into oblivion for many years to come, it has only itself to blame. Just think that rail traffic is already fully run by electric engines. Hissing locomotives are only used for nostalgic outings. It is inconceivable to think that the German high-speed trains ICE or the metro could still run on steam engines!

Experiments with electric drivetrains have been conducted repeatedly over the past decades, both by manufacturers in the form of small series and by independent workshops that converted combustion engine vehicles into EVs.[30] Two decades ago, GM famously tested several hundred electric vehicles—the EV1. And when the company recalled and scrapped them all, their owners held candlelight vigils. Elon Musk remarked in an interview about this that you had to be really pretty tone-deaf to ignore customer wishes in such a blatant manner.

The dynamics changed drastically with the appearance of the American EV manufacturer Tesla on the market. Some of the previous efforts either were made with small numbers in mind and then failed because of lack of funds or were initiated by the big manufacturers but eventually were stopped because of difficulties or a strategy change by the companies. Tesla, by contrast, showed that it was really possible to build a vehicle with an acceptable range, power, and—last but not least—an appealing design. Whereas the Tesla Roadster had still been a niche market vehicle of a rather experimental character that was, however, supported even by skeptics, the Model S represented a breakthrough. This model was convincing in many respects and proved that a startup company could develop a beautiful electric car with high safety standards and novel concepts at an acceptable price—and find customers to buy it, too. A surprising number of customers at that! The Model S was followed by an SUV with the model number X, but the tipping point was the successive model, the Model 3, available from $35,000 and up. As I said earlier, seeing people waiting in line in the early morning hours in front of the Tesla offices was the automotive industry's iPhone moment. Production then started in July 2017.

While the competition and the investors were heatedly discussing and (correctly) doubting that Elon Musk would be able to increase the announced production capacity from 80,000 cars to an incredible number of 500,000 cars by 2018, something else had already happened. Four hundred thousand potential car buyers who made down payments of $1,000 or €1,000, respectively, have dropped out of the customer ranks for German car manufacturers. That is a fact that really hits the industry hard. A *Bloomberg* magazine article demonstrated in a customer flow analysis that Tesla Model 3 customers are particularly interested in German models.[31] Of these customers, 28 percent like BMW, 20 percent like Audi and Mercedes, 12 percent like Porsche, and 10 percent like Volkswagen. In comparison, Tesla buyers would have strong reservations about an alternative car from Kia (8.5 percent). Despite the fact that it was unlikely that all 400,000 preorders actually would convert to Model 3 purchases in the end, you could still make an estimate of the degree of customer migration and compare that later with the real numbers.

Let us just imagine that Tesla delivered all 400,000 preordered Model 3 cars by late 2018. This means that the German manufacturers would have lost 115,000 customers already. BMW would lose 36,220 buyers, Audi 26,247, Mercedes 25,197, Porsche 15,223, and Volkswagen 12,598. This does not include the existing losses to the Tesla Models S and X (see Table 5.2). Those informed guesses and calculations turned out to be true when the actual sales numbers for the Model 3 came in. At the end of 2018, 155,663 Model 3 cars had been manufactured. In the same year, German manufacturers had been selling between 20 and 30 percent fewer sedans in all price ranges in the United States than the year before. That small American car builder sold almost half as many sedans in the United States as all four German manufacturers together.[32]

TABLE 5.2 Overview of the Estimated Production Quantity Loss Incurred by German Auto Manufacturers Owing to Tesla Model 3 Preorders

MANUFACTURER	MARKET SHARE	QUANTITY
BMW	27.6 %	36,220
Audi	20.0%	26,247
Mercedes Benz	19.2%	25,197
Porsche	11.6%	15,223
Volkswagen	9.6%	12,598
TOTAL		115,485

There is some anecdotal evidence from my circle of friends showing the impact. Several people I know had ordered and gotten a Model 3. A friend and her two sisters, all three physicians running their own clinics, all of whom until recently drove BMW X3s and were contemplating a Porsche Panamera as their next purchase, have changed their minds after a test drive of the Model S. As great as the workmanship of German cars is and the interior and exterior look are great, those women saw the *future* and decided that they did not want to drive cars running on "dinosaur juice" anymore. In early 2019, the first one sold her BMW and got herself a Model 3.

This could turn out to be the beginning of the end. The "peak car" moment is just around the corner, and we may have reached the tipping point. While German manufacturers needed the impressive number of Tesla pre-orders to acknowledge the situation, others have made the next step already. Norway, the Netherlands, and Switzerland are among the most successful markets for Tesla—it sells more vehicles there than the premium manufacturers in some segments. Such developments cannot be explained by buyers' premiums alone because Switzerland, for example, does not give incentives to buyers of EVs, neither by means of a premium nor by tax advantages.[33]

Toyota is closing in on Tesla with its hybrid vehicle. The Prius became a best seller in the United States and attuned customers to affordable electric mobility. Until the arrival of the Model 3, Nissan had the top-selling electric car on the streets, the Leaf. General Motors is basing its entire future—which does not seem so very secure after its bankruptcy proceedings in 2008—on the Chevrolet Volt and its successor, the Chevrolet Bolt. Renault launched its Zoë into the race. The only German manufacturer to have serial production of a promising and innovative EV is BMW with its i3, although its appearance will take a little getting used to. Most other EV offerings either are simply not relevant or have been intended to serve as a compliance car as a means of lowering fleet consumption from a regulatory point of view. We will take a closer look at this a little later.

If you proceed rationally and look at all the figures and trends, there should be only one conclusion: by 2030, mostly EVs possibly in the form of robotaxis will be on the roads for both technical and other reasons.[34, 35] First of all, electric engines are five times more energy efficient than combustion engines, partly because they convert energy into motion, not heat. Accordingly, the energy cost per mile is reduced to a tenth. If you then consider that an eight-cylinder engine has 1,200 parts and may even have more than 2,000 parts if you include the transmission and other drive elements, whereas an EV requires fewer than two dozen parts, maintenance cost will be reduced by at least a third or possibly by up to 90 percent depending on various estimates.[36] Fewer moving parts rubbing against each other also means that less energy is converted into heat.

Another element that is not quite as relevant but still important in terms of its acceptance is the acceleration of EVs. Electric motors are much more powerful and produce a higher torque. Consequently, Tesla beats almost all traditional sports cars, including Porsche, McLaren, and Ferrari, in performance. The new Tesla Roadster, which is expected to be available in 2021, has an acceleration so fast that it beats every single super sports car with a combustion engine in existence. Because of the higher torque, an electric car does not require a pinion; that is, no transmission is needed to put the vehicle in motion. This bulky component no longer has to be installed, and thus we have more space available inside the vehicle. Moreover, energy is conserved, and the batteries are recharged as the car brakes. This allows for improved management and more precise status assessments. Combustion engines need much more effort to approximate such scope and quality.

Although traditional manufacturers are aware of these facts, their lifeblood is still with the combustion engine. It hurts to throw all the expertise overboard and set out on a completely different course. Another problem is that management has to take care of the workers who are currently building the combustion engines. This causes some headaches. Too many jobs would be lost if traditional automobile makers decided to swiftly transition to EVs. Labor disputes would be inevitable. Accordingly, people prefer to muddle through as in the past, although we can already see the writing on the wall. At the moment, the system is still working; the companies are still making record profits. We are reaching "peak car" status, and yet we continue to put everything at risk.

As noted earlier, Tesla CEO Elon Musk pointed out the massive difference in the respective leadership cultures when he discussed his reasons for ending the collaboration with Daimler and Toyota on battery technology development in an interview with the German newspaper *Handelsblatt*.[37] He claimed that their efforts to promote hybrid vehicles did not aim to achieve an improved product but just merely to comply with minimum legal requirements. Innovative companies, by contrast, are not interested in picking easily available fruit from the tree, or keeping the authorities quiet, but instead want to master the really difficult challenges. This may not always work from the start, as the first version of Google Glass showed, a Google X project that has meanwhile been partially abandoned. Nevertheless, revolutionary developments force you to take risks.

If you regard Tesla only as a manufacturer of EVs, you miss out on important facets of the whole situation. The innovation the company brings to the automotive market not only concerns the battery performance—although that is already a huge step. Tesla has other offers to make:

- Product system → powerful battery technology, autonomous driving
- Network → charging stations

- Distribution channel → direct sales to the end customer without an external dealer network
- Product performance → acceleration
- Process → fully automated production
- Service → over-the-air updates, free charging
- Business model → upgrades by software activation, preordering
- Customer engagement → hidden "Easter eggs"

Suppose that one day you heard someone say, "From a purely technological point of view, Tesla does not offer any kind of revolution. For example, I cannot discern any extremely advanced technology on the drive train side." If this were a development manager of a large German automotive corporation trying to downplay the danger Tesla poses for the manager's own company with this remark, the reason might be that the manager's company mainly concentrates only on technological progress, not on the multidimensional aspects of innovation.[38] From the engineer-focused, one-dimensional point of view of traditional manufacturers, this is quite reasonable, but it makes them even more susceptible to disruption.

Tesla's mission statement—"Accelerate the world's transition to sustainable energy"—indicates why Tesla acquired SolarCity, set up a network of charging stations, and produces batteries and energy storage solutions. All these activities are necessary to reach the company's goal, whereas traditional manufacturers simply build cars. They do not consider it their problem to think about where the "juice" comes from or who operates the charging station.

Perhaps you have been asking yourself why California manufacturers in particular are so successful with EVs. You might want to point out that American households (in the true sense of the word) usually own two or three cars, often in a combination of an SUV or van and a sedan. Another car with an electric motor is the perfect complement for the city. It is easier for me to install a charging station in the garage of my own home than it is to convince my landlord. It remains to be seen whether these arguments really hold water, because China is also among the leaders of electric mobility, and its conditions are completely different.

How Does the Electricity Get to the Electric Car?
A Quick Guide to Batteries and Accumulators

Putting an electric motor and a battery in a vehicle and using them to drive that vehicle forward are not new concepts. One of the first EVs was a toy car designed in 1828 by Hungarian inventor Ányos Jedlik. As mentioned earlier, around the turn of the twentieth century, it was not at all clear which kind of propulsion would eventually dominate in the automobile sector. For some time, steam-driven, gas-driven, and battery-driven vehicles were all found simultaneously on the streets of European and American cities. The reasons for the end of EVs then are partly the same limitations that are mentioned today, as well as the list of counter-arguments: small range, long charging times, very heavy batteries, short charge duration, the use of rare earth elements only available in certain regions of the globe, insufficient charging infrastructure, and different standards for plugs—the list could easily become much longer.

How does a battery function? A battery produces electric power by converting chemical energy into electric energy. We know similar processes from our daily lives. Imagine a stone lying on a mountain in a little depression. While in the depression, the stone is stable and won't roll away. Push it and move it out of the depression, that is, supply energy to it, and the stone will go from a stable to an unstable condition. When it then rolls down the mountain, it will release that stored energy. Or take the wood for your fireplace: a log won't just start burning by itself. You have to make it burn by supplying it with energy by means of a burning match, thus taking it from an energetically stable to an unstable condition. Then it can release its energy as heat and light.

So now for the battery. The chemical components in a battery are combined in such a way that they are stable. However, as soon as I "poke" them chemically, they react with each other and release electric energy and heat. In some batteries, say, normal household batteries, this process can only occur once. In other cases, involving so-called accumulators, this process can be reversed. You can charge the battery again and "get the stone rolling" once more. Electrolytes in solid, liquid, or dissolved forms provide electrons for transporting energy. The secret of powerful batteries often rests with their combination of chemical compounds, as we will see below.

So what we casually refer to as a battery is, as a matter of fact, much more complex. Indeed, we should more precisely call a battery *cylinder-shaped cells* or, even more correctly, *galvanic cells*. An AA or AAA battery that cannot be recharged is a primary cell with a standard diameter of 0.55 inches (14 mm) and a height of 2 inches (50 mm). A rechargeable accumulator is referred to as a *secondary cell*. Both types are closed systems. You can only draw as much energy from them as they contain, so the quantity is limited.

Laptops use slightly larger and rechargeable batteries. They have a diameter of 0.7 inch (18 mm) and a height of 2.55 inches (65 mm) and are therefore often referred to as *18650-type* batteries. Several billion of these are produced each year. Because they used to be the only mass-produced and cheaply available batteries, they were also used for Tesla's first three models, where the entire production and cooling process was adapted to them. Both the Tesla Roadster and the Models S and X contain several thousand of those batteries. However, there is a tiny but important difference in laptop batteries: the use of electrolyte additives.

A fuel cell, by contrast, receives its fuel (hydrogen) from the outside. Accordingly, it can theoretically provide electric energy without time limits, as long as it has a continuous supply of fuel. Charging is not required in this case, just refueling.

If I use the term *battery* in the following discussion, I am referring to the rechargeable and combinable secondary cells that ensure the energy supply to the electric motor in EVs and thus ensure propulsion. Tesla's Models S and X use exactly 444 cylindrical battery cells to make one battery module. A cooling layer is arranged between the individual battery cell rows designed to guide the heat away from the batteries by means of a cooling liquid (usually glycol). Usually, there are 16 of these battery modules in a Model S or X, housed in a battery box that weigh about 220 pounds. The box has separating walls between the individual modules and small cooling units guiding the heat out of the car. At the bottom is a metal plate designed to protect against mechanic damage, in case something lying on the road penetrates the floor, as has happened before. The central element of the battery box is the battery management system with circuits for controlling the battery modules. This system makes sure that the driver receives the power required and provides information about the charging status of the batteries.

Each 18650 battery weighs 1.5 ounces; consequently, the battery weight alone of a fully equipped Tesla is 700 pounds. If you add the weight of the box, the cooling liquid, and other components, you have almost 1,100 pounds, or half a ton. In comparison, such a battery unit is at least twice as heavy as a combustion engine with transmission and full fuel tank.

Tesla uses slightly larger battery cells for the Model 3. The cells have a diameter of 0.8 inch (20 mm) and are 2.8 inches (70 mm) high, and accordingly, they are referred to as *type 20700 lithium cells*. Given their volume, their capacity is 30 percent higher. However, given the reduced size of the Model 3, fewer cells can be fitted into the car. Tesla will probably need about 3 billion cells for its envisioned annual production of 500,000 EVs. This is almost double today's total production of 4 billion. And here we are looking at the impact of Gigafactory 1: it was commissioned in early 2017. And it can ensure the production of the required numbers.[39]

Not all the batteries for EVs are cylinder shaped. The Nissan Leaf, the Chevrolet Volt, and the Chevrolet Bolt use rectangular cells with coated film instead of wound metals. A study from Sweden regarding "structured batteries" shows that EVs allow experimentation with new battery designs. Instead of simply regarding batteries as an additional element of the vehicle, the vehicle structure itself becomes a battery. If you are using a carbon chassis, lithium and electrolytes can be applied directly onto the microstructures. This would mean that no weight for separate batteries would apply, and the space traditionally occupied by the battery could be used in other ways.[40]

But who knows? Perhaps we will one day be able to do without batteries altogether. There were two companies altogether, Sono Motors from Germany and Hanergy from China, that made electric cars covered with solar panels. The green power provider from Beijing presented four prototypes and expects energy yield from solar power to be sufficient within the next three to five years to charge them.[41] Sono Motors from Munich started a crowdfunding campaign for a low-price electric car that also draws part of its energy from the solar cells incorporated into the vehicle; it should be available for $29,000 upward.[42]

Lithium & Co.: Materials 101

So what are the materials found inside a single battery cell? And how dangerous is such a cell for the environment?

Lithium, Graphite, Nickel, Cobalt, and Aluminum

The batteries used most in EVs are *lithium-ion accumulators*. They consist of lithium compounds for the positive part of the battery (the cathode) and graphite for the negative part (the anode). During charging, electrons are "pulled" from the cathode to the anode and stored there. The process is reversed during the discharge. Just about 11 pounds of lithium is included in the 700-pound battery weight of a Tesla.[43] Half of all known lithium deposits worldwide are found in Bolivia.[44] Both Germany and Austria have deposits that are also mined. This silvery white light metal is very reactive and therefore is only ever found in nature in a reacted form. In addition to lithium, among others, nickel, cobalt, and aluminum are used in batteries.

Polypropylene and Ethyl Carbonate

Between the anode and the cathode, there is a thin separating layer of polypropylene, but tiny holes with a diameter of a hundred thousandth of an inch let the electrons pass through. Ethyl carbonate and other compounds serve as *carriers* for the electrons between the electrodes. If the separating layer is damaged, for example, by local overheating or mechanical damage, the device may overheat, which causes the ethyl carbonate to explode on contact with air.

About Memory and Coffee Filter Effects: What Should a Battery Be Able to Do?

As a driver, you expect your propulsion device to provide sufficient and reliable performance and a range that roughly equals that of a combustion engine vehicle. At the moment, the Tesla Model S is the EV that fits this description best with its range of 800 miles. A lot depends on how the performance density of the lithium-ion accumulators can be improved in the future. This value currently is at a few percentage points per year. LG Chem, the South Korean producer of car batteries, is convinced that it can lower the price for such cells to $100 per kilowatt-hour by 2022.[45] Currently, Asian producers dominate the market for battery cells. Among the leaders in the market are Samsung, LG, and Panasonic.

We know about the so-called memory effect from earlier batteries, where the charging capacity, that is, the amount of energy stored, decreases with every charging cycle. Researchers believe that they have discovered the reason for those losses: parasitic secondary reactions in the electrolytes result in a kind of blockage. On the negative electrode, we discover deposits of solid electrolyte oxidation products that slowly seal the electrode off. Once this has happened, the lithium ions cannot pass through anymore. It is a bit like a coffee filter: if you use the same filter repeatedly, more coffee grounds will accumulate in the pores until nothing goes through anymore. Highly compact batteries, which have low porosity, develop this blockage very quickly after only a few charging cycles. This often leads to a sudden massive loss of capacity. If you use less compact, more porous structures, the effect does not occur, or at least it occurs much more slowly. Electrolyte additives can help prevent such parasitic secondary reactions. Every manufacturer uses its own secret ingredients for this, a secret almost as well guarded as the Coca-Cola formula.

How quickly such electrolyte oxidation products develop and consequently deteriorate battery performance depends on the temperature and the speed of the charging and discharging, not so much on structural changes by expansion. Some battery producers, for instance, say that batteries lose 10 percent of their capacity at temperatures of 140°F (60°C) after 500 charging cycles.[46] The capacity loss is higher, however, if the charging and discharging are slower and the batteries are exposed to high temperatures for a longer time. This is exactly the problem Fisker, an EV manufacturer that had gone out of business and now is resurrected under new ownership, could not solve.

Professor Jeff Dahn from Dalhousie University in Halifax analyzes such capacity loss by means of new techniques that measure the temperatures in the battery very precisely and allow forecasts regarding the loss of performance after just a few charging cycles. This allows researchers to try out different electrolyte additives more quickly. Dahn came up with a rule of thumb: the more electrolyte additives you use in a battery, the longer the battery seems to last. His research was a huge help to the test procedure: in order to simulate realistic conditions, you had to run test cycles over a period of weeks and months, always charging and discharging the batteries as you would do in practical applications. Dramatic loss of capacity sometimes does not occur until you have completed a couple of hundred charging cycles.

In addition to the electrolyte additives, battery heat management is also critical for the performance and durability of the batteries. This starts with production. The batteries are welded in traditional manufacturing processes, which may lead to high temperatures that, in turn, cause electrolyte oxidation. One example of how much potential there still is in the development of battery technology is a technique used by the Austrian company Kreisel Electric. This startup was founded by three brothers who developed a method for welding battery cells using lasers, which shortens the process.[47] This allows for better control of the battery temperature from the start, keeping the hot phase short and reducing production-related electrolyte oxidation. And this is not the only approach Kreisel takes. The company also developed a sophisticated battery cooling system that allows for faster charging and discharging.

The analysis of 500 batteries for the Tesla Model S in practical application showed a comparatively low loss of 5 percent after 50,000 miles (80,000 km), and this process was further slowed down to only 8 percent after 100,000 miles (160,000 km).[48] A group of Tesla owners kept a list of charging cycles and ranges over the course of several years and arrived at similar results.[49] In the Nissan Leaf, by contrast, the capacity sank by 20 percent over the course of three years.[50]

The direction in which battery technology will develop in the future is still unclear. We aim at optimized production processes, improved temperature management, new electrolytes on one side, and new materials on the other. One promising approach involves so-called graphenes, which could be used in anodes.[51] These honeycombed carbon compounds promise a four times increased energy density, much faster charging cycles, and a longer lifetime. Currently, this material, which has a tensile strength that is 100 times higher than that of steel, is still very complex to produce and very expensive. The energy density of the batteries doubled between 1995 and 2005. With the Model 3, Tesla is aiming at another doubling.[52]

Good Things Cost Money

Batteries for an EV cost thousands of dollars. In early 2017, the management consultancy firm McKinsey presented a survey that showed how the price for 1 kilowatt-hour had decreased from about $1,000 in 2010 to $227 in 2016. The cost for Tesla's batteries had actually dropped to $190 per kilowatt-hour since early 2016.[53] With the start of battery cell production in Gigafactory 1, Tesla expected a further cost reduction by approximately 30 percent to about $125 per kilowatt-hour, and it seems to be close to $100 now.[54] As soon as the cost falls to the threshold of $150, the cost equals that of combustion engines. While McKinsey in this study did not expect a price of $100 to be reached before 2025, others are more optimistic and believe that the threshold will be reached somewhat earlier. Professor Tony Seba from Stanford University estimates that the point will be reached by 2022. If you extrapolate the cost decline at Tesla, this could be accurate. CEO Elon Musk even mentioned the year 2020. From that point onward, a vehicle with a combustion engine will be less economical than an EV both at purchase and during use.

Recycling Is Important

The end of a vehicle's life should not be equated with the end of the battery life cycle. Even after a "life span" of eight years, a battery is not completely spent. First of all, it can be repaired if only individual modules or cells are damaged, thanks to its modular construction. Admittedly, you cannot properly use a battery that can only reach 70 or 80 percent of its capacity in an EV because of the resulting reduction in range, but the Association of Electrical Engineering estimates that continued use as a power storage device in houses and companies is possible for another 20 years. Consequently, batteries are not special waste but can be recycled.[55]

Tesla is manufacturing house storage systems that can store excess energy from solar plants. The startup company ReeVolt, from Schwerin, Germany, converts retired accumulators from electric mopeds into energy storage systems for domestic use. In contrast to this, the American startup FreeWire converts those "retired devices" into mobile energy supply Mobi as jump-start support for EVs.

Eventually, however, at some point, the battery cells have to be disposed of. This means that they first must be removed from their housings and then processed either thermally—that is, by melting them down—or by crushing. Prior to that, recycling workers have to ensure that the batteries have been sufficiently discharged, because otherwise there is a great danger of fire if the electrolyte ignites.[56]

The wide variety of battery systems makes automation of disassembly particularly difficult. The way the accumulators are screwed onto each other and wrapped with cooling coils and the like requires disproportionally high manual effort. The shredded and molten parts of the batteries are then split up into their elements by thermal processes at low or high temperatures. However, the effort necessary for the separation of the lithium components can be extreme and therefore is not economical, at least as of now. At the moment, the exact effort required can only be estimated on the basis of scant practical data accumulated so far, because there are not enough recycled vehicles on the market for which battery recycling technology is available.[57]

Generally, the economic efficiency of such recycling processes depends on the current market prices for the elements contained in the batteries. Expensive battery components such as nickel and cobalt would be worth the effort. If, however, the battery manufacturers replace those with cheaper elements because they want to lower production costs, that is good news for anyone looking to purchase an electric car but not good for recycling.

Nevertheless, at present, the disposal offered for retired batteries is mostly of a dubious nature. A Chinese company recently showed a German manufacturer the pit where the batteries were to be buried. The Germans politely declined the offer.

The Charging Primer: Plugs, Standards, and Other Obstacles

How does the electricity get to the electric car? Obviously, by means of an electric plug. So what does such a plug look like? This is where it gets confusing for electric car novices because there are various charging systems. CHAdeMO, Type 2, SAE J1772, direct or alternate current, 50, 130, or do you prefer 350 kilowatts? This is as confusing as the choices available to a new driver having to decide whether to go for diesel, flex-fuel, or gasoline with 87, 89, or 91 octane. What on earth does it all mean?

As happens so often with technologies that are still in their early stages, it is not yet clear what will eventually emerge as the standard. In addition, what becomes the standard is also not necessarily the "best"—depending on how you define that. Betamax and Video 2000 both were considered better technologies than VHS, but VHS still became the video standard for consumers. And thus drivers of electric cars find themselves faced with several competing charging system plugs for charging stations. In a best case, drivers can choose; in a worst case, they cannot find the standard for their vehicle and have to use an adapter or proceed to the next charging station, hoping that the remaining battery charge is sufficient, that they find a plug there that fits, and that the charging gear is not defective or there is no other car blocking it.

The most obvious indication of the various different standards is the plug, because not every shape will fit every car. CHAdeMO is a Japanese standard used by the Nissan Leaf, by the Mitsubishi i-MiEV, and by the Kia Soul EV. The Type 2 plug from the German manufacturer Mennekes is mainly used by German manufacturers, whereas the SAE J1772 is mainly used in North America. There are also adapter plugs for the various systems to allow you a charging session even with the "wrong" plug. The whole setup is reminiscent of the various socket outlets you find on visits to different countries. Plugs with two or three pins at different angles and in various shapes make international travel an electric adventure. How on earth will I be able to shave or dry my hair?

Once you have been at that point, you really understand how important the charging performance is, because that is the value that tells you how quickly batteries can be recharged. Does the power "drip" into your car slowly or rush in like water from a fire hose? Here is a simple formula for you: the higher the kilowatt value, the faster you can get on the road again. Nonetheless, higher charging performance comes with a drawback: the charging cable and the plug will heat up and must be packed in cooling units, which makes both elements less flexible, thicker, and heavier, an extra challenge for delicate hands.

Currently, quick-charging stations are operated with 50 kilowatts. This means that it takes 21 minutes to "fill up" with sufficient energy for 60 miles. The charging time for the same distance can be reduced to 4 minutes if the charging capacity is increased to 350 kilowatts. In July 2017, Porsche installed the first public stations with this charging power to prepare for the launch of its first electric sports car, the Porsche Taycan. Unfortunately, the required power input for such extremely fast charging equals that of an entire city district. And as mentioned earlier, the charging plug must be cooled. At the same time, extremely fast charging is not without risks for the battery, which can be damaged if anything at all goes wrong or even if it is done too often. Once again, heat management is very important.

But what is even more important than public charging stations are those at home. Admittedly, you can plug your vehicle into your normal U.S. 110-volt or German 220-volt power outlet at home (and charging may take all night), but experts discourage this procedure. Sometimes the electric wiring in your home may not be state of the art, and smoldering fires can ensue. The fuses for an electric circuit are designed for a power of 16 amperes, the same as the charging capacity of an EV. If you add other devices or cars to the same electric circuit, the fuses blow immediately. A better way is to have new power lines and a separate charging box installed. A charging box (or wall box) can easily handle long-lasting charging and high currents and ensures speed and safety.

The startup eMotorWerks from San Carlos, Silicon Valley, offers a popular and affordable charging box called JuiceBox, but there are more and more companies with new solutions. The startup ChargePoint, by contrast, has raised venture capital of $160 million already and has more than 57,000 charging points in the United States to show for it. The company expects the number of charging stations to grow quickly, especially in cities that aim to fully electrify their fleets. Anne Smart, responsible for public policy relations at ChargePoint, also mentions problems that arise once charging stations are installed. These problems do not involve acts of vandalism so much but rather lousy drivers who run into the stations when parking and render them inoperable.

The expansion of charging systems is not limited to cars. Trucks, mopeds, forklifts, and even airplanes are being electrified, and there are opportunities for the providers of charging systems to install and operate their products everywhere. Some cities are using electric buses that have a current collector on the roof—like cable cars. The collector can be extracted at every station or only at the end stops and recharges the batteries via a trolley wire.

Another slightly different system uses a hybrid approach: trolley wires are installed on a certain stretch of highway to provide the required energy to the electric truck via a current collector. This procedure allows the truck to travel long distances without recharging. On the stretches of highway without trolley wires, the battery or the diesel engine is used as the propulsion for the truck. By this means it is even possible to overtake other vehicles.[58] The advantage of this technology is the energy savings, which can reduce the cost by 50 percent.

The association CharIn e.V. works to harmonize charging standards. The initiative is supported by Ford, GM, Tesla, BMW, Daimler, and Volkswagen, as well as charging station operators.[59] The first step was to create a unified plug standard, the Combined Charging System (CCS). The next steps include the generation of a specification catalog and a certification process to facilitate use of the CCS standard in the various manufacturers' products. The third field of activity for CharIn is the distribution and implementation of the joint standard, an action mainly designed to counteract the Japanese CHAdeMO standard.

Currently, there is no telling how fast this will happen or whether it will succeed. If you look at the number of charging stations already in operation in Japan, CharIn has a Herculean task before it. In 2016, Japan already had more charging stations than gas stations all over the country: more than 40,000 charging points compared with 34,000 fuel pumps.[60] In the United States, there were only 9,000 charging stations compared with 114,500 fuel pumps,[61] and Germany currently counts more than 6,000 e-stations, Switzerland 800, and Austria more than 2,000.[62] If you look at China, however, all those numbers are merely "peanuts." There were, believe it or not, 270,000 charging stations in China in early 2017.[63]

The oil corporation Royal Dutch Shell surprised everyone in early 2017 by announcing that it would enhance its own gas stations with charging points. The company did not specify the timeframe for this modification, but it did mention the countries it was referring to. I will give it to you straight: Germany was not among them, but the domestic Shell markets in Great Britain and the Netherlands were targeted.[64]

However, even if the number of charging stations still lags behind the number of gas stations, we may confidently expect this to change in the next few years. Charging points with a cost of $3,000 to $7,500 are, first of all, much cheaper and can be installed (almost) anywhere.[65] They do not need any comprehensive environmental impact assessments like gas stations with their liquid fuels that have to ensure that those fuels cannot contaminate the environment or that they agree with other operations in the neighborhood.

Unfortunately, Germany is proceeding somewhat amateurishly in this matter. It is very clear that the German car manufacturers and electricity providers do not believe in electric mobility. How else could you explain the maze of billing systems? Once again, the digital knowledge deficit in Germany becomes apparent: different radio frequency identification (RFID) cards, billing systems whose security standards make experts cry, and public charging stations that can only be used during office hours serve to trip up electric mobility in several ways.

And where there are no amateurs at work, there are saboteurs. The Quick Charging Network for Axes and Metropolis (German abbreviation SLAM) stated as its goal to provide 600 quick charging stations with 150 kilowatts of charging power per charging point on highways and in bigger cities for research purposes by mid-2017. The data are to provide scientists with the basis for models that would indicate where charging stations make sense, how they are used, and what payment systems could be like.[66] The project is financed by citizens' taxes and distinguishes itself by one detail in particular: only two of three charging standards were installed. Or, more precisely, all three standards were included in the delivery from the manufacturer, but the CHAdeMO standard not used by German carmakers, though by, among others, Nissan, was later deactivated. However, nothing much has happened since with the project anyway. By March 2017, just about 50 charging stations had been installed and not many since then. Just as a comparison, by late 2016, Tesla was operating almost 5,000 superchargers worldwide and more than doubled that number by early 2019 with 12,888 superchargers at 1,441 supercharger stations. The combined power of the SLAM project partners—including Daimler, BMW, Porsche, Volkswagen, RWE, EnBW, and the German Federal Ministry of Economy and Energy, that is, all the companies referred to as the "crème de la crème" of the German economy—managed to achieve just about 1 percent of what an American startup can do.

In Vienna, the city management is investigating the option of extending traffic lights and lamp posts to charging points. The idea is to use the 153,000 lighting points with their 3,400 switchboards as well as the 14,000 traffic light installations on the same number of switchboards to provide a basic supply network for charging EVs. The planned cost for a range of 120 miles in 4 hours is $7; at least that is the current plan.[67]

The Open Charge Alliance aims at developing joint open standards for charging stations. Two of the standards so far include protocols for the data exchange between one charging station and a central system and allow a 24-hour forecast of the expected charging volume.[68] Partners in this venture are many different startups as well as research institutions and industrial organizations and energy providers such as RWE. There is also a plan to introduce standards for billing systems on an international level. Hubject aims at a system for Europe that allows drivers to use their providers' payment systems at any charging station.[69]

It would be even better, though, if we did not have to think about charging at all. This is where the so-called induction charging process comes into play. The coffeehouse chain Starbucks already offers a charging system for smartphones in its U.S. stores. There are stands with a selection of ring-shaped plugs on the tables to which you connect to your (older) smartphone; newer smartphones don't even need that ring-shaped plug; then you place the phone on the coupling coil inserted into the table. This is a slow kind of charging—and you cannot use your phone while charging because it has to remain on the coil—but it is still a lot better than having to continue your trip with a dead battery.

Induction charging promises twofold improvements: first, you park your car on a spot in your garage with a coil inserted into the floor, often as a mat with an electromagnetic field that charges without further steps, that is, without any additional cable that needs to be plugged in. The car itself needs to be equipped with a coupling coil as well to act as a receiver. This avoids those desperate morning moments when you realize that you forgot to plug in the car in the evening and now you won't be able to make it to work because your batteries are empty (still talking about the car, of course). Another improvement may come from induction coils inserted into the streets and meant to charge while you drive. In theory, any car traveling on those roads should never have to stop for charging. Which leaves only one problem unsolved: that is, when do you then stop to pee. At the moment, the induction solution is still hampered somewhat by the cost incurred for installing the coils in the tarmac on streets and in parking lots and by the (still) low charging performance. The coupling coils in the car and on the mat ideally also are very close together, just an inch or so maximum, to keep charging loss as low as possible.

Just as in other areas of the new automotive industry, there are several startups busy in this field trying to make progress on induction charging: Momentum Dynamics, Plugless, and WiTricity. These three startups together have raised more than $42 million in venture capital so far. Waymo, the autonomous driving heavyweight, regards induction charging as an element of autonomous electric cars. Both Hevo and Momentum Dynamics have already offered induction pads for Google's (now retired) self-driving Koala vehicles, though the pads still have a very low charging performance. The startups have promised charging performances of 200 kilowatts in the near future, which would bring them level with quick-charging stations.[70]

Great Britain is one step ahead of the European continent: the British already tested the system and want to equip more streets for wireless charging.[71] The startup Electreon from Israel also began its trial run of induction charging for electric buses in Tel Aviv.[72] Charging during use makes it possible to use smaller batteries in the vehicles. This is yet another attempt to reduce the vehicle weight by reducing the weight of today's batteries and therefore increase the range.

Will those wireless charging systems become a new source of electromagnetic radiation with possibly a negative effect on human health? The studies are as yet inconclusive and contradictory. A survey commissioned by the European Union failed to determine any connection between low-frequency electromagnetic fields and leukemia.[73] Another study, however, did establish a relation between the growth of cancerous tumors in mice and their vicinity to electromagnetic fields.[74]

Battery Market and Power Grid: Give and Take

Batteries are becoming big business. According to one survey, this market will grow to $10 billion by 2020. Six manufacturers are likely to deliver 90 percent of the supply, with Tesla planning to put half of the production into its cars, followed by BYD, Volkswagen, Renault, Nissan, GM, and BMW (in this order).[75] The huge demand will be met by several battery producers—among them BYD, LG Chem, NEC, and Samsung SDI—and Gigafactory 1 in Reno, Nevada, operated by Tesla and Panasonic will dominate the market with 46 percent. EVs are the biggest customers of battery manufacturers, amounting to up to 80 percent of the demand for storage solutions.

Thousands or even millions of EVs will also have their effect on the power companies, on the one hand, because the vehicles need power to charge and, on the other hand, because vehicles are considered to be decentralized energy storage units. A huge number of vehicles with their batteries connected to the power grid may help to deal with excess capacity or else pass on current into the network. If a large number of people charged their vehicle at the same time, however, that might lead to an overload, and the power grid could become unstable, just as can happen during peak times of the day when suddenly all the air-conditioning systems or stoves are used at the same time. An intelligent power grid instead communicates with the vehicle and distributes the charging performance over a certain period of time.[76] Such a system is known as a *vehicle-to-grid* system.[77]

Until recently, German manufacturers were characterized by a hesitant to-ing and fro-ing in battery production. Should they do it themselves or better leave it to others? The workers' representatives of some manufacturers expressly voiced their approval of building up competencies in battery production, not least to avoid any loss of jobs in Germany. Daimler has now announced the commissioning of a second battery factory in Kamenz, Saxonia.[78] Volkswagen and BMW wanted to follow suit, but as it turned out, that's not so easy. First, the sourcing of the required rare earth elements and components has proved to be a nightmare. Almost all current sources are running at full capacity and are already contracted to the first movers. Second, even if Volkswagen and BMW secure the supply, it will take some years to get a factory up and running. This means that battery cells for vehicles planned to be launched this and next year need to use batteries supplied by the abovementioned battery companies. And they threatened to cancel the very same supply contracts with Volkswagen if the company was going to build its own battery factory. This is quite a conundrum and a new experience for a carmaker like Volkswagen that is used to dictating conditions and prices.

Regulations and Emergency Solutions: Good Intentions Executed Badly!

Let me say it once again: German automobile manufacturers and the German government are in a predicament. Their extrinsic motivation (the one coming from outside) contradicts the intrinsic motivation of the "young guns." There is a world of difference in whether we get somebody to do something by promising him or her things or providing incentives—such as "win over five new customers and you will get an additional day off!" or "if you exceed the speed limit, you will have to pay a fine of $80!"—or that person is highly motivated to act of his or her own accord: I am attending this lecture because I am truly interested in the topic and in meeting interesting people there, not because someone promised me $100 if I went.

Basically, we humans act very irrationally. The quality of our work deteriorates or we lose interest in a task if we have been lured with extrinsic incentives or are baited. This fact has been proved in many experiments. Adults, children, and even monkeys take less time for the task in that case, they make more mistakes, and their work is less valuable in terms of quality.[79]

Let's say you would like to kindle interest for reading in your eight-year-old. So you promise him a baseball or Pokémon trading card for every book he reads. What, I ask you, do you think will happen? Which books will he read? Of course, the ones that are thin, with large print and the least amount of text. When asked about the story, he will probably be able to give you only a rudimentary outline of the content. And what happens when you stop rewarding him? He will stop reading altogether. In short, we basically achieved the opposite of what we set out to achieve. The intrinsic motivation "reading is exciting" was replaced by the extrinsic "trading card" motivation.

What is the reason for this little foray into behavioral science? Simply the fact that the wrong kind of incentives have taken us down the path to the deplorable current state of traditional carmakers of electric cars and hybrids. We can tell this from the range for the vehicles being offered today with alternative drives. This despite the fact that test cycles such as the New European Driving Cycle (NEDC) and the Worldwide Harmonised Light Vehicle Test Procedure (WLTP) introduced by the European Union and other regulatory authorities are supposed to lower fleet consumption and emissions as long as the fleet meets the criteria. This was a good idea intended to make the automotive industry build hybrids and EVs but was lost in a minimum effort that did not even allow the beginning of a larger vision. Add to this the short-sighted focus on immediate success for the company. The business figures for the next quarter are more important to management and shareholders than securing the corporate future thanks to innovative technologies whose results cannot be expected in less than five years, because by then those people won't be members of the board anymore anyway or have long since sold their stock.

Good intentions therefore do not automatically lead to good results. This is what Elon Musk meant in his assessment of Toyota and Daimler. His intrinsic motivation for liberating the world from fossil fuels is contrasted by the extrinsic motivation of traditional car manufacturers attempting to adhere to test cycles in order to massage and beautify matters. This behavior is by no means limited to the automotive industry. The petroleum industry also has known for years that it is time to leave oil behind. And yet any approach to investments in other kinds of energy sources is pursued rather half-heartedly.

The clocks are ticking somewhat differently in Silicon Valley, and traditional rules do not apply. "Tesla doesn't make any profit at all. They just announced a massive loss!" The smug grin on a German manager's face is soon replaced by sheer astonishment: "And their stock market price still rose to a high level, impossible to understand." It is true: in mid-2015, Tesla lost about $4,000 per Model S sold, as reported by Reuters.[80] In normal companies, this would immediately lead to massive cost cutting, reorganization, layoffs, and possibly the closing of the respective costly business division. Tesla is not a normal company, however, and the product is not standard either—and the automotive market is facing a major disruption. Therefore, the applicable rules are different and hard to grasp outside Silicon Valley. Traditional business models ask a very basic question: "How can I generate revenue?" Silicon Valley is probably the only place in the world where you do not have to answer this question but instead have the luxury to think about how to create value for customers.

German startups traveling to Palo Alto for three months through the German Accelerator Program (for which I served as a mentor) of the German Federal Ministry of Economics often are proud to mention in their presentations that they are already making a profit.[81] That, however, is a bad sign for the local venture capitalists. It means that the startup is not making enough of an effort to make the company bigger and occupy the market before potential competitors make their move. This is what Tesla is doing. Its losses of hundreds of millions of dollars in almost every quarter in past years were mainly due to the expansion of its infrastructure. It costs a lot of money to finance both Gigafactory 1 in Reno, where batteries are mass-produced for the Tesla Model 3 and for household usage, and the global development of the Tesla charging stations and preparations for the production of the Tesla Model Y, the semi truck, and the pickup truck.[82] All of it is an investment in the expected growth. Tesla is preparing to set a de facto standard for EVs, to dominate the market for batteries, and to supply consumers with powerful mass-produced EVs.

Tesla CEO Elon Musk expressed his attitude toward profit and loss very clearly in an interview with the German paper *Handelsblatt* in 2015. He referred to a quote by Daimler CEO Dieter Zetsche, who had said that nobody would make money with EVs, and answered:

> I agree; we cannot be making losses forever. This year we'll be investing a lot into the manufacturing ramp-up of the Model X, and in the long term, the Model 3 as well. So our goal from next year onwards is to be cash-flow positive, but we wouldn't slow down our growth for the sake of profitability.[83]

Like a chess player, Musk places his pieces and is willing to sacrifice some pawns, and if everything goes according to plan, the pieces in the end will be just in the right positions to checkmate his opponents and dominate the market. And that strategy seems to bear fruit. The company turned a profit in the second half of 2018 and started taking market share away from combustion engine cars.

Carl Benz also did not make any money in the first years; as a matter of fact, he seemed to burn it with his crazy idea of a coach without horses. The money for everyday life and that crazy idea of a horseless carriage came from his wife's dowry. She supported him financially and became his "venture capitalist." The real reason for the car trip she took with her sons was that she had become impatient. Benz was very hesitant, and she finally wanted to extend the vehicle testing beyond the limits of their property and show the world what an automobile could accomplish. She set out on her journey, disregarding any prohibitions, ignoring the lack of permits and the scant safety measures. In doing so, she gained public attention, and she also gained important insights for improvements on the automobile. A hundred years later, the CEO of Mercedes counteracts the historical pioneer spirit that marked the birth of his company.

A test report of the BMW i3 published in *Manager Magazin* in February 2014 told the story of how the test-drive failed because of the charging stations, once again highlighting the inadequacies accompanying the introduction of EVs in Germany.[84] The various obstacles included blocked charging cards, charging stations that were only accessible during office hours, the refusal to accept credit cards, and to construction obstacles that restricted the approach and energy fueling of electric cars. The report showed how energy providers simply sleep through new trends and business models, proceed half-heartedly, or just try to find excuses for their inactivity. Add to this another unpredictable element regarding charging stations: the prices charged vary immensely. Providers ask an hourly fee or, alternatively, a basic fee and then payment for the energy volume. At present, driving an electric car is still a business reminiscent of highway robber times. Nobody feels responsible for giving the customer a pleasant overall experience. Car manufacturers maintain that the energy providers are responsible, and they, in turn, do not consider the customers and possible turnover but only the initial cost involved. And in the end, everybody points a finger at the government.

Are Electric Cars Highly Dangerous?

In the winter of 2015, a photo of a Tesla gutted by fire in snowy Norway dominated the news. The car had overheated while charging and caught fire. Was this an example of how unsafe and dangerous EVs really are? A look at the statistics tells us that this would be the wrong conclusion. There are 150,000 vehicle fires per year reported just in the United States.[85] This breaks down to 17 fires per hour. In Germany, the number is just a tenth of that, but that still means 15,000 vehicles on fire every year, or 40 per day.[86]

Combustion engines catch fire at least five times more frequently than battery-driven vehicles,[87] which is not to say that a battery fire is not dangerous. It does indeed pose new challenges for firefighters. A burning battery cannot simply be extinguished with water because the combustion behavior is different in a gas-propelled vehicle. There is no flammable liquid that may spill or blow out spontaneously. In case of an accident, the emergency forces have to render the battery of a destroyed vehicle inoperable, too, and they have to separate specially marked cable connections for this purpose.

After the first fires involving Teslas, the manufacturer reviewed the safety measures. In one case, the floor plate of a vehicle had been punctured by a metal part that damaged the battery. Tesla's reaction was the installation of a reinforced floor plate. Tesla actually was so anxious about reports of battery fires that initially it was very generous to owners, not knowing whether the battery fire was a malfunction—until the fire in one client's brand-new Model S that had burned on the highway just after delivery turned out to have been started by a bullet fired into the battery pack, from the inside, from the backseat of the car. Naturally, fires involving EVs are more likely to make the news than fires involving gasoline engines because this kind of incident is still rare. But it would be a false conclusion to consider all EVs as prone to fire. As Elon Musk said with a wink, "If you want to commit arson, will you bring a jerrycan of gas with you or a battery?"

For a Clean Future: Footprints, Power Mix, Emissions

Arnold Schwarzenegger, action movie star and former governor of California, is also known for his passion for cars. He owns a number of elegant vehicles, a gas-guzzling Hummer, a luxurious Bentley, a Porsche, a Mercedes Excalibur, and a Tesla—anything an automobile lover's heart desires. And he keeps himself busy driving them all. As a cherry on the sundae, he also has a tank in his fleet, allegedly the one the "Terminator" drove during his military service in the Austrian federal army. He bought it from the army and had it sent to California to drive it for fun and use it for charity purposes.[88]

Even Schwarzenegger, who did not necessarily present himself as an environmentalist in the past, given his passion for cars, now takes up the cudgels for EVs, for example, in a Facebook post with the provocative title "I don't give a **** if we agree about climate change."[89] While the rest of the world pretty much agrees that we are experiencing global climate change, the situation is different in the United States, thus explaining the title of Arnold's post. He provides a convincing argument for EVs or vehicles with alternative drives, citing a logical example:

> I have a final question, and it will take some imagination.
>
> There are two doors. Behind Door Number One is a completely sealed room, with a regular, gasoline-fueled car. Behind Door Number Two is an identical, completely sealed room, with an electric car. Both engines are running full blast.
>
> I want you to pick a door to open and enter the room and shut the door behind you. You have to stay in the room you choose for one hour. You cannot turn off the engine. You do not get a gas mask.
>
> I'm guessing you chose the Door Number Two, with the electric car, right? Door number one is a fatal choice—who would ever want to breathe those fumes?
>
> This is the choice the world is making right now.
>
> . . . I just hope that you'll join me in opening Door Number Two, to a smarter, cleaner, healthier, more profitable energy future.

The Austrian Federal Environment Agency supports the eco-friendliness of EVs with hard facts in a survey.[90] In this report, which compared the greenhouse gas emissions with the energy demands of gasoline, diesel, hybrid, and electric cars, the experts took the environmental impact over the entire vehicle life cycle into consideration, from production and operation to disposal.

Compared with diesel and gasoline vehicles, the greenhouse gas emissions for the entire cycle of EVs are 75 to 90 percent lower. With regard to nitrogen oxide, diesel vehicles are the worst, with emissions up to nine times those of gasoline engines. EVs do not cause any nitrogen oxide emissions. Dust emissions are similar for all kinds of propulsion. About 50 percent are incurred in the production of the vehicles; the other 50 percent are generated during the accumulator and electricity production of EVs or the energy provision for combustion engines.

Independently of the kind of propulsion, the energy effort and use of materials during production are similar. Operation of a vehicle requires the most energy. The energy effort over the entire life cycle is three to four times lower with electric cars than with cars with combustion engines. Electric cars require about 50 to 70 percent less energy.

The American Union of Concerned Scientists (UCS) came to much the same conclusion.[91] Although higher energy consumption and therefore higher emissions may be expected during production (approximately 15 percent more for a vehicle with an 84-mile (135-km) range and/or 68 percent for one with a range of about 250 miles (400 km), the environmental impact is balanced again after just 6 to 18 months of operation. Sooner or later, this is amortized, depending on the kind of energy generation. At the end of the entire vehicle life cycle, a combustion engine will have produced double the amount of emissions of an electric car with an 84-mile (135-km) range.

The UCS compared the emission balance of EVs and combustion engines during operation on the basis of the energy mix for the different U.S. states.[92] The specialists pursued the question of how much fuel a combustion engine vehicle would need to use for 62 miles (100 km) in order to equal a similar type of EV in terms of emissions. In the sunny states in the west, a combustion engine vehicle would have to achieve an average consumption of 1 gallon for 97 miles. As an average value for the entire United States, the consumption should not exceed 1 gallon per 68 miles—a dream value no gasoline- or diesel-fueled engine today is even close to in practical usage. You can have a look at the environmental balance of an EV by brand and state using an online emission calculator.[93]

There is such an overview for Europe, too, at least for carbon dioxide emissions.[94] The type of power generation is included in the evaluation. The countries relying heavily on fossil fuels for power generation, for example, Poland, Greece, Bulgaria, Estonia, and Latvia, end up at the lower end of the scale. Those with renewable energy sources reduce their CO_2 footprint; among those countries are Norway, Sweden, Denmark, Iceland, and Austria. France and Switzerland, which put their cards on nuclear energy, also reduce their CO_2 footprint. Germany ranks average: the mix of energy Germany uses today would not lead to a significant reduction in the German CO_2 footprint.

However, it is not enough to measure the emission balance based only on gasoline consumption. Extraction, transport, and production of fuels use huge amounts of energy, and those processes are themselves significant emission sources.

Around 1900, drilling for oil was still relatively simple. The energy required to extract 100 barrels of oil required the energy effort of just one barrel. One barrel is 42 gallons. Today, the easily accessible oil deposits have already been exploited, and you have to drill deeper and deeper or else extract the crude oil from oil shale first, which increases the energy requirements. You cannot obtain 100 barrels of crude oil from one barrel nowadays, but only about 12 to 17 barrels. In the case of oil sands as in Canada and Venezuela, the ratio drops to 5 barrels.[95] It is getting increasingly difficult and expensive to extract "dinosaur juice."

One barrel of crude oil provides about 31 gallons of gasoline or diesel fuel (plus a couple of other petroleum products).[96] Accordingly, refineries with an efficiency of about 85 percent need about 6.4 kilowatt-hours for every gallon of gasoline they produce.[97] Look at this in comparison: 1 gallon of gasoline has an energy amount of 32.1 kilowatt-hours.[98] Twenty percent of the energy amount in 1 gallon of gasoline has been used to actually make it from crude oil.

During the extraction, transport, and storage of crude oil and/or the refined products, you use up energy and emit waste gas. Waste gas and crude oil escape from the conveyor plants, and there are millions of miles of pipes with leakages involving thousands of valves and connectors, with crude oil dripping out or otherwise escaping into the atmosphere. Gasoline and oil tanks are not completely sealed either, as anyone buying property near a gas station can prove with the comments and warnings in the respective documents.

Something that is easily forgotten in the usual calculations regarding the life cycle is the military cost incurred to secure the energy supply. Amory Lovins from the Rocky Mountain Institute believes that a huge part of the annual U.S. defense budget of $638 billion is used for securing the country's energy supplies. The government uses about $507 billion just for military operations in the Middle and Near East, about 10 times more than Americans pay for oil from that region. Even though a lower amount of the money spent on the military is directly attributable to securing the oil supply, a small percentage of that would still amount to a pretty large sum.[99]

The International Energy Agency (IEA) estimates that countries all over the world spend about $500 billion per year to keep oil, gasoline, coal, and gas cheap or to support the industry. According to Fatih Birol, IEA chairperson, the subsidies are triple the amount of subsidies for renewable energy.[100]

We may therefore assume that the hidden environmental cost for the operation of a combustion engine is much higher than is generally known, let alone the humanitarian cost in the fuel-producing countries, where in many cases we support anything but democratic regimes with our fuel money. The conditions in each country determine how fast a switch to EVs will help the environment. China, a country massively relying on coal to produce energy, will have few benefits from the switch to EVs in the short term.[101] EVs improve the ecological balance, especially if the power required comes from renewable sources. Germany also cannot expect any improvement in the emissions balance from the exit from nuclear energy planned for 2022. There will, however, be a decrease in the nuclear waste balance.

While combustion engines cannot change their source of energy—a gasoline engine can, after all, only use gasoline and a diesel only diesel—but EVs do have that option. Electric power does not have a label, so on a medium- to long-term basis, the sources of energy for the generation of the electric power can be changed. Owners of an EV might start out using the usual mix of power sources they get from the power outlet but could then go on to replace that with electricity produced by their own solar panels or make their power mix more environmentally friendly when their utility company adds emission-free energy sources. For the EV, the crucial factor is the mix of power sources, whereas there is not a kind of "gasoline mix" in this sense that a combustion engine could possibly use.

With the recent trend away from particularly polluting fossil energy sources such as coal and the promotion of wind turbines and solar plants, the extent to which the energy production impacts the total emission balance is steadily decreasing. This is why the share of emissions caused by traffic increases relatively. EVs therefore have a comparatively stronger positive impact on the reduction of the emissions balance.[102]

People who own an EV today differ from other car owners in their energy behavior. Sixty percent of them in California alone accepted their power

company's offer of using particularly cheap electricity provided during the
night to charge their cars, despite the fact that they have to pay a slightly
higher tariff during the day instead.[103] They accepted because 32 percent of the
owners of EVs produce solar power for their own domestic use.[104]

Companies: Whose Hand Is on the (Battery Power) Lever?

Incidentally, what about the battery industry in Germany? We have to go
back more than 100 years for that, to the time when the Tudor system accumulator factory Büsche & Müller oHG was founded in 1887. It was later
known as Varta and was a strong competitor of Siemens and AEG. For
many years, German battery technology was the cutting edge. The applications were mainly in the military sector (read submarines), as well as car batteries and flashlights. Today, this industry is just a shadow of its former self.
Manufacturers such as Varta lived through troubled times in the last decades.
Batteries for smartphones, portable computers, and other electronic devices,
as well as cost advantages, moved battery competencies to Asia. Today, the
leading manufacturers are companies such as LG, Panasonic, and Samsung.
A study from Fraunhofer Institut clearly shows that Germany still is not a
leading market for vehicle batteries and has some catching up to do.[105] One
should not acknowledge this only with a shrug, especially because the expertise
would be available. After all, expertise in batteries and battery management
will replace the expertise in combustion engines in the future. However, we
should not expect battery factories to be able to cushion the expected loss of
workplaces in aggregate production. Battery production is a highly automated
business, requiring just about a tenth of the number of employees we usually
find in engine production.

German manufacturers announced the construction of their own battery
factories rather half-heartedly. Mercedes-Benz did, as mentioned earlier, invest
$500 million in the construction of a second battery factory in Kamenz, where
the company has been producing lithium-ion batteries since 2012, but the
company simultaneously announced that it would invest the same amount
into the construction of a new factory for gasoline and diesel engines.[106] This
is the same as Carl Benz putting half his wife's fortune into horse breeding.

Propulsion Technology in Comparison: Motors, Design, and Efficiency

We are talking of battery-powered vehicles compared with combustion engines as if the changes would only apply where the energy for the propulsion is stored or generated, but actually, the entire world around that is changing, too. A vehicle with a combustion engine has a motor with many movable parts, granted, but it also contains all the components needed to store the energy source (fuel tank), supply it (fuel lines), transferring the energy produced onto the wheels (clutch and coupling, transmission, flywheel), and remove and clean the resulting waste gas (exhaust system).

An EV functions with a fundamentally different mechanism. You can do away with fuel lines, fuel pumps, tanks, filters, coupling, and flywheel. Whereas a motor usually requires an unwieldy block-like shape, a battery shape is more flexible. It may come as a block, too, or else be flat and be used as the underbody or be installed in the trunk or even on a trailer.[106] The transmission required to bring the engine power to the wheels in a combustion engine vehicle requires a lot of room. Would you, a nonexpert, have found it immediately in the middle console between the front seats? This is another part that an EV simply does not need. This changes the drive feeling because the torque can be transferred to the wheels directly—no gears—gently and measured. Anyone who ever did a test drive in an EV can confirm this.

The room gained in this way in a vehicle opens up new horizons for body and interior design. Designers can freely play around with the space where the engine, tank, exhaust pipes, and transmission used to be. We can go back to the "VW Beetle feeling" because it is possibly to have the trunk—or as it is called the "frunk"—under the "engine" hood again for storage. And all this liberated space offers many new opportunities, as Italian car designer Andrea Zagato described in an interview with the German *Manager Magazin*:

> EVs do not need a large engine anymore, and the transmission is eliminated. Since electric motors do not necessarily have to be in the front, radiator grill and air vents are obsolete, too. And we don't even need a classic dashboard, because everything is done electronically. All of that allows us to try out a completely different vehicle architecture, and we are free in our designs.[108]

Anybody who has driven in a Tesla will have noticed the huge amount of space between the seats, all of which can be used for passengers and baggage.

Mountains of metal and movable parts are simply gone. While for a combustion engine car you have to ensure that the motor has enough room around it so that it does not move too far into the passenger cabin in case of a crash, an EV designer can plan in a new deformable zone here. In this way, cars become safer for passengers.

Furthermore, the heavy battery installed into the floor lowers the car's center of gravity and stabilizes the chassis. It was therefore almost impossible to make a Tesla Model S or the taller Model X flip on its roof during safety tests in the United States. Such vehicles have much better results in the now legendary elk or moose test.

Electric motors are much smaller and need about two dozen parts compared with 1,200 parts for an eight-cylinder engine. Consequently, they weigh less, can be built in a more compact way, and can be installed directly on the axles. The energy is transferred in a much more direct way and allows for vehicle acceleration that surprises many and allows us to think anew about that fundamental question of what is so special about a sports car. At the same time, the life expectancy is higher, and the engine requires less maintenance, thus further reducing the cost.

Although an electric engine consists of relatively few parts, this does not mean that its technology is simple. There are several types that are driven by direct or alternating current or use strong magnets. Similarly, the performance of an electric motor is influenced by the materials used, the quality of the copper wire, and the way the lamination of the iron core for heat generation was produced. We need research experts for all those issues.

An electric engine can also function like a generator. Electromagnetic fields create a current during braking, and that electric current can be charged back to the batteries. This process is referred to as *recuperation*. When the driver lets go of the pedal, the battery stops producing energy, and the motor does not drive the wheels anymore. And this fact, that the braking effect comes from the electric engine itself, helps to preserve the brake pads. Even more parts experience less wear and tear.

Then there is the efficiency factor of an electric motor to consider. In contrast with combustion engines, where a large part of the converted energy is lost owing to heat and friction and only a tiny part is actually used as energy for movement, an electric motor works loss-free in comparison. The degree of efficiency is up to 97 percent depending on the electric motor.[109] Gasoline engines reach an efficiency of just under 30 percent; diesel and hybrids, up to 40 percent.[110] This cannot be compared directly because you have to take the power generator efficiency into account as well. If we are talking about solar, wind, and water power, we just have to deduct the losses caused during the power transport. If, by contrast, we are using electric current from coal-fired power stations with their efficiency of 35 percent, then the efficiency of the electric motor automatically is reduced in the calculation.[111]

The automotive supplier BorgWarner examined several propulsion systems and their degree of efficiency by eliminating the factors connected with the size of the vehicle, the vehicle weight, and some other elements in order to be able to compare the various types of propulsion. The findings regarding efficiency are similar to those presented earlier. In the survey, hybrids reach an efficiency of 38 percent, combustion engines reach a limit of 25 percent, and diesel engines reach a limit of 28 percent. Compare this with EVs, which are currently at 80 percent efficiency. The efficiency of combustion engines has improved by 14 percent in the last 10 years, but legislators demand more than 30 percent over the course of the next 10 years. This necessitates a gradual switchover to hybrids and EVs.

However, BorgWarner used its results and forecasts to suggest a phased switchover for drive technology to the automotive manufacturers as a means of keeping previous investments safe: a certain percentage of gasoline engines, diesel engines, hybrids, and EVs every year. While this recommendation is perfectly understandable, it is also somewhat dangerous, because it presupposes a linear development in the demand for EVs. Disruptive innovations are, however, exponential; they start out slowly and then seem to explode. BorgWarner's results and recommendations were shredded to pieces by the audience when they were presented in Silicon Valley. Many of those present had lived through similar disruptions in other industries in the past and were therefore very skeptical.

Some suppliers have already noted the trend—away from combustion engines and toward EVs—in their order books. The gear manufacturer ZF announced that it would reduce the gear production at its plants in Saarbrücken and Schweinfurt and instead focus on power trains for EVs. Management states that none of the 8,500 jobs were at risk, but reality will show whether it is really that easy to retrain gear production workers as battery chemists.[112]

Gripping Smart Tires for Heavyweight Cars

EVs require innovation on unexpected components: some Tesla Model S owners among my acquaintances mentioned higher wear and tear on tires not because they demonstrated the accelerating capacities of their car to friends all the time but simply because the rubber and the profile are subject to stronger usage owing to the direct translation of the propulsion energy and the high torque. Tire manufacturers thus have to live up to new challenges; their customers want the tires to be durable, make little noise, and have good road behavior. Unfortunately, you cannot have everything at once. Tires that hug the road are worn out quickly. Tires that are durable are loud. And tires with low tire-road sound emission have inferior grip.

Now, however, companies such as Tesla demand that a tire meet all the expectations: maximum performance and life expectancy together with low tire-road noise and good grip on the road. In order to adjust the limitations, tire manufacturers combine more than 200 variables, from the position of the breaker belts and the sidewalls to the tread with various profile designs and depths. The tires are specifically designed for the individual car manufacturer and vehicle model.[113]

Michelin, for example, the company that produces all the tires for the EV racing series and for Tesla, had to redesign the carcass, that is, the skeleton structure of the tire, as a result of the high weight and performance data of the Model S because this was the only way to ensure that the tire could withstand the extreme strains and still meet all the other conditions as well.

"So How Much Do You Want for It?"

One important criterion for purchasing a vehicle is the price you pay, obviously. The purchasing price, the operating cost, and the resale value are part of the buyer's considerations. At present, EVs are still a lot more expensive in terms of purchasing price because the price for batteries is (still) quite high compared with their performance. But that is going to change, too.

A survey conducted by Swedish researchers analyzed the cost of batteries (from 2007 to 2014) for EVs and noted an annual drop of 14 percent in price. While a kilowatt-hour cost $1,000 on average in 2007, the cost was only at $410 in 2014, and the price was as low as $310 with the market leaders. This trend is expected to continue.[114] A McKinsey survey identified the current battery price at between $270 and $230, so it decreased by 80 percent between 2010 and 2016.[115] With the start of battery cell production in Gigafactory 1, Tesla is expecting another reduction in cost per kilowatt-hour by 30 percent to approximately $125.[116]

As I mentioned earlier, a price of $150 per kilowatt-hour will mean a major breakthrough for EVs, because at that point they become cheaper than combustion engine vehicles. The cost reduction from 2020 to 2025 is expected to proceed in a linear manner.[117]

The cost incurred while charging the battery, by contrast, depends on several factors: the vehicle model, where and when you charge it, and whether the station is public or private. Tesla offers the electricity from its charging stations free for Models S and X and some Model 3 owners; other companies do the same and provide their staff with the possibility of charging on company parking lots. If you want to charge at a public station, you will find different fees depending on the energy provider, and those prices may vary considerably. Some shopping malls or individual stores install free charging stations as a special service to attract more customers. If you charge your car at home overnight, you may be able to benefit from cheaper night rates. The cost is further reduced if you have your own solar plant installed on the roof.

One trend is very clearly discernible in the power industry: electricity fees for end consumers went up over the last few years, while the cost for the generation, transport, and distribution rose only slightly or not at all (see Figure 5.1).

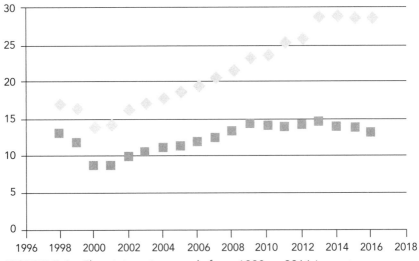

FIGURE 5.1 Electricity price trends from 1998 to 2016 in cents per kilowatt-hour.

At the same time, the cost for the production of solar power is decreasing. A survey by Deutsche Bank compared the price from solar power generation with the cost incurred by power plants for generation, transport, and distribution. From that point onward, power from solar energy will be cheaper than power from power plants. For 2021, the forecast is a solar power price of 5 cents per kilowatt-hour. This means that just the transport from the power plant to the consumer will cost more than that.[118] In connection with a household battery, it will also be technically possible and increasingly economically reasonable to become completely independent from energy providers.

So let us compare the preceding with fossil fuels. The U.S. Department of Energy offers an online calculator so that users can compare the cost for different vehicles.[119] The cost of the necessary kilowatt-hours for a distance of 100 miles is about a third to a fifth of the gasoline cost.

Another factor influencing the decision is the car's resale value. An EV's residual value depends largely on the battery: if it is "done for," the resale value is lower, just as a car with a combustion engine whose owner drove it to the max all the time won't be worth much anymore. Once the durability of batteries increases and the number of charging cycles rises with that, we may expect that EVs will reach a similar or even a better resale value than combustion engine cars. First experiences actually point in this direction. The resale value is furthermore determined by the car's processing quality and the development of the network of charging stations. The Deutsche Automobil Treuhand and the Eurotax Schwacke, both institutions that conduct and indicate possible residual value figures for vehicles, forecast higher prices for EVs with normal previous usage than for combustion engine cars with similar configurations.[120] And there is another factor that could impact resale value, namely, subsidies such as a government buyer's premium. Such premiums, however, are usually given for the purchase of a new vehicle, which pushes more cars onto the used car market, increasing the offer and putting downward pressure on prices. A similar mechanism was apparent in the case of the federal scrapping bonus in Germany a few years ago, which led to a lower general value of used cars.

Vehicles with an electric drivetrain change the price and the cost structure, even before the self-driving technology is marketable. Today, an electric car generally can be said to be worth your money if you use it a lot. Depending on the model and your calculations, an EV becomes more economical due to its low consumption after 30,000 to 60,000 miles. The so-called total cost of ownership (TCO) renders a combustion engine car generally less economical.

Auto repair shop operators estimate that the maintenance for EVs is about 70 percent lower as a result of the batteries and the parts that undergo almost no wear and tear, which is going to lower their business volume to about the same degree.[121] Tesla, for instance, doesn't even give you service intervals. The only things a Tesla owner has to do in the way of service is to check the tires and refill the wiper fluid, so to speak. Another service with a high margin, oil changes, becomes completely obsolete.

All these factors already place pressure on the new and used car prices of vehicles with combustion engines. The Volkswagen emissions scandal and the announced diesel engine traffic bans in many German cities are disconcerting information for customers. They recognize the risk involved in purchasing such vehicles. And because it is more likely for people owning a midmarket kind of car to sell it in order to have a basis for the financing of another new car, then for people owning a premium segment car, the resale value has a massive influence on the decision to purchase. Used gasoline and diesel vehicles might have to be sold at a huge loss later on.

If you now look at all the cost factors and subsidies for EVs and compare them with combustion engine cars, the former are indeed more economical in operation but, as of today, will cost more during the first five years because of the higher purchasing cost (caused by the battery).[122] Should Tesla, however, be able to increase production as planned and build more than 500,000 cars per year, this will have a positive effect on the production cost. Thanks to the scaling effect, the cost for automakers typically decreases by 20 percent for every doubling in production.[123] Accordingly, a Model S today would cost about $50,000, which is still a nice margin at an average sales price of $85,000. The Model 3 is sold at a base price of $35,000 and for about—on average—$42,500 with extras included. The production will have to reach 160,000 cars annually to maintain the base price. If the production reaches 275,000 cars per year, the cost decreases to $27,500. According to *Forbes*, this quantity would indicate the breakeven point if you disregard the cost for research and development, investments in Gigafactory 1, and other expenses.

At the same time, the manufacturers are still hesitant to offer their first EVs at a more affordable price or, for that matter, in a sufficient quantity. The price for the BMW i3 is so high that there was hardly any demand for it initially. The monthly leasing cost then decreased from $800 to $229. General Motors, in turn, limits the production of the Chevrolet Bolt to a total of 30,000 vehicles despite the fact that there are apparently more customers interested. The offer is kept low because some U.S. states require carmakers to provide so-called zero-emission vehicles (ZEV). For example, 14 percent of vehicles sold in California must conform to this standard. If they do not, GM has to buy ZEV credits (basically, penalty points) from other manufacturers, mainly from one competitor: Tesla. Basically, GM consciously accepts losses in the sale of the Bolt just to comply with those regulations.

Both BMW and GM are worried about cannibalizing their profitable models with EVs. However, those who do not keep a slice of the cake for themselves will be outdone. The real indication of the direction in which we are going with EVs has been obvious since the start of delivery of the Tesla Model 3, in production since September 7, 2017. And there won't be any people later who will be able to claim that they did not see it coming.

So let's dare to make a prediction: combustion engine vehicles will lose their economic advantages between 2018 and 2025, and EVs will become the dominant means of transportation. The battery capacities and charging station infrastructure will have advanced to a point that people are no longer worried about the range of their cars.

Cool Porsche, Ugly Electric Dwarf?

Why on earth are so many EVs so singularly ugly or have at least a very strange look? This is a question of interest not only to the customers. The design may have been a factor that scared off a number of potential buyers so far. Tesla was the first to show that an electric car can also look cool without losing any of its efficiency.

The reason why many early and even many current models still have a futuristic and occasionally odd design may be the desire to make the new technology visible to any observer. But that keeps customers away. Many people like change—but only in small doses. If you replace the propulsion technology with an electric motor and a battery, that is enough novelty, thank you. The car does not have to look like a *Star Trek* spacecraft to boot.

Electric "Fleet Maneuvers"

The discovery that some fleet operators were the first to be interested in the new propulsion technologies is not new. Even around the year 1900, public services, postal services, and fire departments, as well as taxi service providers, were the most important buyers of cars in general and especially EVs.[124]

Some customers do not want to wait anymore for the manufacturers to keep their promises, and they take electric mobility into their own hands. Both the German and Austrian postal services are the biggest operators of EV fleets in their respective countries.[125] This went down particularly badly with the German car giants because the EVs of the German postal service are their own development in collaboration with the EV converter StreetScooter GmbH, acquired in 2014.[126]

Neither Fish nor Fowl: Hybrids as an Interim Solution

If Carl Benz had reasoned in the same way as managers in the automotive industry today, we would have had the pleasure of the "Trotter-Hybrid"—a motorized carriage with an emergency horse attached. We would have needed both gasoline and hay, bridles and steering wheel and pedals.

And that is all you need to know when you think of a hybrid, regardless of its form. They are neither one nor the other, neither fish nor fowl. In addition to the motor, there is also an electric drive; in addition to the tank and the exhaust system, there are batteries; and in addition to the tank lid, there is a charging plug. The vehicle thus becomes heavier and technically more complex, which increases the need for maintenance and makes it more expensive. The additional weight runs contrary to the argument for efficiency. And the vehicles cost more initially and at the same time have less room inside because they are filled to the brim with double technology.

The people in the headquarters of the automotive industry thought such an all-purpose vehicle could be sold, but the customers seem to be smarter than those high-paid automotive managers. In 2016, a total of 48,000 such vehicles were registered in Germany. In view of a total number of 3.4 million new registrations, this accounts for only 1.4 percent—small wonder, because hybrids cost about $10,000 to $17,000 more than "pure" combustion motor vehicles.[127]

So where did the automotive managers get the idea of offering hybrid cars? There are two reasons for this. The first is connected with whitewashing, or glossing the fuel consumption over, as, for example, in the New European Driving Cycle (NEDC). Short driving cycles are simulated under laboratory conditions, and fuel consumption and emissions are measured. The trick is starting the test cycle with the (fully charged) battery and continuing with the motor as soon as the battery is empty. Most hybrid vehicles offered cannot exceed an electric range of 30 miles (50 km). This figure is based on a practical test BMW conducted with electric Minis in 2008. The company found that 90 percent of the daily trips people take are not longer than about 60 miles (100 km) in total, 30 miles (50 km) one way and the same back. This is the reason for the current dimensioning of the electric distance range of hybrid cars.

And as if by magic, this makes the average fuel consumption and emissions look very low. There are other "optimizations" to assist the figures: the heating and air conditioning are off during this test cycle, as are all other elements using electric current, such as wipers and headlights. Slight overpressure in the tires makes for a decreased roll resistance and a top speed of 80 miles per hour. These vehicles are nothing but *compliance cars*, built exclusively to appease the lawmakers.

The second reason is the amount of investment required. A full, abrupt switch to EVs would swallow up billions for battery technology and electric motors, whereas at the same time investments already made in the manufacturing of combustion engines would be lost, an approximately similar amount. Simultaneously, a third of the workforce would become redundant, even if the workers could be retrained in battery chemistry.

These are the reasons why hybrid cars appear so beautiful to the beholding eyes of German automotive managers. You can manage a soft transition from one technology to the other without risking any labor disputes or giving too much offense to the owners. And all that would work wonderfully if there wasn't that tiny little flaw: we are not just talking about Germany and German manufacturers or the traditional carmakers all by themselves. Tesla is a company that comes to fill the gap exactly in that premium segment, offering a vehicle that has sufficient range, closely followed by, on the lower end of the price range, Renault-Nissan, Lucid Motors, NIO, and Byton, as well as any number of Chinese manufacturers. They are all occupying a market that almost all traditional manufacturers continue to ignore because they do not believe in it.

Fuel Cell Development Under Fire

When we are discussing alternative means of propulsion, we cannot disregard fuel cells. The reason is not that this technology will have its breakthrough in the near future and will soon be available for customers, but rather that it has been among the favorite projects in the automotive industry for years, mainly because it offers a model that car manufacturers understand. Instead of an engine burning gasoline or diesel fuel in a controlled reaction, transforming the fuel into motion energy, a fuel cell burns hydrogen in a controlled reaction and transforms it into motion energy. Liquid hydrogen is saved in tanks and pumped into the car just like a liquid fossil fuel.

The drop of bitterness with this method is that even a fuel cell does not need as many components as a combustion engine, and you therefore need fewer workers. But fortunately, this vision of the future is still very far away, and nobody in the automotive industry has to worry today about what should happen with all those workers that nobody will need anymore.

Nevertheless, the principle behind the fuel cell is quite promising. When you burn hydrogen, you get water. Accordingly, the exhaust from the car is nothing but pure water vapor. Sadly, reality thwarts us with a series of practical problems. First, pure hydrogen is highly explosive, showing a "strong exothermal reaction," as a chemist would put it. The tragic fate of the *Hindenburg* blimp shows clearly just how explosive it is. At the time, dirigibles were filled with hydrogen. Because it is lighter than air, this let the dirigibles rise. Germany did not have any reserves of helium, which is a noble, or inert, gas and does not burn on contact with oxygen or react with other substances. Hence, the German Zeppelin *Hindenburg* had no choice but to use hydrogen. We all know what happened.

So we understand that hydrogen must be stored with appropriate safety precautions. It is first synthesized (i.e., generated by using energy) because it does not exist in a pure form anywhere on the planet. Refining of fossil energy carriers is the most frequently found way of manufacturing hydrogen; it is separated when hydrocarbons are split. And this does not make this process any more environmentally friendly. Another type of hydrogen synthesis is electrolysis, which allows you to obtain hydrogen directly from water, but requires way more energy.

The efficiency is no more than about 60 to 70 percent. Almost half the energy is lost during electrolysis, and because hydrogen must be kept cool, you need further energy for storage and transportation. Because the efficiency of the fuel cell itself is only 60 percent as well, and it takes some time before the hydrogen is finally translated into motion energy, the entire chain of processes takes up three times as much energy as the hydrogen pumped into the car will be able to provide.[128] In contrast to this, the degree of efficiency of EVs is more than 80 percent, and combustion engines may achieve 40 percent in an ideal scenario only resulting from loss of heat and friction.

Another shortcoming of fuel cells is the supply infrastructure. As for today's fossil fuels, an entire network of fuel stations and pipelines has to be provided for hydrogen, a network that has yet to be implemented, starting from zero. This is the aspect that Professor Hans-Peter Lenz from the TU Vienna also perceives as one of the greatest obstacles. The cost for a hydrogen fuel station was estimated at about $2 million in 2010.[129] If these stations existed, and hydrogen was not so expensive, vehicles could be refueled just as quickly as they can be with the liquid fuels we are using today. Traditional fuels have a price per mile of one-third of the price of hydrogen. While the fuel cell infrastructure is completely missing, we do, however, have power for EVs in every household today. The network exists and is extended step by step. The introduction of quick charging stations is less expensive because there are not as many environmental regulations to consider as for fossil fuel- or hydrogen-based energy carriers. In addition, electric mobility is gaining on charging speed.

All this leaves us in the paradoxical situation that the same automobile managers who are telling us that EVs are not suitable for daily use because there are no charging stations seem to go "blind" in both eyes when they talk about fuel cells and do not want to acknowledge the same problem there. A survey done by the consulting company KPMG showed just how widespread those ideas are among managers of car manufacturing companies. The survey found that 78 percent of the participants consider fuel cells to be the real breakthrough for electric mobility, whereas 62 percent believe that EVs with battery cells must necessarily fail because of problems of infrastructure. The geographic differences could not possibly be any greater. In Europe 70 percent of the managers in the automotive industry expect battery vehicles to fail, an opinion shared by only 34 percent of Chinese managers.[130]

More challenges lie in the way, however. Fuel cells emit energy steadily, but a vehicle requires more peak energy when accelerating. Additional batteries have to be installed in the vehicle to deal with energy peaks. This makes the car heavier, more complex, and more expensive. There are several reasons why automobile managers still hang on to fuel cells. These managers may actually believe in them. Another reason may be that the cells are a technically interesting and challenging technology with great promise. Or else it may be that the automotive managers have already put in so much money and effort, that their own reputations are at stake now because ceasing work on this project would be interpreted as failure.

However tempting and promising the fuel cell model may be, there are too many open questions. For example, it seems unlikely that our Max or our Sofie will already be riding around in vehicles with such a propulsion system by 2030. Small wonder that Mercedes now wants to abandon the project—just a few weeks after the company made a grand statement confirming the work that went into the development process. Meanwhile, EVs have surpassed fuel cells both technically and economically.[131]

How brittle this technology is was demonstrated in 2019, when an explosion in a chemical plant in the San Francisco Bay Area created a hydrogen fuel shortage, and an explosion in a hydrogen fuel station in Norway had all fuel stations shutting down and Toyota even temporarily halting sales of its fuel cell vehicles.

Too Expensive, Too Unreliable, Too Quiet: Some Reservations Should Remain

Educated and less educated guesses and opinions about EVs distort the image of the technology's usability and potential. Some issues are whitewashed; others are presented as worse than they are. Some arguments are regarded as being disproportionately important because the experience with and technical implementation of combustion engines are translated exactly to EVs.

This dilemma is to no small extent due to consciously false information provided by the automotive manufacturers themselves, which so far have been toying with this topic somewhat listlessly. The authorities were rather timid themselves, despite all the lip service about how important the topic is. With Tesla's success, it gets increasingly difficult to plausibly make EVs sound like a bad and undesirable idea.

The many years of hesitant progress and disinformation certainly are in part to blame for the fact that German customers have not yet warmed up to electric mobility. The car manufacturers talked ill about electric mobility for many years, and now they complain that car buyers are skeptical about this development.[132] A survey among 7,000 consumers conducted by McKinsey in the United States, Norway, China, and Germany showed that there are reservations regarding the price, the availability of models, and the range of EVs.[133]

Let's take a look at some of the arguments presented again and again and analyze them in detail.

1. "The Range Is Insufficient!"

Ninety-nine percent of all daily trips by car add up to less than 75 miles (120 km). The 43 million cars on German streets move 381 billion miles (611 billion km) every year, which breaks town to 8,875 miles (14,200 km) per vehicle per year, or 24 miles (39 km) per day.[134] Cars are actually rolling for only 38 minutes per day, at an average speed of 38 miles (60 km) per hour. The situation in the United States is not much different. The 260 million vehicles registered are driven a total of 3,200 billion miles (5,125 billion km) per year or a distance of 12,300 miles (19,700 kilometers) per year, or 34 miles (54 km) per day.[135] The average speed is the same as in Germany, and every car spends 54 minutes per day in motion. In other words, American cars are just standing around for 23 hours and 6 minutes every day.

One could say that it is a "stand-o-mobile" rather than an automobile. Even the tiniest EVs today are able to cover that distance without being charged in between. Basically, the range anxiety is unfounded.

2. "The Network of Charging Stations Is Insufficient!"

This is true! The network is neither nationwide, nor are there as many charging stations as gas stations, but you only need them if you are planning to go on a long trip. The installation of charging stations on transit routes and in public places has begun, perhaps not quite as fast as one would wish for, but the overall situation is improving. Bertha Benz was not worried about the lack of gas stations on the way when she set out on her first journey.

The discussion about the availability of charging stations is slightly over the top. This is understandable because German manufacturers especially offer electric cars whose range has been calculated very tightly. One aspect that has not yet been considered appropriately is that EVs are no longer mainly charged at public stations but at home or at work. At the workplace, we need charging stations that are more than a simple power outlet. This allows the car to do its job while you do yours.

Technical progress allows us to expect an annual range increase of 5 to 15 percent. The Tesla Model 3 puts a medium-sized car with a range of more than 300 miles on the market, and other manufacturers aim at still longer ranges.

3. "Electric Cars Are Too Expensive!"

This is also true. But it won't be true for long. Today, electric cars are still more expensive at purchase but cost less in operation. Translated to their average life span, EVs will therefore become cheaper from a certain point onward. Some models have already reached that tipping point, where the purchase of a comparable combustion engine vehicle comes at the same price or may even be more expensive. By 2022 at the latest, midsized EVs will be cheaper than combustion engine vehicles, and it will be economic nonsense to buy a car with a gasoline or diesel engine.

4. "EVs Are Not Environmentally Friendly Because of Their CO_2 Emissions!"

If the electric power used comes from plants with fossil fuels, this is correct. However, the more we switch over to environmentally friendly sources of energy to generate power, the lower are the CO_2 emissions and the less harmful are EVs to the environment. One element that can be eliminated immediately is the share of the fine dust pollution and other pollutants resulting from exhaust gases. And yes, fossil fuel power plants can deal with energy and emissions more efficiently as a result of the use of exhaust gas purifying systems and filters and because exhaust heat is monitored and maintained by experts, which means that utilities are far superior in their results to thousands of smaller combustion engines run by nonexperts.

5. "The Batteries Are Bad for the Environment!"

Battery production most certainly has an impact on the environment. So does the production of engines and fuel. During operation, more substances that are dangerous to the environment come from both gasoline- and diesel-powered engines, such as oil and oil filters that must be exchanged or that may leak and pollute the environment. And let's not forget that batteries have a much longer life cycle and that the materials can be recycled. Not so much gasoline, which is just burned and dumped into the air.

6. "Where Is All That Electric Power Supposed to Come From?"

Two scientists from Berlin did some calculations, determining the mileage of cars today and contrasting that with the energy input for EVs.[136] The annual 381 billion miles (611 billion km) of all the 43 million cars on German roads correspond to almost 115 billion kilowatt-hours if you use 29 kilowatt-hours for 100 miles of travel on average. In 2016, Germany produced 648 billion kilowatt-hours of power, which is to say that there is a need for 17.6 percent of today's electric power production for 43 million electric cars with the same annual mileage.[137]

We have this much electric power available even today, and not all vehicles will draw from the power grid at the same time. On the contrary, EVs help to take up electric peaks because the batteries can be used as interim storage and can be charged mainly at night. Using intelligent controls such as a SmartGrid, charging can occur at times during which it is already difficult to use the power that coal or fossil fuel power plants, solar power plants, and wind turbines produce. In this way, electric cars become energy buffers for the power producers.

And let's not forget in this context that the refining of gasoline also requires electric power. Refineries with an 85 percent efficiency need up to 6.4 kilowatt-hours for every gallon of gasoline they produce.[138] Compare these data: 1 gallon of gasoline represents energy in the amount of 32.2 kilowatt-hours.[139] This means 19.2 kilowatt-hours for 3 gallons of gasoline, which, in turn, is the mileage of an EV, almost 60 miles. This is the amount of energy we save because we do not have to refine the fuel anymore.

Germany furthermore exports about 50 billion kilowatt-hours of electric current to other countries nowadays. Considering these figures, Germans would have to increase power production by 16 percent if they suddenly replaced all vehicles with electric cars. However, that won't happen overnight; the process will occur in several stages.

So what conclusions can we draw from this? Not only do we already produce sufficient power for EVs, but we could actually use them to solve problems that we have with our power supply.

7. "Electric Cars Are Too Silent!"

It is hard to believe this could be considered a disadvantage, but it is. EVs are definitely quieter than combustion engine cars. The only things you hear are the rolling noise and the drag. This silence is often used as an argument, ostensibly because people with visual impairment are exposed to a higher degree of danger. As a matter of fact, such people often have much better hearing than non–visually impaired people. In an initial phase, it is more likely that people with eyesight who rely on their hearing and do not observe the traffic would be in danger.

Believe me: we will get used to those almost silent cars. Just like most of us do not feel any nostalgia for loudly clattering typewriters or hissing steam locomotives, we all shall not shed a tear when the roar of combustion engines and the rattling of diesel engines are gone. We already know what really constitutes a danger to our health: noise. Street noise increases the stress level and is responsible for the premature death of residents nearby. The risk of dying 7 years earlier is increased by 10 percent for over-75-year-olds exposed to noise.[140]

8. "Fuel Sales Account for a Lot of Tax Income."

Given the ingeniousness of legislators, they will always find ways to tax citizens. "What good is this 'electricity'?" asked William Gladstone, British minister of finance in the nineteenth century. "One day sir, you may tax it" was the answer Michael Faraday gave.

Obviously, the revenue from fuel taxes and oil taxes will decrease as more EVs are found on our roads. In Germany, the state earned more than $44 billion with its energy tax in 2015.[141] In Austria, the oil tax amounted to just over $4.6 billion in 2014, and in Switzerland, the amount was a total of $4.86 billion in 2015.[142] And in the United States, the federal fuel tax resulted in revenues of $36.4 billion in 2016.[143] A large part of the tax income is earmarked for road maintenance and other traffic infrastructure projects.

One way of handling this loss of income might be a different kind of tax calculation for energy. The EU Commission presented a draft proposing a tax based on energy content instead of energy volume.

Brief Conclusion

In 10 years, taxis will be fully electric, although the time frame for buses will be slightly longer, at least in Europe. This is so despite the fact that there are offers from companies such as BYD from China, Proterra from the United States, and the German company Sileo even today. However, the European traffic agencies are still hesitant. In contrast, London has already ordered BYD buses, and the 16,400 electric buses in Shenzhen, as well as the shuttles used at Stanford University, are made by the same manufacturer. BYD produces both in Hungary and—from 2018—in Hauts-de-France in France.[144]

Germany continues to straggle behind with regard to electric mobility, even if some German people try to whitewash the figures on the hybrids. China is at the forefront with almost half of all electric cars on the streets, with the United States a close second. A quarter of all battery cells and 37 percent of all electric motors are produced in China today.[145]

CHAPTER 6

The Future Comes Rolling In

Autonomous and Self-Driving Vehicles

F*** self-driving cars. Self-cleaning apartments, that's what we need!

—Hearsay

I have to make a confession: one of my secret pleasures is watching videos of car crashes on YouTube. I can spend hours watching those blurred, fuzzy videos of accidents, always hoping that nobody was really hurt. At the same time, I am surprised at how an apparently normal traffic situation can lead to an apocalyptic scenario in just a few seconds. Most accidents are very obviously caused by driving errors—someone ignoring the right of way, turning into oncoming traffic, driving too fast, passing on the wrong side of the road, or attempting to still make it while the traffic light is turning red. Depending on the season, add ice and snow to this equation and the trend among young people such as the "drifters" in Arab countries to enjoy making their cars skid on purpose, and it seems as if we are asking for accidents to occur.

This impression, gained from the videos, is not wrong at all. Every year, 1.2 million people die in 10 million traffic accidents worldwide, and 50 million are injured.[1] According to the National Safety Council (NSC), more than 40,000 fatalities occurred in the United States. As a result of traffic accidents in 2016 and 2017, which means 2,000 more than in 2015 and 5,000 more than in 2014, more than 1.3 million persons were injured.[2] Traffic accidents are the main cause of death for Americans up to the age of 39. They are still among the top five causes of death for the older generations, after cancer, heart attacks, unintentional overdosing or poisoning, and suicide.[3] According to the statistics, 1 in 112 Americans dies in a traffic accident. And every person living in the United States—this includes men, women, and children—pays $784 including taxes and insurance fees for the medical treatments necessary after road incidents.[4]

The economic and social costs of car crashes are estimated to amount to $1 trillion (yes, indeed, $1,000 billion).[5] In addition, 94 percent of all the crashes are due to human error.[6] And the drivers are rarely called to account. As the *Wall Street Journal* discovered by research in 2014, 95 percent of all fatal incidents did not have any criminal consequences.[7] An analysis for the state of Oregon came to similar conclusions.[8] Car accidents are apparently seen as inevitable strokes of fate, and the drivers manifestly are not at fault as long as they did not act grossly negligent (i.e., driving under the influence or talking on their cell phones).

In 2014, 1,854 persons were killed in the United Kingdom, and 185,540 were injured. In Germany, 3,377 people lost their lives in traffic accidents, and 374,142 were injured, about twice as many as in Britain[9] (see Table 6.1). Just to give you an idea of how many fatalities this is per year, it is as if a full passenger aircraft crashed over Germany twice a month or one every day in the United States. The probability of being fatally struck by lightning, assuming an average life span of 80 years, is 1:13,000.[10] The probability of dying in a car accident is 1:112. The data show that 60 percent of the victims die on country roads, 29 percent within city limits, and 11 percent on highways; 70 percent of the injuries are sustained in urban areas, 24 percent on country roads, and 6 percent on highways.

A 12-month study in 2006 looked into the distribution of accidents and near misses. The researchers counted 69 accidents, 761 near misses, and 8,295 "incidents."[11] For each accident, there were 90 incidents and 8 near misses. Half of all accidents occured within a radius of just a few miles from home. Knowing the area makes people inattentive—because you know your route like the back of your hand—and so this is where you are most in danger if something unexpected happens.

The practical driving experience of drivers contributes to the probability of accidents. Novice drivers differ from experienced drivers by their different patterns of observation. They tend to mainly check the nearby area at the front

TABLE 6.1 Traffic Fatalities by Country and Year

COUNTRY	2015	2016	2017
Austria	479	432	413
Belgium	732	637	620
Bulgaria	708	708	682
Canada	1,860	1,898	1,841
Croatia	348	307	331
Cyprus	57	46	53
Czech Republic	737	611	577
Denmark	178	211	183
Estonia	67	71	48
Finland	270	250	223
France	3,461	3,477	3,448
Germany	3,459	3,206	3,177
Greece	793	824	739
Hungary	607	597	624
Ireland	162	186	157
Israel	322	335	321
Italy	3,428	3,283	3,340
Latvia	188	158	136
Lithuania	242	192	192
Luxembourg	36	32	25
Malta	11	22	19
Netherlands	620	629	613
Norway	117	135	107
Poland	2,938	3,026	2,831
Portugal	593	563	624
Romania	1,893	1,913	1,951
Serbia	599	607	579
Slovenia	120	130	104
Slovakia	310	275	276
Spain	1,689	1,810	1,827
Sweden	259	270	253
Switzerland	253	216	230
United Kingdom	1,804	1,860	1,783
United States	35,485	37,806	37,133

Sources: https://etsc.eu/12th-annual-road-safety-performance-index-pin-report/ and
https://www.statista.com/statistics/191521/traffic-related-fatalities-in-the-united-states-since-1975/

of their cars and the curb. They disregard the exterior mirrors, even during maneuvers where mirrors are absolutely essential (e.g., when changing lanes).[12] Experienced drivers, by contrast, know what they have to take into consideration. We know this from our professional fields: experts have a more comprehensive view of matters than newcomers.[13] One surprising figure is related to the drivers of company cars. For them, the probability of being involved in accidents is increased by 49 percent, even if you consider the higher number of miles driven as well as other factors.[14]

A study conducted with 1,700 teenage novice drivers—with cameras mounted in the car interior—was devastating. In 89 percent of all the accidents in which those young drivers ran off the road and in 76 percent in which they crashed into other vehicles, they were not paying attention. Why? Surely you have an idea: they were looking at their smartphone, talking to other passengers, or looking anywhere but in the direction of travel.[15] In 8 percent of the incidents, the drivers were singing along with music, and in 6 percent, they were applying makeup.

A striking example is the case of Courtney Sanford. Driving to work, the 32-year-old American veered into the opposite lane and crashed into a truck. She died at the site. Friends hearing about her accident became suspicious when they compared the time of the accident with her last Facebook post. The selfie showed her at the wheel with the text, "The happy song makes me so HAPPY." Just seconds before her fatal accident, she had uploaded several Facebook posts and caused the crash herself.[16]

The frequency of accidents is also influenced by activities around driving. "Well yes" and "of course" are the typical reactions. It seems reasonable to say that a lot of distraction negatively impacts our driving skills. Loud music on the radio, animated discussions with passengers, and dense city traffic with a wide range of activities seem to just be "accidents waiting to happen." However, boring stretches of road where nothing happens are also surprisingly high on the list of accident frequency. This is confirmed by the Yerkes-Dodson law, which says that cognitive performance depends on the level of stimulation and/or activation.[17] The curve between low and high stimulation levels has the shape of a U. If you are not stimulated or are overstimulated, performance will be lower than expected. The ideal performance level is between the two extremes. Traffic that challenges you too little or too much has an equally negative impact on traffic safety.

After tiredness, the second leading cause for accidents is voyeurism.[18] Surely we all have had our experiences with that. An accident happened in the opposite lane. There are emergency vehicles, desperate or injured people are walking on the road, debris is lying around, and vultures are circling the body parts. And invariably, traffic slows down in the direction you are going. Many people look across curiously. And this is when rear-end crashes occur. An accident in one direction becomes a double crash with traffic jams in both directions.

Some people are convinced that the color of a car influences its accident frequency. Even 30 years ago, a friend's mother who had just been hit by another driver mentioned that she could not help but feel that her new silver-colored car was frequently overlooked. She had had a lot of close calls since she exchanged her former bright red car for the silver Renault. Her feeling was spot on. An evaluation of 16,700 taxis in Singapore whose color is either blue or yellow showed 6 more accidents per 1,000 taxis per month at the same mileage for the blue compared with the yellow taxis. Over the course of the study, a period of three years, this added up to an increase of 9 percent in the number of accidents for the blue cabs. Yellow taxis are more conspicuous and easier to see by other road users.[19]

An alarming trend in recent years led to an increased fatality rate for accidents in the United States: the higher car weight. This is not due to the increasing weight of drivers or passengers, but because people have developed a liking for heavier and bigger vehicles. The paradox is that bigger and heavier cars are usually bought because people want a higher level of safety. This is actually true, but it only applies to the passengers in that particular car. Their chance of incurring major injuries in case of an accident—regardless of whose fault it is—decreases by 29 percent. Unfortunately, this is offset by a 42 percent increase in the injury rate for the passengers of small cars.[20] If we also consider other accident victims, each fatality in traffic with an SUV or pickup truck equals 4.4 dead passengers on the part of the smaller road user, pedestrians, cyclists, and motorbike riders. For an increase in weight of 1,000 pounds, the risk of death on the part of the other road user increases by 47 percent.[21] What this boils down to is that those allegedly safe heavy vehicles are extremely deadly for the rest of the population. If we were to ban them, the effect on safety would be equal to that obtained when safety belts were first introduced, according to the calculations of the authors of that study.

If you drive along San Antonio Road in Mountain View, California, crossing the bridge across the local railway track, you have a good chance to see one of the Google-Waymo self-driving vehicles out by itself. This is where Google has its research center on self-driving vehicles, an office building with a double-level garage housing more than 60 cars. In the past, you could see the Lexus SUVs, all with a selection of different sensors mounted on all sides and the roof and then a series of smaller cars the media dubbed "koala cars," made in cooperation with a company called Roush. Those two-seaters meanwhile have been retired, but they looked cute, and they did without steering wheels, pedals, and other controls we expect to see. Today the fleet consists of several hundred Fiat Chrysler Pacifica minivans and soon should reach several tens of thousands of minivans plus Jaguar i-Pace electric vehicles and even semi trucks. Consequently, Google is now testing the fourth-, fifth-, and sixth-generation of its self-driving vehicles.

When the *New York Times* first ran a story on Google's self-driving vehicles in 2010, the news hit the public like a bomb.[22] Nobody had expected an internet company such as Google to be working on something like that, and nobody had a clue that the technology was already so advanced. At that point, Google vehicles had already covered more than 100,000 miles.

The initiator behind the scenes of the project was the Defense Advanced Research Projects Agency (DARPA), which has conducted research projects for the American military for decades and occasionally initiates challenges. One of them, incidentally, brought us the internet. In 2004 and 2005, there happened to be two competitions on autonomous driving.[23] In 2004, a prize of $1 million was offered for a vehicle that could cover 150 miles of desert terrain (about 240 km) without human interaction. The first competition remained without a winner, and only during the second competition in 2005, offering a prize of $2 million, did a team from Stanford University with project leader Sebastian Thrun manage to complete the task. The team members took a Volkswagen Passat model that they named Stanley, equipped it with sensors, programmed it, and set it off on its journey.

For some of the participants in the DARPA Grand Challenge, autonomous driving is a very personal concern. In a TED conference, Sebastian Thrun talked about one of his friends who died in a car crash as a teenager.[24] And Anthony Levandowski's unborn child was almost killed when another car crashed into the one his pregnant fiancée was driving.[25]

David Stavens was a student then and a member of Sebastian Thrun's team, and he told me about the difficulties they encountered. Thrun had just become a professor of artificial intelligence at the Stanford AI Lab when he heard about the first competition, which nobody had won. He manually drove along the route through the desert to apply for the second challenge. And that route was extremely hard, even for a human driver. For the next 19 months, Thrun worked on the vehicle and the algorithms with a team of 5 to 10 students, including Stavens—with the result we all know and the realization that focusing on the software was the most promising approach. The other teams continued to regard the task mainly as one of hardware problems. Thrun's victory and the fact that in the second competition there were already five vehicles that managed to cross the much longer distance and made it to the finish line are considered the "Kitty Hawk moment" (the place where the Wright brothers took their first motorized flights) for autonomous driving.

In 2007, this was followed by the DARPA Urban Challenge, for which cars did not drive just in the desert but also through the built-up area of a former military barracks on the deserted George Air Force Base in Victorville, California. The vehicle from the Cornell University team, however, was involved in the first-ever collision between two autonomous vehicles. The MIT and Cornell cars touched lightly. Despite this incident, the Cornell team, which had only very limited financial means, finished fifth and earned some bragging rights.[26]

After the Urban Challenge, Google quickly hired Sebastian Thrun and let him pick the best people from the competitors' teams. Anthony Levandowski's company, 510 Systems, a company that developed sensors and light detection and ranging (LiDAR) systems, was simply integrated.[27] For some years, Google secretly researched and developed self-driving vehicles until the *New York Times* made the project public. The search engine giant's interest in the automotive sector not only inspired several other startups but also served to suddenly force all the established automotive manufacturers to intensify their efforts or, in some cases, take their first steps in this direction. Today, the leading protagonists in this struggle for the realization of the dream of autonomous driving are mostly not the carmakers, but many well-financed digital companies and startups. Currently, several thousand companies are involved in the development of components and solutions for autonomous cars.

The vision of self-driving vehicles is not new. Both the German Bundeswehr and Carnegie Mellon as well as Daimler conducted first experiments as early as the 1980s. A team from the Universität der Bundeswehr (the German University of the Armed Forces) started its first experiments in 1985 and is considered a pioneer in this field. Under the direction of Professor Ernst Dickmanns, a Mercedes passenger coach was remodeled as a "test vehicle for autonomous mobility and computer vision" (German abbreviation VaMoRs) in 1987 and set to drive along a blocked stretch of highway.[28] Carnegie Mellon conducted tests with the Terragator in 1983 and with the NavLab

1 in 1986.[29] In 1994, Mercedes started its Prometheus project, packing two Mercedes SEL 500s full of cameras and computers. One of the vehicles drove over 600 miles (1,000 km) in public traffic on the highway to Paris, followed by a trip of over 1,000 miles (1,700 km) from Munich to Stockholm, reaching top speeds of up to 110 miles (175 km) per hour.[30]

Despite the success of these experiments, research was subsequently restricted. The automotive manufacturers were reluctant to put this nascent technology into the hands of their customers, and the technology proved to be too expensive, too difficult to handle, and not tested sufficiently. Autonomous vehicles for the masses only got another chance when progress in computer technology, software, and sensors and the related reduction in cost made them seem an obtainable goal again.

So what counts as an autonomous, or self-driving, vehicle? The term *automobile* refers to a vehicle that moves as if by itself, without being pulled or pushed by an animal or humans. It is only *controlled* by humans. In a self-driving, or autonomous, car, this task is executed by a computer, control elements become obsolete, and human beings are relegated to the role of passengers. On the way to self-driving vehicles, the Society of Automotive Engineers (SAE) International differentiates between six levels of automation:[31]

- **Level 0: No Driving Automation.** The human driver completes every driving activity, even in the case of warning signals.
- **Level 1: Driver Assistance.** The vehicle may assist steering or speed adjustments under certain conditions, but the driver still maintains full control.
- **Level 2: Partial Driving Automation.** The vehicle may take over steering and speed adjustments under certain conditions, while the human driver remains fully responsible for overall control. Short formula: hands off!
- **Level 3: Conditional Driving Automation.** The car steers, adjusts the speed, and observes the road. The human drivers have to take over when the system needs assistance. Short formula: eyes off!
- **Level 4: High Driving Automation.** The vehicle manages to make practically every decision, even if the human driver does not respond to a request for assistance from the system. Short formula: mind off!
- **Level 5: Full Driving Automation.** The vehicle takes over and replaces the human driver without any expectation that a passenger will respond to a request to intervene.

The highest level assumes that the vehicle would not expect the passenger to take over control even in the case of technical defects, such as a blown tire, an important system failure, or similar circumstances. The vehicle must be able to come to a safe stop by itself.

Brad Templeton, who was working on self-driving cars with Google, criticizes SAE's level system and suggests a different kind of classification for autonomous vehicles.[32] His examples are autonomous shuttle buses, because they are used on company premises or university campuses. Such vehicles can do without a steering wheel or a driver, but they are limited to certain streets or segments of roads. This means that despite the fact that this vehicle therefore technically belongs to level 4 (high degree of automation), it is still a far cry from actually being one because the traffic scenarios on the routes traveled are too protected. Templeton therefore suggests a different subdivision, namely, according to the area in which the vehicle moves and the required/permitted speed at which it travels. It would be possible to create subcategories and classifications of streets and locations in which certain types of self-driving vehicles may travel. This categorization might include the following criteria:

- Only allowed to move on designated streets and intersections
- Driving on streets with a speed of about 19 miles (30 km) per hour
- Driving allowed on highways only
- Driving allowed only at night and/or times of little traffic
- No traveling in the vicinity of schools between 8 and 9 a.m. or when school is out
- Driving allowed only in areas with sufficient telecommunications and network coverage

Whatever the classification will look like in the end, some of us cannot wait to have the new technology implemented. The shuttles used by Facebook, Google, and other companies show that members of the younger generation really are not at all interested in sitting behind the wheel themselves. Every day, hundreds of those usually white, unmarked double-decker buses travel along Highway 101 and take employees to work and back home. Thanks to the WiFi connections on board, they can work on the bus or dedicate themselves to various types of entertainment.

Currently, there is no provider that even remotely manages to come close to Google's efforts. In 2016, Google cars drove more than half a million miles on public roads—mainly in urban traffic—and another 1 billion miles in a simulator. With this, Google was responsible for 97 percent of all test miles driven in California.[33] In 2018, the distance traveled had reached more than 10 million miles, most of that in city traffic. At present, "real" 25,000 miles per week are added to the simulated almost 3 million miles per day covered by self-driving vehicles.

In July 2016, this amounted to 88,000 miles (slightly more than 140,000 km), all of them in busy city traffic including all kinds of complications. The reason for these tests is less the pure number of miles driven but rather the experience and analysis of a large number of different driving situations. From pedestrians crossing a street, to trucks backing out from a dead end, to construction sites with unclear signposting and improvised traffic signs, the vehicles have to "test-drive" everything themselves. Once, even an elderly woman was spotted by the sensors of a Google vehicle. The woman was riding around in the middle of the street in her electric wheelchair. Why she did that? She was shooing away a duck with her broom, as Chris Urmson, former manager of the Google Self-Driving Project, recounted with a chuckle in a TED presentation.[34]

As of August 2019, 63 companies received permission to use test vehicles on public streets from the California Department of Motor Vehicles (DMV).[35] In addition to well-known manufacturers, there are also several that are not generally known:

- Volkswagen Group of America
- Mercedes Benz
- Waymo
- Delphi Automotive
- Tesla Motors
- Bosch
- Nissan
- GM Cruise
- BMW
- Honda
- Ford
- Zoox
- Faraday & Future
- Baidu USA
- Valeo North America
- NIO USA
- Telenav
- NVIDIA
- AutoX Technologies
- Subaru
- Udacity
- Navya
- Renovo.auto
- Plus.ai
- Nuro
- CarOne
- Apple
- Pony.AI

- TuSimple
- Jingchi
- SAIC Innovation Center
- Almotive
- Aurora Innovation
- Nullmax
- Samsung Electronics
- Continental
- Voyage
- CYNGN
- Roadstar.Ai
- Changan Automobile
- Lyft
- Phantom AI
- Qualcomm Technologies
- SF Motors
- Toyota Research Institute
- Apex.AI

- Intel
- Ambarella
- Gatik AI
- DiDi Research America
- TORC Robotics
- Boxbot
- EasyMile
- Mando America
- Xmotors.ai
- Imagry
- Ridecell
- AAA NCNU
- ThorDrive
- Helm.AI
- Argo AI
- Qcraft.ai
- Atlas Robotics

These companies have more than 600 test vehicles in California and over 1,400 in total across the United States on the streets, not steered by more than a thousand registered nondrivers, also known as safety drivers. Nondriving is actually more tiring than you might think. Ford noticed that some people doze off while observing the vehicle.[36] Both Google and Ford have installed bells, buzzers, vibrating seats, and warning lights to keep nondrivers focused. A second engineer on the passenger seat is also there to ensure that the non-drivers do not fall asleep. This phenomenon is the reason why manufacturers such as Ford aim to simply skip level 3 (conditional automation), at which the car "expects" the human present to be able to take back control at any point—after a warning. Tests showed, however, that humans need up to 20 seconds or even more as a warning period, and even that may not be enough to ensure that they are fully aware of the situation, which means that they may be in additional danger.

In addition to the companies just listed that hold California test licenses, another dozen or so manufacturers run autonomous vehicles in California. Because they are using testing grounds, private roads, and sometimes federal lands (which are not subject to California legislation), they also have per-mission to test autonomous vehicles.[37]

In April 2018, the California DMV issued a license that allows vehicles even without a driver to be on public roads.[38] Shortly thereafter, Waymo was the first company to receive such a license, which allows the company to operate more than three dozen vehicles without a driver in an area including Mountain View, Palo Alto, Sunnyvale, and Los Altos, even in light rain conditions.

The tests, however, are not just limited to the ground. For example, an autonomous flying Black Hawk helicopter can be found on the NASA Ames Moffett Airfield near Mountain View. The U.S. Air Force is teaching it flight patterns for combat action.

In Nevada, the state that issued the first official test license to Google in 2012, we can find five big corporations involved in such projects. The number of registered autonomous vehicles there currently is 30.[39]

- Google
- Continental Automotive Systems
- Volkswagen Group of America
- Delphi Labs
- Daimler/Freightliner

It helped that Nevada implemented a law as early as 2011 to the effect that testing autonomous vehicles was permitted under certain conditions.[40] One condition is that a liability insurance policy for $5 million must be held, minimum safety standards must be observed, and the vehicles may only be used in special zones designated for testing. One of the safety measures is the presence of at least two people in the car who have completed a training program in operating the vehicle. The vehicle must have a switch-off function that allows exiting the autonomous driving mode and a system that can cede control of the vehicle to the driver. Furthermore, the vehicle must be able to record the sensor data of the last 30 seconds and before they get a license for driving on public roads, the have to prove that they drove 10,000 miles in autonomous mode with a vehicle on a test track.

Other startups include AIMotive (formerly called AdasWorks) in Hungary, Nauto in Palo Alto, and nuTonomy in Boston.[41] In Singapore, two manufacturers, Delphi and nuTonomy, are testing self-driving vehicles. Singapore, by the way, was the first place where left-hand traffic with autonomous vehicles was tested. Nissan started testing in London in February 2017.[42]

The public discovered in August 2016 just how far the development had progressed. This was when nuTonomy started testing self-driving taxis with customers in Singapore for a short period. Only four months later, these "robomobiles" were used in Boston, too.[43] Uber received the first batch of Volvo XC90s equipped with self-driving technology that were also available for customers in Pittsburgh at the time.[44] Because of a fatal accident involving an Uber vehicle (more about this later), Uber has drastically reduced its efforts and taken a much slower approach.[45]

Anybody who would like to ride in a self-driving vehicle can do so in several places, at least in limited versions with autonomous shuttle buses being tested on mostly private land such as office parks, college campuses, and hospital campuses. These include Bishop Ranch in California, Sion in Switzerland, and the Charité in Berlin.

Chinese internet giant Baidu works with BMW and is conducting tests in Beijing.[46] BMW was planning to start work on autonomous driving with 40 vehicles in Munich from 2017 onward but seems to have postponed the project.[47] Baidu is testing an autonomous fleet of taxis in Wuzhen. Last but not least, Volvo started testing an autonomous taxi fleet in Goteburg, Sweden, in 2017. But nothing compares with the efforts of the over 60 companies with a test license in California. Silicon Valley is the world's hotspot for this technology.

The manufacturers' approaches vary: some develop their own system, and others are waiting to purchase a complete solution to integrate into their vehicles.[48] Companies do not always comply with the authorities' regulations; one example is Ot.to, a producer of autonomous trucks in Nevada that was subsequently acquired by Uber and after the fatal accident shut down. Apparently, the first tests were conducted there without ever having asked for permission.[49]

While the traditional automotive manufacturers take one step after the other and start with driver assistance systems, the disruptors plan to go all in at once. Google, Uber, and nuTonomy have been working on full autonomy from the start. Google's former head of development and now cofounder and CEO of autonomous drive startup Aurora, Chris Urmson, compared the car manufacturers' approach with an attempt to learn how to fly by trying to jump a little higher every time.

The reports the manufacturers need to provide to the DMV indicate the extent to which the companies are involved in the development of driverless vehicles.[50] The immense advantage Google has compared with other companies is immediately visible. There are now four reports available from 2015 to 2018. In those years, Google-Waymo cars accumulated at least 10 times more miles than all other manufacturers together managed to drive. And it has more cars, full stop.

Various data are requested in those reports, and the number of disengagements is one of the most important figures. According to the DMV, this figure indicates the following:

DMV regulations define disengagements as the deactivation of the autonomous driving mode for the following two situations:

(1) "if a malfunction of the self-driving technology is determined" or

(2) "if a safe operation of the vehicle demands that the driver switches off the autonomous mode of the autonomous vehicle and takes over manual control of the vehicle immediately."

And as an addition: "This clarification is necessary to ensure that the manufacturers do not report any other or any routine deactivation."

The first disengagement report is dated January 1, 2016, with a total of seven companies providing data about how the total number of 71 vehicles had performed until November 30, 2015. The report for 2016 includes 11 companies. In 2017, there were 12 companies, and in 2018, a total of 28 companies had logged autonomous miles in California (see Table 6.2).

The 2018 disengagement report shows that over 400 cars were used by these companies to develop and test autonomous driving technologies on public roads in California. Google's sister Waymo drove 1,271,587 miles in that period, almost triple the miles that its closest rival, GM Cruise with 447,621 miles, had managed. Since 2009, when Google started developing the technology, its cars have accumulated more than 10 million miles in autonomous mode.

When we compare the number of disengagements with the miles driven, we get a glimpse of how often safety drivers had to intervene. And the results are striking. Waymo's safety drivers had to interfere once every 11,154 miles, doubling the miles from 2017. If an average car owner in the United States drives 8,000 to 12,000 miles per year, Waymo's cars would drive about a year without human intervention. GM Cruise, the San Francisco–based startup Zoox, Nuro, and Pony.AI also charted disengagement rates that were beyond 1,000 miles.

The contrast with the bottom end could not be starker. Uber manages less than half a mile before a human has to intervene, Apple intervenes every 1.1 miles, and Mercedes, the pioneer in autonomous driving, has a human take control every 1.5 miles.[51] Basically, these companies don't even manage

TABLE 6.2 Manufacturer Disengagement Report to DMV

2018	DISENGAGEMENTS PER 1,000 MILES	MILES PER DISENGAGEMENT	DISENGAGEMENTS PER 1,000 KILOMETERS	KILOMETERS PER DISENGAGEMENT	NUMBER OF VEHICLES
Waymo	0.09	11,154.3	0.06	17,846.8	111
GM Cruise	0.19	5,204.9	0.12	8,327.8	162
Zoox	0.52	1,922.8	0.33	3,076.4	10
Nuro	0.97	1,028.3	0.61	1,645.3	13
Pony.AI	0.98	1,022.3	0.61	1,635.6	6
Nissan	4.75	210.5	2.97	336.8	4
Baidu	4.86	205.6	3.04	329.0	4
AIMotive	4.96	201.6	3.10	322.6	2
AutoX	5.24	190.8	3.27	305.3	6
Roadstar.AI	5.70	175.3	3.56	280.5	2
WeRide/JingChi	5.76	173.5	3.60	277.6	5
Aurora	10.01	99.9	6.26	159.8	5
Drive.ai	11.91	83.9	7.45	134.3	13
Plus.ai	18.40	54.4	11.50	87.0	2
Nullmax	22.40	44.6	14.00	71.4	1
Phantom AI	48.20	20.7	30.13	33.2	1

(continued on next page)

TABLE 6.2 Manufacturer Disengagement Report to DMV (*continued*)

2018	DISENGAGEMENTS PER 1,000 MILES	MILES PER DISENGAGEMENT	DISENGAGEMENTS PER 1,000 KILOMETERS	KILOMETERS PER DISENGAGEMENT	NUMBER OF VEHICLES
NVIDIA	49.73	20.1	31.08	32.2	7
SF Motors	90.56	11.0	56.60	17.7	1
Telenav	166.67	6.0	104.17	9.6	1
BMW	219.51	4.6	137.20	7.3	5
CarOne/Udelv	260.27	3.8	162.67	6.1	3
Toyota	393.70	2.5	246.06	4.1	3
Qualcomm	416.63	2.4	260.39	3.8	2
Honda	458.33	2.2	286.46	3.5	1
Mercedes Benz	682.52	1.5	426.58	2.3	4
SAIC	829.61	1.2	518.51	1.9	2
Apple	871.65	1.1	544.78	1.8	62
Uber	26,08.46	0.4	16,30.29	0.6	29

Source: https://www.dmv.ca.gov/portal/dmv/detail/vr/autonomous/testing.

getting their cars to go the distance to every other intersection without human intervention.

Although many manufacturers working on the development of autonomous vehicles are missing from the disengagement reports and the interpretation of a disengagement leaves some room for interpretation, California does offer first insights into the state of the art, thanks to the number of testing companies and the public reporting duty, and allows us to compare the progress that the manufacturers are making. It would only be possible to have a better overview if the manufacturers working in other states and other countries also had to publish such data. We should also consider that some companies that keep making bold statements about their autonomous driving features (such as Tesla) do not (yet) appear in the report.

Another aspect that becomes obvious is that we need to develop standards for technology comparisons. The DMV indications are still too vague to allow more exact comparisons, and the manufacturers are not obliged to provide raw data. Companies use different definitions of disengagements and could call for disengagements to occur at an earlier stage. Because autonomous vehicles are not available on the market yet, testing agencies and authorities cannot yet purchase such cars and conduct independent tests to verify the figures provided by the manufacturers.

There is another factor distorting the DMV disengagement reports: they only cover the tests in California. Many manufacturers, however, also test in other states or countries where no annual report is required. It is conceivable that manufacturers, in an attempt to play their cards close to their chest and not give clues to other manufacturers about the state of their technology, only test relatively simple or tried scenarios in California, whereas the more experimental tests causing more frequent disengagements are set up in other regions or in test areas not subject to reporting.

One solution for creating more transparency could be the installation of an independent consortium to define key figures and scenarios and obligate the manufacturers to use standardized measuring procedures and reports instead of limiting themselves to state or even just federal specifications. Despite the limitations just outlined, we can nevertheless say that Waymo is clearly the leader in this technology—with a huge advantage at that—although many companies have increased their engagement. It will become very exciting indeed when manufacturers such as Tesla release autopilot functions, Uber and nuTonomy disclose a little more and start reporting, and Waymo starts working with its own fleet of Fiat-Chrysler minivans, more than doubling its number of vehicles in a single stroke.

If you compare the testing activities with the number of patent applications submitted by the companies, there does not seem to be a connection. According to patent statistics for autonomous vehicles, traditional manufacturers are in the lead in purely numerical terms.[52] Toyota, followed by Bosch, Denso, and Hyundai, is at the top of the list; Google comes only twenty-sixth. However, the number of patents is no indication of the degree of innovative power and progress in the development of new technologies in a company.

Google also lists "simulated contacts" in its report to the DMV. Google reports one such contact for every 74,000 miles. Contacts are defined not only as actual physical contact with other vehicles but also as events such as driving over the curb. Human drivers have one such contact every 500,000 miles, according to accident reports, although this only includes the collisions reported to the police. The number of unreported incidents is estimated to be at least twice as high, and contact with curbstones and similar occurrences are not included at all.[53] Considering the accident statistics, we probably should be less worried about protecting humans from robocars than about protecting robocars from humans.

Near the Bay Area, we find a unique testing area called *GoMentum Station*, part of the former Concord Naval Weapons Station. This area, still declared to be a restricted military area, features a 19-mile network of roads and highways that all the manufacturers have taken into consideration for vehicle testing. Honda, Acura, Ot.to, and Easymile have been testing in this area.

The hopes for self-driving vehicles are very high. Even if only a small percentage of the traffic fatalities and injuries caused by human error could be avoided, that would still be worth using this technology. However, it is not clear yet whether this technology can live up to such expectations. A study by the University of Michigan states the hardly surprising conclusion that there will always be crashes. The researchers furthermore were unable to determine that a self-driving vehicle automatically drives better than an experienced human driver. Particularly during the interim phase, that is, when automatic and manual vehicles have to share the roads, there may even be an increase in the number of crashes.[54]

Surveys regarding the advantages of automated driving resulted in the following: 43.5 percent of those interviewed stated that searching for a parking space would be practically eliminated, 39.6 percent answered that one could do something else while traveling, and 53 percent liked the idea of switching between automated and supported driving modes. Two-thirds of the people interviewed saw self-driving cars in the context of new vehicles such as hybrids and EVs.[55]

A combination with ridesharing services can definitely reduce the number of vehicles on the roads, and a more efficient style of driving allows you to save on fuel. Fewer cars on the streets also leads to a decreased demand for expensive traffic infrastructure, including parking. In addition, such vehicles make it possible for hitherto disadvantaged groups of people to once more participate in mobile, social, and economic life. This includes elderly people and those with limited eyesight, as well as children. Such a development, in turn, would provide relief for those who so far have had to provide these services to compensate. Voyage, a company spun off from the online university Udacity, does exactly this. The Palo Alto startup developed a robotaxi service for gated communities with a large number of retirees. The average age of the passengers is 76, and while Voyage CEO Oliver Cameron expected this demographic to be more reserved about self-driving cars, they instead love them. Making self-driving mobility available to this demographic group means being able to participate in a social life.

Traditional manufacturers and newcomers view the future of the car from different perspectives, a fact that is also mirrored in the definition of the terms they use. The managers of Volvo and Renault Nissan use the term *autonomous* for vehicles that will still be delivered complete with steering wheel and pedals in the future but nevertheless comply with level 4 vehicle standards. Traditional manufacturers believe that customers in the future will continue to want a private car that they occasionally want to steer and drive. They refer to cars as *self-driving* if they have no manual control elements, such as those mainly used by taxi fleets and ridesharing companies.[56] The newcomers on the market, by contrast, do not differentiate. They expect automobiles not to require any control elements in the foreseeable future because those elements would increase the vehicle cost and create an unsafe situation if a passenger has to intervene in the control of the car.

I will discuss the impact on other economic sectors and our society a little later. First, let us take a look at how a self-driving vehicle actually works, what it needs, and what kinds of questions we have to find answers for. And also, what a fatal crash involving a self-driving car by Uber in 2018 meant for Uber and the industry.

Seeing and Being Seen:
Cameras, Lasers, and LiDAR Systems

It's impossible, it's difficult, it's done.
—FREQUENTLY USED SAYING IN SILICON VALLEY

Although the dream of driverless cars has existed for decades, a real break-through seems possible only now, a fact that is due to the technological progress in several areas: computing power and data storage capacities, machine learning, robotics, algorithms, broadband networks, and sensor technology. Normally, self-driving vehicles are equipped with a series of sensors that allow the car to "see" its environment. Not all the automotive manufacturers rely on the same technologies; some skip certain technical steps altogether or attempt to obtain similar results by combining others. The following list of items that I discuss is based on Waymo's self-driving vehicles, which are considered the currently to embody the most advanced and far-reaching approach: LiDAR, radar, camera, graphic processing unit (GPU).[57]

The most striking feature is the LiDAR system attached to the roof, protected by a glass dome the size of a small church bell. Inside the dome is a group of 32, 64, or even 128 revolving lasers that measure the distance to other objects and generate a three-dimensional (3D) map of the surroundings in a radius of up to about two-tenths of a mile. *Surroundings* in this case refers to a 360-degree circle around the car. The basic principle of this function is similar to radar, except that laser beams are sent out instead of radio waves to optically determine distances to the objects around the vehicle. Thirty complete 3D images are generated per second.

The data from the LiDAR system allow one to become aware of objects, categorize them, and calculate their speed and direction by comparing the position of the respective object for every measurement. This allows the computer to predict where, for example, another car is headed, and it can react accordingly. In this way, the autonomous vehicle can grasp the essentials of a normal traffic situation. The intensity of the reflected laser light also permits the computer to read and interpret street signs.

The LiDAR system is used not only for detecting the environment and other objects but also for creating high-precision 3D maps. The amount of data collected with a LiDAR system is huge, because many objects in a regular street setting can be recognized. The complete capture of 300 miles of highway lanes around Palo Alto provided 1 terabyte of data for mapmaker

Civil Maps.[58] The company was able to reduce this huge amount of data to 8 megabytes by intelligent filtering. Still, those 3D maps require a lot of work to keep them updated. These are the challenges the LiDAR technology poses and the reason why several startup companies and corporations try to manage without a LiDAR system.

The first LiDAR systems, resembling a spinning popcorn bucket, were sold for several hundred thousand dollars a few years ago, but prices have dropped for some newer versions to a tenth of that amount, and the expectation is that the price will drop still lower. While you had to pay $400,000 in 2007 to purchase such a system, you could get the same system in 2015 for an "affordable" $40,000.[59] A real bargain! But as long as LiDAR systems still cost more than a couple of hundred dollars and have not become much smaller, they are not suitable for the mass market. Simple static and therefore very inexpensive LiDAR systems are found, for example, in every robot vacuum cleaner and in those laser distance meters from do-it-yourself (DIY) stores. In the previous models, the laser arrangement revolved around its own axis several times per second, but now so-called solid-state LiDAR systems do not have any movable parts.[60] Admittedly, their field of vision and range are more limited, but they require less maintenance and are cheaper.[61]

Waymo CEO John Krafcik mentioned the drop in prices at the Detroit Automobile Show. Waymo, for instance, developed its own LiDAR system, which costs only a tenth of the asking price several years ago. From $75,000 for the first systems, the price has decreased to less than $8,000. This price is still too high for installation into mass-market vehicles, but companies such as Innoviz from Israel have already announced LiDAR systems that will be available for $100 or even around $10.[62]

Because several automotive manufacturers plan to launch autonomous vehicles on the market in 2020 or 2021, the demand for LiDAR systems is definitely expected to grow. The San Jose–based LiDAR manufacturer Velodyne estimated that it will sell about 12,000 units in 2017, 80,000 in 2018, and 1.7 million in 2022.[63] Many new startup companies with huge amounts of risk capital are positioning themselves for this expected boom. One of them is Luminar Technologies, founded by Austin Russell, age 22.[64] Rumor has it that capital providers have invested $150 million with an evaluation of $1 billion. Quanergy received about the same amount of money. A lawsuit Waymo filed against Uber (which was settled in 2018) shows just how highly competitive the fight is for dominance in the LiDAR market. Uber was accused of having accepted more than 14,000 documents about LiDAR technology as well as supplier lists that former Google employees took with them when they left to work for Uber. At the center of the controversy was former Google employee Anthony Levandowski, who was involved in the first line of LiDAR technology development. Self-driving technology is so immensely important for the future success of ridesharing programs that any defeat in court would make the technology deficit almost impossible to overcome.

The continued development of LiDAR systems with thousands of laser pulses per second in every direction is, in any case, progressing rapidly. Waymo's long-distance LiDAR system can see for up to 700 feet and—as Krafcik explained—recognize and identify a football helmet at a distance of two football field lengths. "Short-sighted" LiDAR systems, by contrast, recognize hand signals and gestures (e.g., from a police officer or a cyclist), as well as a pedestrian's angle of vision, and therefore are able to estimate the expected direction of travel or motion and make the vehicle react accordingly.[65] Today's leading companies shifted from recognizing and categorizing objects to the next stage of interpreting intent of other traffic participants.

All LiDAR companies are working on even more robust and powerful LiDAR systems. Several scenarios and road conditions pose greater challenges for the developers than others. Direct sunlight, raindrops, and laser light from other cars with LiDAR systems can interfere with the signal. Raindrops confuse the LiDAR system because they reflect the laser signal and are superimposed on the reflections from any other object. To the sensors, the water droplets from the preceding vehicle appear to be a solid object, and so the vehicle applies the brakes. This reaction must be filtered out using very advanced algorithms. This is already possible for light rain, small raindrops, and snowflakes.[66] Snow on the tarmac, by contrast, covers up the markings on the road and therefore renders the cameras useless. This is why the MIT Lincoln Laboratory has been working on a radar system—a localizing ground-penetrating radar (LGPR)—that can record the reflections from the road surface with a high-frequency radar and thus "see" the road markings.[67]

Another potential problem for LiDAR systems is the color of cars. As we saw earlier with the study of blue and yellow taxis in Singapore, humans recognize vehicles in bright colors more easily and have fewer collisions with brightly colored and more visible vehicles, and the same is true for sensors. Dark colors absorb more laser light and therefore reflect fewer useful signals. Plastic parts and composite materials also absorb more light. By contrast, metal paint blocks ultrasound, so such sensors installed in the car cannot penetrate the metal. Last but not least, radar also can be blocked by certain paint components.[68] This can also extend to structures. It won't simply be a question of aesthetics in how street equipment, carriers, and other objects on the street are presented. We will also have to consider the extent to which they impact "visibility" for sensors. For all of this, there are solutions that are able to filter certain wavelengths and angles, emitting multiple laser pulses, calculating clusters, and separating scattered light from signals.[69]

In addition to LiDAR systems, we will need cameras that are also arranged to provide a 360-degree view around the vehicle. The cameras are looking for pedestrians, cyclists, and other vehicles, as well as for street signs, traffic lights, and street boundaries. The resolution of cameras today with fewer than 2 megapixels per 36 images per second (i.e., 60–70 megapixels altogether) is better than that of LiDAR systems, which only reach up to 2 megapixels per second. Furthermore, cameras are an essential part of the selection of available sensors because they can recognize both the form and the condition of other objects. Is that object on the street a wooden plank or plastic packaging? A camera image is normally two-dimensional, but multiple cameras together can form stereovision and thus provide the computer with a 3D image. Just get the right cameras and algorithms to do this. This is why some companies developing self-driving cars believe that ultimately they won't need LiDAR at all.

Radar-based distance meters are installed in the bumpers to recognize other vehicles as well as objects crossing directly in front of and behind the vehicle. Of course, these radar systems are not the rotating radar antennas we are used to seeing but are instead mounted on chips the size of a stamp.[70]

However, before a vehicle with a LiDAR system is really able to move autonomously, the surroundings must be captured in a 3D map. Mind you, not just once. The environment is constantly changing, and this has to be taken into account. An antenna station in the rear of the vehicle receives geo-localization data from GPS satellites. And a sensor is installed into at least one of the wheels, tracking the movement of the wheel. The sensors are completed with gyroscopes (measuring the acceleration) and a speedometer. And there is one more thing: the condition of the roads in California, right in the largest test area, is extremely poor. Missing and faded road markings, potholes, and traffic signs obscured by overgrown vegetation increase the challenges for the manufacturers.

Cameras are used for the view not just outside but also inside the vehicle. The Fraunhofer Institut is conducting tests in collaboration with Volkswagen,

Bosch, and other manufacturers to find out how many persons are traveling in a vehicle, who they are, what their position is, and what they are holding in their hands. From these data, the researchers attempt to draw conclusions regarding passenger activity.[71] It is also possible to measure whether a driver is upset or bored.

The MIT Lincoln Laboratory is working on discovering where Tesla drivers look when they have an autopilot and what their emotional state is.[72] It was a surprise to find that a smile on the driver's lips is not a sign of contentedness but actually a sign of frustration and dissatisfaction: with the navigation system, with their own car, or with the other vehicles around them. While a satisfied driver tends to look rather bored, frustrated drivers often point out the absurdity of their situation with a smile.

You need a lot of computer power to evaluate and interpret sensor data. It is a small wonder that the manufacturers of computer chips, also located in Silicon Valley, have identified this as their next big opportunity. Because the geographic distance between the experts in the respective companies is so small, the development proceeds quickly. One corporation offering specific processors for autonomous vehicles is NVIDIA. Many elements that are installed today on boards no bigger than your palm and for a price of a couple of hundred dollars were sold for millions only 10 years ago and were so big as to fill an entire room.

The first step is to know why the cars need this computer power. Let me give you an example to illustrate. The Google image database includes tens of thousands of shapes of moving objects: an adult, a person in a wheelchair, a person with a cane, a person with a dog on a leash, a child, a person lying on the ground or crawling, even the duck pursued by the old lady in the wheelchair because it happened to a Google vehicle. On the basis of these kinds of data, the car has only fractions of a second to assess what is in front of its sensors.

Can the car safely drive over such an object to be identified, or would it be better to stop? However, a more frequent task for the autonomous vehicle is to correctly identify streets, road markings, traffic signs and traffic lights, buildings, other vehicles, people, trees, and so on. For this purpose, the algorithms discern between a semantic segmentation and actual object recognition.[73] *Semantic segmentation* is the categorization and reference of individual image pixels to the so-called object classes. Is this thing that I see in the object class of tree or human, or is it just an indication on the road? Object recognition is more detailed and tries to understand whether this is a static or moving object and, if moving, where it is moving to and whether it is necessary to react to its movement.[74] Some things can be predicted and programmed by software developers, but other things simply cannot (e.g., that duck). The vehicle has to experience all of reality with all its sensors. This is where artificial intelligence and machine learning come into the equation.

CHAPTER 7

Artificial Intelligence

America Invents, China Copies, Europe Regulates

When asked about artificial intelligence, Tversky replied, "We study natural stupidity."

—Amos Tversky

Our brain is a natural miracle. No computer today can even approach the computing skills a human brain exhibits, and at a fraction of the energy consumption of a computer, too. Human brains need only about 50 to 100 watts, a quantity corresponding to the power of a light bulb. You might want to take it as a compliment if someone says that someone else does not "shine very brightly."

But before we delve into artificial intelligence and understand why the car of the future needs it, let us first take a look at what exactly the car needs to know and learn. The first two big questions are, "Where am I?" and "Where did I drive?"

GPS is great for finding out where on the globe you are right now (with a precision of 5 to 30 feet), but this would by no means be exact enough for a self-driving vehicle. Secure navigation in traffic needs information that is exact down to a few inches. In order to obtain this level of precision, the vehicles can use stationary objects for orientation. This may be a door, a special building, a tree, or something similar. Comparison with the GPS signal then allows the vehicle to understand where it is. As soon as the vehicle starts moving, other stationary objects come into its "line of vision," and the calculation of where it is and where it is going to becomes more precise.

Think of it this way: aliens abducted you one night and then put you down again on the ground a little while later. But just where on earth are you now? Somewhere in the distance, you see a light illuminating a parking lot in front of a supermarket. The writing is in German, so you may assume that you are in some German-speaking country. Some cars parked outside have German license plates, so the probability of you being somewhere in Germany increases. An advertising poster for a Bavarian beer brand above a restaurant sign tells you that you probably are in Bavaria. In this way, you narrow down your location using the available reference objects. Admittedly, there is still a small chance that you are in a Bavarian village in the middle of China, copied by the Chinese like they did with Hallstatt and Paris, but every additional detail you identify makes that possibility a little less likely.

It is easy to see that an autonomous vehicle requires adequately detailed and processed maps in order to be able to answer such questions. This is why Google and Apple, both offering map and navigation solutions, undertake their immense efforts, and it shows why a consortium of German automobile manufacturers bought Nokia's map service HERE for a sum in the billion-dollar range.

Once the car knows where it is and where it is going, more questions need answers:

- Are there other objects around me?
- What kinds of objects are they?
- Do they move?
- How quickly do they move?
- Where do they move?
- What about stationary objects? Are they giving me crucial information such as signs, traffic signals, or lane markers?

Compared with these, the first two question of where the vehicle itself is and where it is going seem almost laughably simple. The car now has to interpret the objects. Which ones can be simply and safely ignored, which ones have to be observed, and which ones are critical?

The different sensors in the car receive various depictions of reality, and erroneous information is possible. A camera may be blinded by sunlight, the

LiDAR system may be useless because of raindrops and snowflakes, and the GPS might show the wrong location because of metallic or large structures in the vicinity, let alone a potential simple defect, for example, caused by a loose connection. The vehicle now faces the huge challenge of having to make sensible predictions that allow safe driving from the existing measurements. The sensor data are fused and interpreted, a process referred to as *sensor fusion*.

Once the vehicle has solved all these problems and knows where it is, where it is going, who and what are around it, and whether all those objects are moving, it can turn to planning its own route. Which possible route is the most efficient one? Not all the ways are equally convenient, and the assessment may vary during the drive. Anybody who has used Google Maps knows that the system displays the shortest available route and may suggest an alternative during the drive that leads you away from the normal main roads to side roads and dirt roads. For example, routes that require a lot of left turns are not efficient in the United States because you have to wait for the oncoming traffic and the journey therefore takes more time. The navigation systems used by delivery services such as UPS and FedEx take this into account by suggesting routes that mainly let you make right turns, although the distance traveled may be longer.

The most efficient route may not depend on the direct route in other cases either. Sometimes traffic in the direction of travel does not allow you to choose the ideal route. A truck that simply will not let you take a turn might force you to make a detour that effectively is more efficient in the end. Or a route that is faster may be very inconvenient for human passengers; for example, a bumpy or winding road may make passengers carsick.

So how does the car drive? Well, turn on the engine and get going, right? This does not work, not even on straight stretches of road. The street may be inclined, one wheel has slightly less air pressure, and voilà, your car is slowly drifting off course. This is when countermeasures need to be taken. In a curve, the vehicle should not just slam in and turn right but should reduce speed and gently follow the road. The motion planner has to plan in gentle movements. Rocking, jolting driving that may even make the rocking motion on a horse look soft and even is not a pleasant experience and may be accompanied by negative side effects.

Sebastian Thrun, winner of the DARPA Grand Challenge, makes a detailed analysis of the basics of programming for exactly these questions in a Udacity online course.[1] Anyway, this just covers the general basics. There must be a lot more adjustments made and situations experienced to create a detailed database of all the possible situations a vehicle may be exposed to during a trip.

The time needed to generate this data treasure trove is immense. But as soon as even one car made a certain experience, that experience is available to all the other vehicles almost immediately. In contrast to humans, with every single new license holder having to start at zero and gaining driving experience over time, each new self-driving vehicle can immediately access that database. This is why the several hundred test vehicles Waymo has driving for 25,000 miles every day are so vastly important. Waymo's driving has long since grown out of Mountain View, California, with its usually sunny and clear weather conditions. In addition to the hot weather in Austin, Texas, and Phoenix, Arizona, and the rainy weather in Kirkland, Washington, the next destinations to be "haunted" by self-driving vehicles may be London with its left-hand driving and several snowy areas.

Other car manufacturers meanwhile are still struggling with some rather basic problems concerning the way to safely steer vehicles. One strategy to diminish Waymo's lead is the one Tesla currently employs, namely, its over-the-air update procedure. Some of the sensors built into Waymo's self-driving vehicles have already been installed in the several hundred thousand Tesla cars already delivered to customers. When 15,000 Teslas had semiautonomous driving functionality installed overnight in an over-the-air update in late 2015, hundreds of videos were uploaded in just a few days showing cars changing lanes and traveling on highways without any interaction from the driver.

Since October 2016, Tesla has been putting into every manufactured car the Autopilot Hardware Kit 2 comprised of eight cameras, ultrasound and radar sensors, and powerful GPUs. As soon as the software reaches a state that allows fully autonomous driving, it can also be installed via an over-the-air update. In one single stroke, all Tesla cars that have the hardware kit installed (which comes at a price of several hundred dollars) will turn into self-driving cars—so far several hundred thousand vehicles altogether, including the Models S, X, and 3. Driving data from those cars are already passively captured and communicated to Tesla, which uses the data to train its machine-learning system with driving scenarios and street maps that will enhance all Tesla vehicles with its aggregated data. Even without any visible road markings on a snowy surface, the vehicle can benefit from the combined driving experience of the entire Tesla fleet.

Basically, Tesla combines the experience Google has to painstakingly implement on its own by linking all the Tesla car owners in a crowdsourced effort. The vehicle learns from the driving behavior of many human drivers. In this way, results are obtained more quickly, and more different scenarios can be generated, although it all has to be monitored carefully. For example, one of the results from such a trial was that there were problems with incident light, because most drivers in California were driving while the sun was out and therefore were mostly exposed to this kind of problem.[2] Nevertheless, we are likely to see an exponential increase here, which will make the tentative steps of some traditional manufacturers look futile and very old-fashioned. After all, most manufacturers so far have not even decided how far they want to go, let alone selected any hardware to be implemented in their production vehicles.

Now, the Tesla and Waymo approaches—as I call them—are two different philosophies. Waymo, with the majority of autonomous technology developers, always has LiDAR as an important component in its sensor suite, whereas Tesla (and other companies) believes that it can do without LiDAR and instead have cheaper sensors such as cameras and radar fill in the missing LiDAR information through algorithms. Tesla is so sure about this approach that it is reflected in Hardware Kit 2, which has no LiDAR. Some developers believe that you may be able to do without LiDAR, but will the algorithms and camera technologies improve quickly enough that the price decreases of the still expensive LiDAR systems cannot catch up? Who will be right time will show.

If you put something unusual in front of a baby, the child will look at it for a longer time. This is the kind of experiments researchers use to determine whether ideas like moral behavior are inherent in a baby or the result of socialization, or whether babies have a sense of humor.[3] And indeed, they found that babies tend to look mainly at their parents when there are funny things happening (e.g., loud noises made with an object by mommy or daddy). They do that so that they can judge their parents' reactions and intentions. If the parents start laughing, so will the baby.[4] Developers use a similar strategy to teach self-driving vehicles.[5] A camera observes the driver and recognizes whether he or she checks the interior or exterior mirrors or looks over his or her shoulder. This usually signals an imminent maneuver such as changing lanes or passing. The data from the sensors that are pointing outward then help the vehicle to learn the conditions under which such a maneuver occurs, and then the vehicle proceeds to realize that maneuver itself.

You can surely imagine how much computer power processes such as this require. Once again, specialized hardware comes into play and takes us straight back to suppliers such as NVIDIA, which introduced processors on the market that have the power of 150 MacBook Pros or more.[6] NVIDIA now uses four such processors on one computer unit for self-driving vehicles and currently works on this platform with the amazing number of 80 customers.[7] Tesla even develops its own graphic chips to handle the data and AI algorithms specific to automotive requirements.

Today, the manufacturers use a lot of so-called electronic control units (ECUs) in their cars, small processors that, for example, control the sensors, brakes, and electronic entertainment devices on board. Practically every sensor has its own ECU, and the quantity used quickly reaches three-digit numbers. Although the manufacturers have been debating the use of fewer and more central processors for years, this topic really has been pushed onto the main agenda since the development of self-driving vehicles took off. Not only are central ECUs cheaper, but they are also simpler to program, and installation of software updates is easy. Because automobile production is based on a deep vertical integration of supply chains, which means that the car manufacturers are very closely linked to their suppliers, tasks that used to be handled locally by the component the supplier produces are now handed over to the head developers of the central ECUs. Once again, this means a lot of change, because decentralization so far has always supported pronounced insular thinking of departments within the manufacturing corporations and led to the existence of many redundancies among the workforce. This is another reason why previous attempts to centralize ECUs were not very successful. Now this is forced by the development of autonomous cars and therefore meets with quite a lot of resistance in the respective business divisions.[8]

Chip manufacturers such as NVIDIA, Qualcomm, and others are ahead of Intel, which seems to have missed the boat and is attempting to retain some relevance in this profitable future industry by forming partnerships with producers such as Delphi and Mobileye.[9] In a logical act of liberation, Intel therefore bought Mobileye from Israel for $15.3 billion in March 2017. It is interesting to see how technology shifted from predominantly central processing units (CPUs) for computers to mobile CPUs for smartphones to now graphics processing units for autonomous vehicles, and with this the dominance of different chipmakers.

But what happens if sensors malfunction? Will other sensors be able to provide the missing information and ensure the safety of the vehicle? Researchers in Cambridge followed up on some solutions to this question.[10] The task was to find out whether a simple $200 camera installed behind the windshield would be able to provide sufficient information about objects in front of the vehicle. We have encountered the relevant questions before: which objects are ahead of me, are they moving, and if so, where are they headed? The scientists managed to do exactly that with the camera only. Thanks to clever algorithms and machine learning, it is possible to compensate for lack of data from malfunctioning sensors with just a camera. Tesla demonstrated that as well, when at the Tesla Autonomy Days in early 2019, Elon Musk showed examples of algorithms accessing multiple onboard cameras for creating 3D-point clouds of the surroundings. The field of application is not just limited to autonomous vehicles. Such routines can also be used for household robots to recognize objects in the house and classify them appropriately.

The technical term for this task is *semantic pixel labeling*. The technique uses algorithms that attempt to find meaning from image pixel clusters. The program tries not only to recognize the individual objects using changes in lighting, shadows, outlines, and texture but also to classify the objects correctly. While this is quite a challenge for individual pictures, it gets really difficult when you have videos with a handful or as many as dozens of images per second to analyze.

AI is a key technology for future automobiles and many other industries. Companies from the software industry understood this at an early point in time. Uber, Tesla, Google, Apple, NVIDIA, IBM, Baidu, and Microsoft invest billions in AI, and robotic startup companies are quickly bought for huge sums, or AI experts are lured away from universities and research institutes with astronomical salaries. For example, Ford invested $1 billion in the startup Argo AI in early 2017, a company founded by former Google and Uber employees. Uber allegedly acquired Ot.to with its 90 employees for a sum of $700 million. Aurora.ai, by contrast, was cofounded by former Google manager Chris Urmson, who already raised $620 million in venture capital for it (and rejected an acquisition offer from Volkswagen), and Nuro.ai is the baby project of two other former Google employees.[11] Intel's acquisition of Mobileye with its 600 employees, among them 450 engineers, cost $15.3 billion. The purchasing price corresponds to a per capita price for engineers with self-driving technology expertise of a staggering $33 million.

The market for specialists will already be depleted before the established automotive companies fully realize why AI is important. The widespread ignorance of many governments with regard to AI is illustrated by a summary by Fabian Westerheide from the German Federal Association of Start-up Companies. In a hearing of the Digital Agenda Committee of the German

Bundestag (Parliament) in March 2017, the questions centered on issues of how to regulate AI instead of focusing on the chances and opportunities.[12] America invents, China copies, and Europe regulates.

Practice Makes Perfect!
Machine Learning and Deep Learning

A friend in my youth came toward me with a swagger: "What do you say. Let's play Ping-Pong! I read everything about it in that book. Now nobody can beat me!" I admit, I was not the best table tennis player in the world, but that friend never had a chance. It was, after all, the first time he had held a table tennis paddle in his hand.

For us, it is glaringly obvious that there is a difference between theory and practice. Knowing everything about the theoretical aspects, how to hold the paddle, which angle is best, how to predict the trajectory of the ball, how to move—all of that is of no importance once you really have to hit the ball with exactly the force and speed that lets the ball cross the net just far enough so that you opponent cannot get to it while it still hits the table.

And this insight holds true for other disciplines as well: you learn how to walk by doing it. And one crucial part of learning to walk is definitely falling. Coordination of movement, application of theory, and special intricacies have to be learned by practice.

Self-driving vehicles are no different. Even if you program in all the rules and regulations, the many nuances of real traffic, as well as the situations that no engineer could ever predict, make it almost inevitable that the machine has to drive itself and has to learn while doing so. And that is what we call *machine learning*. AI makes its experiences just as humans do, and practice makes perfect.

Anca Dragan, researcher at the University of California Berkeley, works on the interaction between humans and machines. Instead of programming a behavior into cars, for example, what to do at a four-way stop, a frequent situation in the United States, she lets the cars find out for themselves what they need to do. The robot knows that it has to stop at a stop sign, but the challenge is knowing when to start again without endangering any other traffic participants. Human drivers often do not come to a complete stop, which they actually should do to comply with the law. Google's experience was as follows: a robot car would wait until the other vehicles have come to a complete stop and then depart again after a short period of waiting. Human drivers interpret this as hesitating and simply take the right of way. Consequently, a robot intent on safety would never be able to leave the intersection.

Dragan therefore had the vehicle calculate the best moment to take off safely using algorithms. She discovered that the vehicle started to display a different behavior. One action was to reverse a few feet to indicate to the other vehicles that they should take the right of way. Waymo's cars have also displayed surprising behavior. Human drivers learn to change direction in a narrow street with a three-point turn. The driver first drives forward and left, then reverses to the right, and then completes the maneuver, driving forward and left again and then straight. The robot car found different possibilities for such a turn. It first reversed or turned right into a driveway. Not everything the machine learns is actually useful. What is important is that it shows a behavior that does not confuse humans or make them feel uncomfortable.

As the car continues to be able to learn, it is better prepared for new driving situations and can find solutions for them. But what happens if there is a situation for which it cannot find a solution or way out? That is exactly what happened during the DARPA Grand Challenge, during the desert competition for robot cars. A robot car got stuck on a slope and continued to spin its wheels until the tires burst into flames.

Nissan believes that there will always be situations in which an autonomous car needs help. Instead of trying to find a way out by itself and potentially getting stuck, the car calls a call center where a human observer assesses the situation from a distance and steers the car clear again. The startup company Zoox and Toyota have submitted patents describing the remote control of vehicles in unexpected travel situations.[13] The startup Phantom.auto offers such a solution, with remote control of autonomous vehicles when they get stuck. Waymo, which is the first company in California to have received a license to test completely driverless vehicles—without a driver or even a person in the car—had to provide a solution to the California Department of Motor Vehicles specifying as to what the car will do to get out of such situations. In such cases, a call center operator will communicate with the car and provide instructions to guide the car out of the difficult situation.

Some Words About AI

There is an alarming fact connected with AI: researchers do not fully understand how a machine does something and what exactly it learns in the process. This had already become obvious with Google AlphaGo, the Go computer that not only beat Korean Go champion Lee Sedol four games to one but actually humiliated him. Observers were particularly astonished by the way AlphaGo played. The computer made moves human players had never seen before. While the IBM chess computer Watson still used massive computing power—a so-called brute force approach—to calculate as many moves in advance as possible and determine its chances of winning, AlphaGo used machine learning and something we would tend to call *intuition* or *instinct*. Go requires rather simple moves in contrast to chess, but there are a lot more permutations that cannot be predicted with brute force.[14]

AlphaGo first played against Fan Hui, triple European Go champion for several months, and then against itself to prepare for the competition. In the thirty-seventh move of the third game against Lee Sedol, the machine made a move that irritated all the Go experts and even upset Lee Sedol to the extent that he first took a 15-minute break and then eventually quit the game. The probability of a human player making that move was 1:10,000. Contrary to the experts' opinion, AlphaGo made that move in a manner we would call *intuitively*.

This was the first occasion of an effect we refer to as a *singularity*. Machines become so intelligent (in a domain) that they will teach us. If humans teach a monkey or ape, they can bring the animal to a higher cognitive level than when monkeys teach monkeys. Researchers demonstrated this with Koko, a female gorilla who learned sign language and acquired a vocabulary of more than 1,000 words.[15] Today, humans are taught by humans. But what if a higher intelligence were to take us under its wing, so to speak, and taught us? We could rise to a previously unheard-of cognitive level.

This is what happens with advanced AI systems. AlphaGo showed human players alternatives they did not know existed until then. Fan Hui, who ranked 600 in the international Go ranking list before his matches against AlphaGo, has since progressed to position 300. Similarly, Lee Sedol did not lose any more in the months following his losing to AlphaGo, but he actually discovered new joy in playing Go because the computer had shown him more options.

Philosopher Nick Bostrom from Oxford describes several experiments in which AI systems were confronted with certain tasks in his book *Superintelligence*. The scientists initially regarded the alternative solutions suggested by the system as wrong and as system errors, until they eventually proved possible when investigated in detail and disclosed some surprising approaches.

This effect, which some of us may see as a unique opportunity, makes other people worry. Our understanding and acceptance of decisions often are based on our ability to understand the conclusions drawn. What are the rules, which decisions were taken, and what leads us to which kind of final result?

In other words, AI is a little bit like the game "Chinese whispers," also known as "telephone." You whisper something in your neighbor's ear. The neighbor whispers the words, or what she understood, to the next person, and so on until finally the last person says the words he thinks he heard out loud. Everybody is delighted when the final result is often completely different from the original message. When you add AI to the game, it gets a little more complicated. Each participant in line receives a whispered message not only from one other person but also from several people at the same time. The recipient now has to combine these messages into something sensible and decide what it is he or she wants to pass on and to whom. Increase the number of people involved in the game to 1,000 or even 1 million, exchanging information in several long lines, and it will be almost impossible to predict what might come out at the other end.

A system that uses machine learning makes decisions based on probabilities—on many probabilities! Each node has a small but possibly important influence. In a similar manner as for the butterfly effect in chaos theory, which says that the beating wing of a butterfly on the other side of the globe may cause a hurricane here, we humans cannot understand how the individual probabilities at the nodes affect the end result. Humans find this difficult to accept because our entire system of science, as well as public awareness, is based on transparent decisions.

But how will we react if an autonomous vehicle makes a decision we do not understand and one that might put people in danger? Ethical questions such as the *trolley problem* (I will say more about this later) are difficult enough for us as is, and we expect clear rules and criteria for decisions to ensure our legal position. The government could pass laws for foreseeable cases allowing for unanimous decisions, but as we discovered with the example of the duck-chasing old lady in the electric wheelchair, reality is much more complex and surprising than a set of rules. An AI-controlled autonomous vehicle must be able to react safely in any situation.

These concerns keep traditional car manufacturers from pushing the development of self-driving vehicles. After all, a car is the largest movable and, following one's house, usually the second most expensive purchase people make. So now imagine a two-ton robot traveling autonomously and getting out of control. Because of potential legal action and potentially high claims for damages, traditional manufacturers are extremely careful, given their past experiences with faulty transmissions and accelerators that, for instance, Audi and Toyota had gone through.

This philosophy becomes apparent in other ways as well. A machine built by traditional manufacturers is already mostly tried, tested, and perfected. Following this logical train of thought, it is therefore unnecessary to provide vehicles with technology that will only be marketable and usable at a later point in time. While the storage space of a smartphone is almost empty when you buy it, a car's computer memory is already full. Updates for corrections are just about possible, but updating to include new functions is not.[16] This has less to do with the cost of free memory and a lot to do with the thought that an app ecosystem would result in too many questions and make people worry about safety, reliability, cost, payment services, and external developers. They do not have any experience with many of these elements and regard the concept as a risk that is too high at too little benefit them.

Cars are usually regarded as self-sufficient machines that are able to reliably operate independently of others. Everyone for himself or herself. Self-sufficient systems do not need any network; they do not need to communicate with other systems. Even if linked systems initially require more effort to be ready for use, they pay off very quickly. Removing a tiny software bug in Volkswagen's e-Golf or the Jaguar iPace is a costly undertaking if thousands of individual cars have to be ordered back to the mechanic so that they can upload the software patch via a USB port (not to mention the frustration on the part of the owner who—once more—has to make an appointment at the service center).

Tesla and Google no longer play this game. Their roots are in the digital industry, where an uncomplicated procedure to deliver improvements and enhanced functions as they become available is perfectly normal. We already saw this in Tesla's over-the-air updates. The surprise and obvious enthusiasm can be measured from the sheer quantity of demonstration videos on the internet. However, the travel data transferred to Teslas and visible in the videos also show critical behavior on the part of the driver, for example, someone falling asleep or dangerous situations like a Tesla crashing into the rear of another car. The company adjusted some of the autopilot functions or switched them off entirely.

Such cases are red flags for traditional automobile manufacturers. The fear of installing defective components and having to face litigation for damages that hurt the company's reputation and finances is too great. Whatever you deliver must be perfect; modifications are a costly measure. Exchanging an airbag or ignition switch involves a massive logistical effort, not to mention the inconvenience for the car owners. From this perspective, delivering unfinished software seems like a sacrilege. Testing is shifted to the client, and the culture of *zero errors* is abandoned. Reid Hoffman, founder and CEO of LinkedIn, put it like this: "[You] jump off a cliff, and assemble an airplane on the way down."

The approach adopted by Tesla and other (software) companies should not be mistaken for carelessness and recklessness. Horizontally integrated systems cannot be fully tested under laboratory conditions the way vertically integrated systems can. Even if this were possible, the effort would be astronomical. The advantage of the Tesla approach is that it is incredibly versatile, and feedback from customers about the improvements to be made is much faster. In vertical integration, a manufacturer controls every element. By the time everything has been tested, the components that are finally released in a fast-paced world such as ours may already be hopelessly out of date. The almost unmanageable (from the point of view of the owners) iDrive menu button in BMW cars is a perfect example to illustrate a laboratory development that the engineers thought was perfect and one about which quite obviously nobody ever thought to ask the customers for feedback.

Christoph Keese, author of *Silicon Germany: Wie wir die digitale Transformation schaffen* (*Silicon Germany: How We Can Manage Digital Transformation*), describes the difference in culture between digital and traditional companies as follows: risk management instead of preventing mistakes; discovery instead of maintaining the old; continuous improvement instead of perfection on delivery.

We can turn to the video game Elite Dangerous to see that the entire issue is not quite as abstract as you might think and to show that difficult situations with AI systems have already occurred. In this game, the AI system suddenly had enhanced functionality, allowing it to develop new skills and weapons and specifically target the spaceships of human players for destruction. The AI started to wipe out humans controlling those characters in this game, until the administrators shut down the game and reset it.[17]

Waymo vehicles drive more than 25,000 miles per week, and not only in California. Waymo is also testing its vehicles in Texas, Washington state, and Arizona with different road and weather conditions. Tesla, by contrast, has already collected 1 billion miles of autopilot experience from Tesla owners since it released the autopilot function in October 2015.[18] Because all data are reported, the company can draw conclusions from the behavior of both drivers and cars and thus continually improve the autopilot. Tesla actually offered the entire data collection of 1 billion miles already recorded back then by Tesla drivers to the U.S. Department of Transportation (DOT).[19] This would allow the DOT to draw its own conclusions from these data and use them as a basis for decisions regarding legislation on autonomous driving.

In the second part of his "master plan" published in July 2016, Elon Musk mentioned that he expected the approval of autonomous vehicles by the authorities at a point when, first, the road safety of the systems had improved by a factor of 10 compared with human drivers and, second, the authorities had test system data for 10 billion miles driven.[20]

In the meantime, there has been a heated discussion about the sense and nonsense of driver assistance systems and autonomous driving in the general public, triggered by the fatal accident of a Tesla Model S driver in May 2016. He had the autopilot activated at the time. Forty-year-old Joshua Brown was traveling in his Tesla in autopilot mode when a semi truck crossed the highway and Brown's car crashed into and under the trailer. The autopilot system was not designed to recognize a vehicle standing at a right angle and probably could not distinguish the white-painted sidewall of the truck from the sky behind the truck.[21] A subsequent National Highway Traffic Safety Administration (NHTSA) investigation, however, concluded that the human driver was at fault. Brown had been warned repeatedly by the system and should have resumed control of the vehicle, according to the operating instructions. NHTSA also decided that not only was it not Tesla's fault but that, according to the submitted data, the autopilot had already resulted in a general 40 percent decrease in accident frequency.[22] Since then, several other fatal accidents with the autopilot engaged have been reported.

The traditional automobile manufacturers with their experience from the past, when they were often sentenced to huge recall actions and penalty payments as well, have a story or two to tell about this. Accordingly, General Motors and Audi, in an effort to compete with Tesla's autopilots, announced driver assistance systems for 2017 and 2018 that also included a camera looking inside the vehicle. It is there to see whether the driver is looking at the traffic or is distracted. If the driver appears distracted, the system reminds him or her to concentrate again. In addition, these assistance systems are to function only on segments of road approved in the navigation systems.[23] Anything that enhances the safety of such vehicles should be welcomed.

In addition to the automotive manufacturers and the internet giants with their full "war chests," there are also smaller startup companies trying to get a piece of the action. They usually concentrate on hardware and software to be installed into standard vehicles to enhance the car with an autonomous drive mode. Cruise Automation in Mountain View and Comma.ai in San Francisco are just two of those startup companies. Whereas the former has already been acquired by GM (for about a billion dollars), the *enfant terrible* of the industry, George Hotz from Comma.ai, initially set out to create a build-it-yourself kit that turns vehicles into self-driving cars for less than $1,000.[24] Hotz, who made a name for himself at the age of 17 for a hack letting him unlock the iPhone, was able to collect several million dollars of venture capital for his enterprise. His approach to the generation of software: crowdsourced machine learning. Testers are to record their behavior while driving with a smartphone app. The data are then input into the AI system and included in the DIY kit. Comma.ai published the first data sets in the summer of 2016 to accelerate joint efforts.[25]

The German startup Kopernikus.auto is also working on developing a kit to refit manually controlled cars with technology that on the turn of a switch makes the car self-driving. In fact, the company made a proof-of-concept experiment with Porsche by having the car drive from the parking lot into the workshop without the mechanic having to jump into the car, start it, drive it, and then park it. Similar applications of such a refit kit could be used to move cars from the end of the production line to the parking and delivery lots.

Traditional manufacturers start sweating when confronted with this kind of approach by startups. Ralf Herrtwich, former manager of driver assistance and chassis systems with Daimler and now Senior Director of Automotive Software at NVIDIA, found problems with the individual learning of vehicles:[26]

> We currently feel that it is too early for our vehicles to learn by themselves in the sense of modifying their own algorithms. So we are talking about the individual software entity levels. This would, after all, present a problem: whenever a vehicle indicates an error, we would have almost no way of reproducing it because we would not have the individual level of knowledge of that vehicle. That is the reason why I said at the time that I did not expect each vehicle to learn individually.

We can, however, imagine that we would aggregate the experience made by the fleet at the back end. We could then develop training for all vehicles from that and update them accordingly so [that] all the vehicles would drive using the same logical structure. It is, after all, a very important factor for us that our vehicles continue to behave in a deterministic fashion, namely, that we are able to reproduce it. If each vehicle adapts to its own data experience, every single one will behave slightly differently. I know, that does sound very humanized, and would actually be sort of cool, but for us, it would mean we are not really able to guarantee which behavior our vehicles offer and which one they do not.

This is a clash of worlds. On the one hand, we have traditional corporations proceeding extremely carefully and therefore slowly because they do not want to risk their reputations as the makers of safe vehicles that they spent years building. The prejudice against autonomous driving propagated from top management does not help either.[27] On the other hand, we have the industry mavericks Google, Tesla, and Apple backed by fantastic financing and using huge resources as well as small startup companies such as Comma. ai. They all use unconventional methods that carried them to industry leadership straightaway. There are also several newcomers from China that try to score points with huge financial backing and political support and can accelerate the entire process considerably. In China, prohibitions or penalty taxes are quickly provided for combustion engines, or companies are obligated to install charging stations. Chinese Google rival Baidu and automotive manufacturer Geely (which bought Volvo in 2010) are among those competitors.[28]

So what about the expectations the authorities have? NHTSA lists the reduction of fatal injuries by 50 percent as the first criterion for the approval of self-driving vehicles. Mark Roseking, administrative director of NHTSA, commented:[29]

> The goal I want to set is start with two times better. We need to set a higher bar if we expect safety to actually be a benefit here as opposed to just bare compliance with requirements. While no one wants to say, "How good is good enough?," I'd actually say, "Start at two times and then let's work from there."

Authorities and experts agree that we need to find new ways to evaluate the safety and efficiency of autonomous vehicles. Their idea is to take inspiration from the aircraft industry. One example to be followed is a network for the anonymous exchange of safety data, allowing pilots, air traffic controllers, and so on to exchange confidential information about systemic issues and near misses, with the aim of correcting mistakes and avoiding accidents.

As a matter of fact, the safety standards in the aircraft industry have been written in blood. Each plane crash is investigated and the reasons for the crash determined; then the authorities react accordingly. My brother, an airline pilot in Austria, was responsible for safety standards for part of his professional life. Each plane crash anywhere in the world is analyzed in the trade journals, and the results of the investigations are passed on to all airlines. This allows them to identify the safety issues that are relevant for their own fleet and introduce new standards. By the way, based on my experience, I would not recommend reading such reports with the knowledge of an amateur. Once you have burrowed through the technical terminology, they include details that would make you sick. And sometimes accidents happen as a result of human error that will leave you with your mouth open.

We should expect something similar with self-driving vehicles. The successive improvement of the vehicles is going to be founded on existing scenarios from fatal accidents. The analysis of Joshua Brown's fatal accident already has led to several improvements on the autopilot despite the fact NHTSA did not blame Tesla in the least.

After a visit to Google's offices, the NHTSA representatives also understood that the current requirements from the authorities are not sufficient and that, thanks to autonomous vehicles, we understand driving behavior much better:

> As [several NHTSA representatives] took a test drive near several parked vehicles, one NHTSA employee along for the ride opened a door, causing the Google vehicle to lurch to a halt. Had the incident occurred on public roads and not in a closed test environment, current rules would mandate that Google report the hard-braking stop to the California Department of Motor Vehicles.
>
> It would have gotten labeled as hard braking and used against them as "Oops, hard braking," as opposed to "Crash avoided," said Rosekind, who was on the trip.
>
> So we need new safety metrics.[30]

In order to accelerate the learning process of autonomous systems, the analysis includes billions of miles in simulators as well as millions of miles driven on actual streets in the real world. Waymo's CEO John Krafcik reported at the 2017 Detroit Auto Show that his vehicles had covered a billion miles in the simulator in 2016. He said that those virtual miles were invaluable for the further development and the predominant form for the improvement of the vehicle's drive performance.[31]

The simulations are based on the data from actual test drives. The simulator then allows for a modification of those data to test various scenarios. Would the car, for example, have recognized the old lady in her wheelchair at night, too? What if a particular sensor had malfunctioned? Would the remaining ones still have provided the entire image and interpreted it correctly? What if it had been raining? Real test drives are the foundation on which hundreds of test cases could be played out in the simulator. The almost 25,000 miles driven in the real world every day, plus the approximately 100 million miles simulated every month, provide about 6,000 different simulation scenarios that can be created out of one real test drive. The AI behind the scenes learns from each one and generates an enormous database as well as deep-reaching decision trees.

Other manufacturers and research institutions also use simulations to develop and enhance autonomous vehicles. Udacity offers drive data and simulators in the context of its online course. Intel Labs and the University of Darmstadt use data from the popular video game Grand Theft Auto V and then annotate them and superimpose data from the real world to create training simulations for autonomous driving. The University of British Columbia adopted a similar approach.[32] László Kishonti from Almotive, a Hungarian startup company for self-driving technology, used a car racing game for the Microsoft Xbox game console for his first software training. The software had been given the task of steering the racing car. At the beginning, it kept crashing into the wall or hurling off the road, but the vehicle learned with each try. After innumerable attempts, the software then was able to let the racing cars drive on the virtual track fast and without errors.[33] And more recently, Chinese internet behemoth Baidu has joined the race of autonomous operation systems and simulators with its open-sources Apollo initiative.

Other manufacturers offer entire virtual driving simulations for autonomous vehicles. The Computer Vision Center in Barcelona, for example, has drawn up an entire range of driving scenarios in a city.[34] The simulations software Synthia allows researchers to quickly check whether their software reacts appropriately to different situations.[35] While driving on a highway is a fairly easy lesson for an AI system, city scenarios and the kinds of situations that do not occur very often are difficult. How should a self-driving vehicle correctly react in the case of a traffic accident, the arrival of an emergency vehicle, or construction vehicles maneuvering at a construction site or on adjacent streets? How does the same traffic situation present itself in different weather conditions and seasons? A simulator allows you to re-create such scenarios in a better way and adapt them to the frequency of occurrence so that the AI system can encounter such situations as required and learn from them.

There may also be some unexpected effects if traffic rules have to be violated or a certain driving behavior affects the well-being of passengers. A car parked on the side of the road but protruding into the lane or a garbage truck blocking the lane can irritate a self-driving vehicle. Should it wait until the other vehicle frees the lane again, or should it cross the double line and avoid the obstacle? Human drivers are quick to grasp whether the obstacle is likely to remain there for some time and will react accordingly. An autonomous vehicle must be able to react in the same manner.

Next, how should the vehicle deal with potholes? If the vehicle recognizes a pothole, should it drive over it, jolting its passengers, or would it be better to drive around it? What if the road is full of potholes? It turns out, in fact, that some of the most challenging scenarios for autonomous cars involve driving in a parking lot in front of a shopping mall. With no real rules, lots of vehicles parking, cars backing up, shopping carts rolling around, lots of pedestrians, and variable wait times as people unload their carts, this is probably what autonomous cars have nightmares about.

This brings us to another stage that self-driving cars have to master— understanding the intent of other traffic participants. Sometimes the self-driving cars can deduce intention from the other objects' trajectories. A vehicle driving in a certain direction will probably keep doing so. A pedestrian crossing an intersection will probably try to make eye contact with the driver, so the pedestrian's face will turn to the car. This is what an autonomous vehicle tries to spot: where is the face looking? All this gives us a glimpse at the stages of development self-driving cars go through:

1. Controlling and directing the vehicle
2. Identifying position and route planning
3. Recognizing and categorizing objects
4. Behaving safely around those objects
5. Identifying intent of other participants
6. How humans use self-driving vehicles
7. Comfort

Looking at the current stage of autonomous vehicle development, the first stages are basically solved. Identifying intent is one of the large tasks today, with potentially hundreds of millions of scenarios that cars need to be able to interpret, from two ladies talking at a crosswalk with no intent to cross the street, to predicting the next move of a school bus, to the expectations of a police car parked in the right lane with the flashing lights turned on, to the duck-chasing lady in the wheelchair.

At the same time, the cars are supposed to drive comfortably. We already talked about driving styles for challenges such as avoiding potholes or not driving in too many stop-and-go motions. But comfort also encompasses where passengers want to be dropped off or picked up. As it turns out, when shopping, passengers want to be dropped off at the supermarket entrance but picked up at the shopping cart return. Considering these details in the vehicles' overall experience, and which will distinguish future robotaxi fleets, makes using autonomous vehicles more pleasurable, comfortable, and smooth. And these details can only be determined by driving the fleets for tens of millions of miles in real life with real passengers.

This variety of options in real life is a huge challenge for AI systems and machine learning. There is a lot to do, and the work requires staff with the respective expertise. Such experts and researchers are, however, in hot demand because of the wide range of applications for AI, so we currently experience what could be called a *talent auction*. Carnegie Mellon's robotic research division lost more than three dozen specialists at the same time in 2015 when Uber made them an offer they simply could not refuse.[36]

And when Uber acquired the eight-month-old startup company Ot.to in August 2016 for allegedly $700 million, the price broke down to $7.5 million for each employee in the company. Intel was ready to pay $25 million for each employee when purchasing Mobileye—a total amount of $15.3 billion. Ford invested $1 billion in the startup company Argo.ai. These are the huge sums companies today are ready to pay to get engineers for the continued development of autonomous driving, and the annual salaries also seem astronomical: self-driving technology engineers earn between $232,000 and $405,000, an average of $295,000. Google offers $283,000, not including the starting bonus of $30,000 and other benefits, and at times is prepared to pay $348,000 per year.[37]

Any subjects not yet offered at the universities are taught online. The learning platform Udacity, founded by Sebastian Thrun, has offered what it calls a *nanodegree* as an engineering certificate for programming self-driving vehicles since late 2016.[38] This course is quite something. Starting out fairly easily with recognizing road markings, it quickly fans out into deep learning and neural networks, TensorFlow, classification of street signs and recognition of other vehicles, transfer of human behavior to robots, and decision trees. And this is just the first part of the three-part course.

The first students started out in October 2016, and another 100 from all over the world joined them each month. I know how much work this course involves from my own experience as a participant. Knowledge in several programming languages and dealing with large amounts of data is required, as is access to the required computer hardware for calculations. Most course participants easily spend 30 hours per week to be able to solve the respective tasks.

Some traditional manufacturers set up partnerships with startup companies or the competition and invest in or acquire the entire package in order to be able to keep up with new trends. For example, GM invested $500 million in Lyft.[39] Fiat, by contrast, initially provided Google with 100 vehicles.[40] Volkswagen invested in Gett, and Apple put billions of dollars into Didi Chuxing. And German manufacturers bought the map service HERE from Nokia.[41] Between 2013 and 2017, more than $80 billion had been spent on developing the autonomous car, according to the Brookings Institution.[42]

The Trolley Problem:
How Will Self-Drivers Solve Ethical Issues?

Whenever the discussion is about self-driving vehicles, sooner or later there is usually the question of whose fault an accident is and how the software would decide in the inevitable ethical conflict of whom to run over or kill. Matthias Müller, former chairman of Porsche and Volkswagen, was quoted in the German magazine *Auto Motor und Sport* as saying, "I always ask myself . . . how a programmer is to decide in the course of his work whether an autonomously driving car should, when pressed, verge to the right into a truck or to the left into a compact car."

The same question was asked in an edition of the German weekly *Der Spiegel* in January 2016 under the headline, "Lotterie des Sterbens" (The Lottery of Dying).[43]

One day, this is going to happen, or something very similar to this:
A self-driving car is racing along a road; the computer is driving.
The driver is sitting comfortably, reading his newspaper.

Suddenly, three children jump out on the road, and to the left and right there are trees. The computer now has to decide instantly. Will it do the right thing?

Three human lives depend on that.

Exactly! Who will live and who must die? Let us take a closer look at this question and the related ethical problem, and let us also analyze the implications the question has for the mind of the person asking. Namely this: the person either knows very little about self-driving vehicles and accident statistics or knows the facts and is not, as a matter of fact, asking an honest question.

So let us first ask a few counterquestions:

- Do you drive a car yourself?
- If so, how long have you been driving?
- If you replied "Yes" to number 1 and "For several years or decades" to number 2, have you ever had to make a decision on whom to kill with your car? Have you ever experienced this dilemma while at the wheel of a car? Do you know somebody who faced the conflict of either running over a person or persons or risking his or her own life by crashing into a tree? Did you have to go through such scenarios in driving school? Did you train for that?
- Whom do you trust more to make the right decision—if that is possible to determine? A driver who has to decide on such an ethical issue in a fraction of a second, never having faced that situation before and never having learned about it in driving school, or a software developer who had the time to think about this problem for hours, days, weeks, or months; do simulations; learn from them; and work out algorithms for it?
- Did you know that in the end, it is not the programmer making the decision but the car that learned to decide the issue through machine learning and human support?

Such a dilemma actually occurs so rarely that it's purely hypothetical for most of us. However, this is called the *trolley problem*, and it is regularly presented as an example by critics and people concerned and quoted as an argument against this "immature technology." Researchers have been acting out various scenarios of this type in an attempt to demonstrate ethical conflicts and recognize behavioral patterns.[44]

The original trolley issue is about the following problem scenario: a railway car with no brakes rolls down a slope. At the end of the slope, there are several workers on the tracks. From the position of the person interviewed, those workers cannot be warned in time. However, there is a railroad switch that would allow you to guide the car away from the workers onto a different track. Unfortunately, there is a person on the other track as well. The question is this: would you activate the switch, thus risking killing only one person, or would you not react, which would probably result in the death of several people?

This surely is an intellectually stimulating and ethically interesting question, but it (almost) never occurs in real life. Accidents are far more frequent at level crossings when people cross the tracks even though all the warning lights are flashing and the barriers are closed or the train engineer was not paying attention. These scenarios are more relevant because they occur more often and more people are injured, but let us take one step at a time.

In a utilitarian model, an action is assessed according to its greatest benefit for society, even if that sounds cruel. Could it, for example, be a greater benefit for society if the car runs down the grandmother, not the baby? From a utilitarian point of view, one would argue that a senior citizen has already lived most of his or her life and now is mostly a "burden" to society, whereas a child still has yet to live out his or her life and could contribute a lot to the common good. But is this always true? What if the child is seriously ill and will only cause a lot of medical bills, whereas the grandmother is just about to publish her first best-selling book? How can we and how can the car know things such as this, especially in a situation that does not allow for any other option? Install a facial recognition system, connect it to a huge database with private information about each and every individual, and make decisions based on all that? Certainly not!

Other variations of the trolley problem point out the availability of other objects or persons allowing for direct or indirect interaction. There is a fat man we might consciously push in front of the railway car to prevent the tragedy or a rod you might want to make the fat man stumble over inadvertently so that he will fall in front of the car and slow it down. Version 1 is rejected by the trial participants because nobody wants to actually be personally guilty. Version 2 is more palatable because the rod acts as an intermediary so that we are one step removed from our own action. It was not me but the rod that brought the fat man down. In one case, you would be directly responsible; in the other one, the responsibility is indirect. In the first case, you kill someone;

in the second, you let someone die. But the end result is the same every time: the fat man falls down in front of the railway car, which lets us zoom in a little closer to the moral dilemma. Applying the trolley problem to autonomous vehicles usually involves suppressing the context. It is hypothetical and artificial, developed by researchers who intended to keep as many parameters constant as possible. There is no third alternative, no acknowledgment of an alternative. For instance, we withhold the information that autonomous vehicles always maintain a 360-degree view of their surroundings; can see for about 200 to 300 yards into the distance thanks to the combination of LiDAR, radar, and camera sensors; and therefore can react more quickly than a human being.

There is always another option, as we saw in the example of the Go game—and as Captain Kirk of *Star Trek* proved in one episode. Mr. Spock had developed a program for the exam at Starfleet Academy that asked candidates to make decisions that were destined to result in their failure. Kirk hacked into the training simulator system to modify the parameters and thus became the only candidate who ever passed the test. Sorry, this is a *Star Trek* fan digressing . . .

Even if we theoretically think through a (moral) decision, this is no guarantee that we would really decide in that way in the moment of truth.[45] But what if we understand that our own life will be sacrificed for the lives of others through no action of our own? Should an autonomous vehicle endanger its passengers to save pedestrians? While we are willing to sacrifice ourselves in certain circumstances for the sake of others—think of firefighters and soldiers—we want control over that decision and not have a machine make it for us.

Chris Urmson, former head of the Google X self-driving technology group, said that Google had equipped its car with a very defensive driving behavior and had a kind of priority list of who and what it should consider with special priority. Accordingly, the autonomous car first tries to protect the most vulnerable road users, namely, pedestrians and cyclists. They are followed by larger movable objects such as other cars and trucks. Immobile objects come later.[46]

The vehicle has to experience many of these behavioral patterns itself, and this brings us to one of the misconceptions many people foster: it is not the engineer who makes the decision and programs the system accordingly; it is the system deciding on the basis of its machine-learned knowledge. The engineers initially provide a set of rules to the vehicle and help it with situations in which it encounters difficulties, but with millions of miles driven, the AI will develop its own behavior.

Google made this plausible using the example of cat pictures. Naturally, a programmer can try to define all the criteria determining a cat in a software program. How can you recognize from the photo that it depicts a cat? The fur, the shading, the eyes, the ears, the paws, the nose, the teeth? But how do you describe the paw to the computer? With claws retracted or extended? So you see the paw from the top or bottom, from the side, extended, or pulled close to the body or even under it. Immediately, we run into problems. There are too many things we would have to include in the algorithm, and even that would not guarantee that the computer could recognize a cat. Instead, the programmer gives some framework indications and then lets the system "look at" millions or images of cats. At the beginning, the computer is going to be lousy at this. It won't recognize the cats or may identify something as a cat that actually is not. Now the human comes into play. A human checks the things the system learns and how, changes the algorithm, and adds criteria and parameters. Over time, the system improves its success level in determining whether an image shows a cat or not. Each additional picture increases the score.

This is exactly what the developers of self-driving cars do. They enter algorithms and rules as framework conditions. Then they let the car drive around, first in the simulator to identify the most basic mistakes, and then they carefully start with the first ventures into the real world. The machine will make any number of mistakes. Those situations are introduced into the simulator and are repeated, parameters are modified, and new scenarios are added, repetition upon repetition. Slowly, the machine approaches ever more complex traffic situations, learns from them, and successively improves. Just as the AlphaGo computer refined its Go play until it was better than the Go champion, the car (and with it all the other cars of the same type) improves its driving skills until it eventually outdoes the average human driver.

So this approach demonstrates that the fundamental question is a very different one. The question should not be "Whom will I decide to kill?" but "How can I avoid killing anyone?" If you take a look at traffic accident statistics involving passenger vehicles, you will see that human error is the main cause of accidents, heading the list at 94 percent. The resulting global economic damage is $500 billion each year.[47]

Experts suggest that 55 to 80 percent of motor vehicle accidents in the United States are not even registered.[48] Half the accident victims in Western countries are actually the driver and front seat passenger, whereas in countries such as Kenya and India, the number is just 5 to 10 percent. More than 80 percent of the traffic accident victims there are pedestrians, cyclists, and motorcyclists.[49] India in general has the most disastrous traffic statistics. In India, 400 people die in traffic every day: more than 140,000 fatalities every year.[50] And more than 11,000 people die there every year simply because of badly installed speed bumps that are actually meant to prevent accidents.[51]

Most crashes in the United States happen on Saturdays and Sundays, between midnight and 3 a.m. More people die in those intervals than at the same time of night on other days of the week.[52] In Europe, the number of traffic fatalities is distributed very unevenly. Russia alone accounts for more than two-thirds of all European traffic fatalities.[53] Accidents often are also the result of corruption. Police and other people responsible for traffic safety often turn the other way if the price is right. In Mexico City, this behavior led to a drastic measure: in 2007, the last male traffic officer was retired and replaced by a female.[54] The number of accidents decreased, and simultaneously, the number of traffic tickets rose sharply by 300 percent.

Another method for reducing the number of accidents is the *girlfriend effect*.[55] Female passengers work as a corrective measure, especially for young men. They have a calming influence on male drivers and make them drive more carefully and slowly. The Israeli military uses this effect by assigning trained female comrades to soldiers returning home, thus commanding them to be their male comrades' "guardian angels."

All these facts are true, not just hypotheses. Today, many people lose their lives in traffic, and the main reason for this is human error. Eventually, therefore, the question remains: whose fault is an accident? Who pays for the damage, and who pays compensation to the injured party? The question of who is guilty is receding into the background in the case of self-driving vehicles, however. Volvo announced as early as 2015 that the company would compensate any accident caused by its autonomous vehicles.[56] Anders Kärrberg, vice-president of Volvo and head of compliance, said at the time:[57]

> Carmakers should take liability for any system in the car. So we have declared that if there is a malfunction to the [autonomous driving] system when operating autonomously, we would take the product liability.

We can imagine this as being similar to another civil law case: a divorce. Until 1976, divorce in Germany was only possible once the question of guilt had been established. This was necessary to decide the support obligations of one spouse toward the other. Since 1976, a marriage can be dissolved without determining whose fault it was that the marriage did not work out.

This might be a model for self-driving cars. Fleet operators or manufacturers of such a vehicle would then automatically be responsible. The expected low accident frequency would ensure that any persons suffering damage could be compensated more quickly. Sensor data would allow for a settlement on site, and insurance companies could transfer the compensation sum directly from the site of the accident.

Let us take a quick step back to review our starting point: how many accident scenarios really confront us with an ethical dilemma like the one presented by the trolley problem? Practically none. So why is this question so popular, and why does it get so much attention? Brad Templeton, technology expert and former consultant in the Google X Self Driving Car Project, provided a detailed statement that clearly shows the real questions one should be asking if one is honestly interested in the chances offered by this new technology.[58]

In questions concerning robots, we are more Catholic than the pope. We simply cannot accept unethical behavior in robots; they must be perfect. We grudgingly accept such behavior from other people, and if we ourselves behave unethically, we rationalize the thought and eliminate it.[59] Behavioral scientists such as Dan Ariely from Duke University research human behavior, especially irrational aspects such as cheating. Which conditions will make test persons make decisions that are less than ethical?[60] And these include doing it for somebody else, being separated several degrees from money (such as the experiment where researchers put six Coke cans and six $1 bills into a dorm fridge and after 72 hours observed that the Coke was gone, but the money was still there. "I am not stealing money, because then I would be a thief. I am drinking the Coke that would go bad anyway, so I am actually doing something good"), maybe not being watched, if everyone is doing it, if you are not being reminded what is ethical and what is not, or simply because smarter people think they can rationalize it better.

Humans regard robots as artificially created machines that should have been adequately tested and programmed without errors by their engineers. This approach remained valid for a long time based on a mechanistic view of the world. However, robots become more like people with AI and neural networks. They learn (like us), they have to gather experience (like us), and they make decisions in a way that seems almost like human intuition. The one move that AlphaGo made and that made the Go world go into a frenzy was a move Go experts considered plain instinct.

Consequently, asking ethical questions might also go in another direction: How should we classify robots and therefore autonomous vehicles? Should they become independent legal entities? Do we accept that they may make mistakes and show frustration, that they may hold certain rights, and that they can be punished? This is exactly what the European Parliament believes to be true. One draft for a civil law includes very specific proposals for the classification of robots, namely, as legal entities with all rights and obligations.[61]

Because we are in new territory with robots, our previous legislation might be insufficient. *"Nullum crimen sine lege"*—no crime without a law. Wherever laws are not available, we turn to ethics. Ideally, legislation, morality, and ethics will always converge, but reality is different, as always. What is the "right" decision, and why? Although this question is most interesting, it is not very helpful when we try to come to terms with the positive potential of self-driving vehicles: fewer traffic accidents, more mobility for previously disadvantaged groups of people, fewer vehicles on the roads, and consequently a reduction in traffic infrastructure.

However, we first anticipate the dangers emanating from new technologies. The media's tendency to mainly emphasize negative aspects and potential dangers is inherent in the system. A Facebook site I manage provides a collection of articles and arguments from the German-speaking press and impressively shows how the dangers and risks are exaggerated.[62] The strategy of playing with people's fears sells better and gets more clicks and more advertising money, but in the long term, it has a devastating effect on an objective and honest handling of new developments.

In an article for *Spiegel Online* on the topic of AI, the AI scientist Jürgen Schmidhuber was asked right at the start whether we should be worried about AI being out to get us.[63] Schmidhuber immediately called a spade a spade:

> I can see that today you don't want to talk with me about how our AI and artificial neural networks already can help billions of people, for example, by smarter smartphones and automatic early cancer diagnosis; you are more interested in potential dangers in the mid-term future.

This, in a nutshell, is the problem societies and politics are facing. We block ourselves by only discussing the dangers of new technologies but never the potential. Often we eventually discover that risks are no longer relevant because the idea has long since been implemented in a different way. We appear more intelligent if we present ourselves as critics/admonishers/

negativists than if we are idealists who want to change the world with our creativity. Harvard Professor Teresa Amabile explored this kind of thinking and proved its existence. She presented students with two book reviews. One was more favorable, and the other one was critical. The students were then asked to evaluate the writers' intelligence. They estimated the intelligence of the author of the critical review as being higher. The students did not know, though, that Amabile had written both book reviews herself.[64] So Amabile had to find an answer to the question of who was more intelligent—she herself or she herself?

Naturally, such cases as the trolley problem have to be analyzed carefully, and appropriate measures must be implemented. However, it is dangerous and irresponsible to make it the main argument against self-driving cars. Such discussions usually stem from less than honest motives.

The really important debates should be about, for example, the distance an autonomous car should maintain when passing a cyclist. Is it acceptable for a car to pass a cyclist at a lateral distance of only 3 feet, although the probability of an accident is higher than if a distance of 6 feet is used? But then we are in the opposite lane and endanger the oncoming traffic. Or else, is it acceptable that the traffic flow stalls because the car had to swerve slightly into the oncoming lane?

Such an experiment might already fail because of lacking permission from an ethics commission trying to save human lives. It is impossible for an engineer to simply program the "right" behavior. The width of the street, traffic conditions, obstacles, and the weather might make it impossible to respect the required distance. A human being is not able to forecast all the possible combinations, so the machines in the end must be able to find a safe combination of distance, speed, passing maneuvers, and staying behind the other vehicles by themselves or deciding simply to stop. The example of the cyclist is, admittedly, much less "sexy" than the trolley problem, but it is definitely more frequent and is therefore more important.

Speaking of ethics, there is an ethics commission for automated driving in Germany, and it presented its report in June 2017. Mind you, Germany was already deeply immersed in the discussion of ethical problems before there was a single executable legal act on autonomous driving. Thankfully, the commission, consisting of, among others, representatives from business, universities, and religious and legislative bodies, did not fall into the trap of the trolley problem but instead saw the positive effects of a functioning future self-driving technology. In the United States, most political leaders immediately think of progress. Senators occupy themselves with fundamental measures for allowing autonomous driving on the streets as soon as possible while still ensuring road safety.

After evaluating the accident statistics, the ultimate ethical problem we should discuss is whether we should not completely prohibit humans from controlling vehicles. Because human beings are mostly responsible for accidents, it is ethically not acceptable to allow this situation to continue once we have better technology available. People who object to this and does not want to hand over what they think is their control should talk to the surviving dependents of traffic fatalities about what is "loss of control."

Anyway, there is at least one category of (unwilling) traffic participants on which the developers of autonomous cars already focus, and no, I am not referring to the construction workers from the trolley problem. In Australia, neural networks for autonomous driving were developed with the express purpose of recognizing kangaroos and evading them on the road.[65] Hop along then!

Only the Others Make Mistakes

The greatest challenge for autonomous vehicles today is not so much the traffic rules or lousy road markings but instead other road users' mistakes and traffic violations: vehicles still moving when the traffic light has changed to red, cyclists going against the direction of travel, cars changing lanes without signaling first. The list is endless. People act irrationally. If they did not, we would have been using autonomous vehicles for years, and the world would be a simpler place—probably much less exciting, but safer.

Consequently, the British traffic authorities are expecting a deterioration of traffic situations during a transition period in which self-driving and manually controlled cars are (ab)using the streets together.[66] The reason for this is the behavior of autonomous cars: some are enthusiastic, others seethe with rage. As a pedestrian, I want a robot car that drives carefully and defensively, one that is helpful and attentive to me. However, as the driver in the car behind the robot car, I want it to finally shift up a gear because I am in a hurry.

The simple fact that autonomous vehicles in the first phase of their introduction will move particularly carefully in traffic is the reason why the British authorities are expecting an increase of 0.9 percent in traffic congestion. Because the surroundings influence the driving behavior of self-driving cars, the share of robot cars will only result in the expected traffic efficiency increase when robot cars reach a critical mass of 50 to 75 percent of all traffic participants. In other words, in the beginning, there will be more traffic jams. Hence it must be in the best interest of traffic planners to keep the transition time as short as possible. This may, for example, be ensured by a step-by-step transition in certain districts in which defined groups of vehicles are not allowed to drive anymore, whereas autonomous vehicles may move freely.

Human Asshole or Google Terminator?
Transgression of Rules and Other (Cavalier) Offenses

Imagine that you are at a pedestrian crossing and a black sports car slows to a stop. The driver revs the engine. What do you think? Sure enough, an asshole driver. But if the car in question is an autonomous vehicle, we immediately think of the movie *Terminator* and start to feel uneasy. Waymo employees used this to point out how important the design is for the acceptance of self-driving vehicles. The rather cute (and admittedly ugly) Koala cars Google was testing are designed to look harmless and to behave in a friendly and courteous manner so that people are more willing to use them.

We can now start to understand how important it is that robots—and autonomous vehicles are robots—demonstrate a behavioral pattern that inspires trust. One element of this is the physical appearance. Helen Greiner, cofounder of iRobot, spoke about testing customers' reactions to the first prototypes of vacuum cleaner robots.[67] The clients initially rejected the idea disgustedly. They imagined a humanoid robot with a vacuum cleaner in its hands. When Helen then presented the Roomba with its appearance of a slightly oversized Frisbee, they suddenly could all imagine using it. The harmless-looking Roomba actually gets nicknames from its owners, and there are even entire shops specializing in Roomba clothing. Yes, that is not a typo, clothes for a vacuum cleaner robot, because it looks so sweetly clumsy that people tend to anthropomorphize, or humanize, it. Helen Greiner also reported that some customers refused to send their defective Roombas in for repairs and instead expected an "ambulance" to come and pick it up.

Even military robots are not sent to the Joint Robotics Repair Facility, which is the official name of the workshop, but instead admitted to the "robo hospital." Soldiers even said goodbye to robots that could not be repaired with military honors, including gun salutes as if at an official funeral.[68] This is crazy, right?

And it shows that the sleek designs of traditional automobiles will have to undergo a profound change. The crucial elements will no longer be a streamlined body and showy details, but friendliness and harmlessness. Once self-driving vehicles are part of the sharing economy, in which many classes in society will no longer own a personal vehicle, it will generally be negligible what a car looks like as long as it does its job well. Even today, there is hardly any passenger in a taxi who chooses the vehicle on the basis of its design or color. Honestly, do you?

In addition to its cute exterior, the car should also be courteous: yield to pedestrians, go with the flow in traffic situations currently still dominated by human drivers, and convey its passengers swiftly and yet in clearly perceived safety. Google had to learn this lesson the hard way. It is not enough to simply comply with traffic regulations. There are fluent transitions in which, for example, traffic in total travels slightly faster than the speed limit, and anybody hesitating will simply be stuck in the same place forever. And what is even worse, behaving contrary to the majority of drivers exposes you to other drivers' wrath and incites them to become a traffic hazard as a result of their own (over)reactions. This is the reason why Google's test vehicles allow a manual speed increase on highways over the stated maximum speed limit.

Another danger for the Google vehicles at the moment is drivers who—admittedly, I am one of them—immediately whip out their cameras for unique shots and videos of such a car and forget that they should be watching the traffic instead of the Google car.[69] So it is high time people like me were removed from the steering wheel.

Traffic participants also notice other participants' subtle gestures. Audi, for example, discovered that drivers planning to change lanes often move closer to the road markings before switching on the turn signals.[70] Copying this behavior programs a more "human" driving behavior into the car. It remains to be seen whether this is actually the better driving behavior.

Another example is waiting for a traffic light to turn green. Google noticed that cars still pass an intersection from the opposite direction (where the light is red) with surprising frequency in the first two to three seconds after the light has changed to green for the others. In order to avoid accidents with such "red light runners," Google's cars are set to wait 2 to 3 seconds before they start moving again. This, in turn, makes the drivers waiting behind them nervous: they sound the horn, flash their headlights, and initiate risky attempts to pass just to finally get going again. Don't forget: 2 to 3 seconds . . . which seems like an eternity.[71]

Then there are certain situations in which the law must not be obeyed. One example is the double line that cars really should never cross. If, however, a car has to break for another vehicle stopping at the side of the road—say, a garbage truck—then it makes no sense to wait until such a vehicle gets moving again. This implies that an autonomous vehicle must be able to assess this situation correctly and then be able to make the decision of crossing the double line.

But what if the vehicle is ordered to disregard traffic rules because its owner instructs it to do so? If a parking space is only available for a maximum period of 2 hours, I might have to find a new spot. Now I can order the self-driving vehicle to find a new place to park every 2 hours. So, is this morally acceptable? Instead of deciding to park in a paid parking lot where I can stay for an unlimited amount of time, I circumvent the orders issued by the city or municipality with my command to the car.

Not everyone trusts the companies when they say that their autonomous vehicles will be able to make ethical decisions by themselves. Christopher Hart, chairman of the National Transportation Safety Board (NTSB), thinks that the federal agencies should draw up guidelines for ethical questions and safety systems, similar to what today is normally provided in the aviation industry.[72] Other experts doubt that it is at all possible to control this because the situations an autonomous vehicle could potentially be exposed to are much too complex and unpredictable for that.

On the whole, road users today expect an autonomous vehicle to drive like they do, without endangering passengers and people outside the car and without taking risks or causing nausea because of its driving style. It should be safer on the road than a human being and still find the fastest and most efficient way to its destination. The idea is to be pregnant, but just a little bit.

Skeuomorphism, Siri, and Symbols: How Do We Interact Today?

When my then five-year-old son first encountered Apple's iPhone voice assistant Siri, we could not stop laughing when he began asking Siri questions like, "How many days to go till Christmas?" and "When is Halloween?" He never gave up, despite the fact that Siri could not provide any helpful answers at the time. In the end, he asked her, "Why are you so stupid?"

While we were amused, his approach followed the same patterns we used to apply during times of technology change ourselves. Is there anybody around who still has a landline phone with a rotary dial? The hand raised to the ear with two fingers extended, intended to convey the message "Call me!," is hardly comprehensible for the "generation smartphone": they mainly use their fingers for texting. Equally, the finger placed on the wrist to ask for the current time of day is incomprehensible to the generation that lives without a wristwatch. Why would anyone wear a wristwatch when time is indicated on the smartphone anyway?

And younger people cannot even relate to the symbol for "save" used in many applications with the disk shown as the icon. Floppy disks disappeared from use in the early 2000s. Many of the digital applications we use imitate the physical object from which they were once derived: the notebook app looks like a notebook, and the e-book seems to turn pages. This sort of thing is intended to facilitate the transition from the physical object to a digital application for the user. The idea behind it is to present the users in times of technology change with small chunks of change that do not overtax them. This design principle is called *skeuomorphism*.[73]

Too many changes all at once are too much for most people. They reject the product. This is a partial explanation for Tesla's success, because the company did not fall into the trap of deviating too much from today's standards of automotive design to demonstrate that it is using a completely different drive system. Other manufacturers trying to underline their innovations with daring new designs often fail. As soon as the users have gotten used to this step, the designers can take the next one and overwrite the former language of form and image.

Apple's operating system for the iPhone changed its outward appearance only a few years after its launch and discarded in the application images of many elements that included similarities to physical objects from the real world. Leather or paper backgrounds were only dumped after a number of years, when users had become comfortable with the new options. This also allowed the developers to leave the limitations of analog technologies behind. A digital calendar and diary, for example, are much more versatile than paper-based versions.

One of my friend's daughters—the same age as my son—uses her smartphone in a completely different way than adults do. While I personally prefer writing longer texts on my MacBook, she, as well as many of my acquaintances and especially the younger generation, communicates with voice assistant Siri, dictates e-mails to her grandmother in Spain at breakfast or in the back seat of the car, listens to her e-mails being read to her, listens to the search results, and launches apps with a voice command. Whereas her mother uses Amazon's voice assistant Alexa only for setting a timer or an alert, the child uses it for many applications. All that is perfectly normal for her and my son, just as we think it perfectly normal to use a keyboard, whereas our parents and grand-parents use(d) the "eagle method" to find the characters on a typewriter: the index finger is poised above the keys, circles round until the right character is found, and then dives down on it. And some would definitely prefer putting an ink pen to paper for any text.

The way we communicate with machines is changing. Input devices such as punch cards, rotary knobs, switches, keyboards, joysticks, voice commands, and motion sensors came and went in a time frame of just three decades. And now several tons of machinery approach, move among us, and interact with us. These machines have to understand our intentions, and we need to be able to comprehend theirs.

Driving a car is a social event. This may be a surprising statement if you look at people in cars behaving like the streets are their personal property. Somebody who would normally be courteous and polite and hold the door open for others does not think twice about ignoring other people's rights of way or cutting in directly in front of them. Traffic rules are one thing; one's own behavior quite another. There is a huge gray zone when it comes to driving a car. You just have to travel out of your own city or country to see that in other places, they may have the same traffic signs, but the pre-vailing norms differ substantially. In Turkey and Italy, using your car horn is perfectly standard, and a car with a broken horn can easily be regarded as a write-off. In India, drivers sound their horn when they pass another to announce their intention. For the four-way stop frequently required at inter-sections in the United States, you need to observe carefully who arrived first and therefore is allowed to enter the crossing. Break your middle finger in Austria, and you are unfit to drive because you lost an important element of "communication."

Often the nuances are very subtle indeed: a little nod, a glance, a gesture may decide the priority of road users. It is somewhat reminiscent of a tango. At *Milongas*—tango events—in Buenos Aires, you ask someone to dance with you by establishing eye contact all the way across the dance floor. For dancers from other regions of the world, the dance floor is full of land mines. The question and acknowledgment or rejection is so subtle that anybody who is not Argentinian constantly commits social blunders. Non-Argentinian tango forums are full of embarrassing *Milonga* stories in Argentina. A dance on four wheels on a street is a similarly complex social activity, and now we have machines messing around with us as well.[74] Researchers from Sweden and England have put together an entire video collection that shows social traffic interactions between people and cars in autopilot mode as well as the resulting misunderstandings.[75]

How do we as pedestrians, cyclists, and drivers today communicate with an autonomous vehicle? How do I recognize that the car has seen me and leaves me the right of way? In traffic today, we usually try to establish eye contact with the driver of the (other) car. If someone averts his or her gaze, we are more careful; after all, this person may have overlooked us. Some drivers use this strategy to insist on their right of way.[76] But the question is how can we establish eye contact with a self-driving vehicle? Where should I look? As it turns out, the trend of presenting the front of a self-driving car in the shape of a friendly face is more than just a design trick. Similar to the cars in Pixar's film of the same name, such front parts can indeed serve to allow communication between humans and these machines: they can display text or symbols, or the vehicle talks to the pedestrian, indicating its intentions. Audi tested an LED display behind the windshield in an autonomous A7 model. The display told pedestrians that the car had noticed them—a kind of digital equivalent to a hand sign from a human driver.[77] A vehicle might also project a zebra crossing on the ground to demonstrate to pedestrians that they have the right of way. A more playful version would be for the car to smile at pedestrians to show that it is safe for them to cross the road in front of it.[78]

The Silicon Valley startup company Drive.ai (founded by alumni from the Stanford Artificial Intelligence Lab and acquired by Apple in 2019) is concerned with exactly this kind of thing. A mix of text and emojis—universally understood symbols—that, as a first step, is designed to help pedestrians learn to understand a vehicle's intentions.[79] Pedestrians are at the center of this interaction because most people will probably come into contact with autonomous vehicles from the outside or as drivers of a manually controlled car, not as passengers in an autonomous vehicle. Drive.ai uses sensor equipment on its test vehicles that includes a display. The text there indicates to pedestrians and other drivers what the car plans to do next.

Waymo has already included in its cars a sound as a means of communication, and it is none other than our old-fashioned car horn.[80] A computer algorithm decides when it is to be used. Is this a dicey situation? The horn is sounded if another vehicle changes lanes too close to the Waymo car and therefore poses a potential danger. Or if a car is exiting a driveway or side street to merge into traffic and it is not clear whether the driver has seen the Waymo car. Or if there is a vehicle approaching in the wrong lane. The car sounds the horn twice briefly if another vehicle reverses in front of the Waymo car.

Originally, any horn sound was only indicated in the interior of the car itself so that the test driver could tell the vehicle whether the signal was appropriate or the car had misinterpreted the driving situation. Waymo eventually wants to simulate an experienced and patient driver as regards the horn sound. However, the sound also has another purpose. Google's Koala cars are electric vehicles and therefore do not emit any engine noise. They are easily ignored. A polite little horn sound can warn pedestrians and cyclists.

The car not only has to communicate its intentions to the other road users, but also has to interpret the others' intentions and react accordingly. One example is the hand signals used by bikers to indicate that they are about to change direction and/or lanes.[81] We should remember that it is not enough to simply recognize the hand signal; the comprehension must be followed by an appropriate maneuver.

Cyclists, for example, are slower in comparison, but they are very agile traffic participants. Their movements are often difficult to predict, especially in the context of the surroundings. One of Waymo's examples shows cyclists passing a row of parked cars and having to swerve around an open passenger door. The algorithm must recognize the open passenger door and predict that the cyclist might make a sudden movement without making a hand signal prior to the maneuver. The Waymo car has to leave sufficient room for this movement to be completed. In another case, the hand signals might not be executed correctly, or they might be sloppily done. Sometimes all the indication cyclists give is one look over the shoulder; then they initiates their change of lane or turn.

Hand signals may also be received from a police officer directing traffic, a construction worker keeping one lane free for a construction vehicle, a passenger/hotel staff member signaling for a taxi, or a driver stranded by the side of the road and in need of help. Or they may come from somebody planning a robbery. The vehicle also has to understand whether it sees a hand signal that is intended for it or just a casual gesture. Let me tell you about an incident a friend experienced. Back then, he worked for Waymo. On our way to a brewery in Mountain View, we wanted to cross the road when one of the vehicles he works on passed us. His colleagues were inside, and he waved to them from the side of the road. The car interpreted this wave as a hand signal and braked to a stop.

A group of English researchers developed a language for autonomous vehicles called Blink that uses hand signals.[82] At present, all the system can understand are the signals for stop and continue, but thanks to machine learning, it is currently being trained to recognize hundreds of other signals, among them some that are particular to a certain culture.

Gestures shape human behavior, as scientists from the Berlin Center for Gesture Research found out, and not just because we seriously communicate with autonomous robots. Making swiping gestures on a smartphone, pointing to our wrist to ask what time it is or to demonstrate urgency, and making a telephone receiver gesture with your hands to ask someone to call you later—all these are just some of the gestures we use to communicate or used to use to communicate.

A first step for providing robots with some understanding of gestures is to classify them. The Ars Electronic Future Lab in Linz in collaboration with Mercedes Benz has created categories for almost 150 patterns of interaction and hand signals. Which of these gestures are ones that we use every day? And which of them can sensibly be used in communication with robots?[83] Do we have to invent new gestures, or can we use the ones already available? Furthermore, which gestures can be misunderstood? Depending on the country you are in, the same gesture can be confirmation or something nasty. An "O" formed with your thumb and index finger signals okay in our country but is seen as a symbol for an excretory organ in Brazil—and is hence an insult. A "V" made with your index and middle fingers and shown with the back of your hand facing the other person is the same as showing just the middle finger for anybody in Great Britain. This sort of thing is not easy for us, and there are innumerable cultural blunders that the controls of an autonomous car can make as well—all of which may well cause the people to reject the new technology. And even in the same part of a city, gestures can vary during the day. Maya Pindeus, CEO and cofounder of London-based startup Humanising Autonomy, is creating a catalog of gestures and tagging them with their locations and times of the day. Businesspeople crossing an intersection in the morning signal differently than do inebriated pub crawlers at 11 p.m.

Another company working on deciphering intent and behaviors of people around autonomous vehicles is Perceptive Automata. Predicting human intent in traffic makes everyone safer.

China is one step ahead in this matter. Scientists at Nankai University are doing research on how to control self-driving cars with your thoughts. A passenger wearing a headset with 16 sensors to communicate instructions to the vehicle via electroencephalogram signals was quite successful in the first round of tests. This solution may offer great opportunities for disabled persons.[84]

However, as one Waymo vehicle had to painfully experience, the old adage of "Might makes right" also applies to self-driving vehicles. The car brushed against a public bus that simply would not behave as the system predicted. The car had expected the bus driver to leave the right of way to it in this particular traffic situation, but the driver did not. Fortunately, the only damage incurred was a bent fender and a torn-off sensor.[85] The learning experience was that a larger vehicle is less likely to give up its right of way than a smaller one. A smaller one better bends than breaks, and the stronger one makes its own laws. Consequently, machines have to understand and be able to accept human behavior. You cannot just program them with a fixed set of rules and have them adhere to those rules at all times. People break rules or interpret them flexibly, and a machine accordingly has to know when it can do that too and when that is actually what is expected of it.[86]

How far do people trust self-driving cars? A survey on this topic brought to light a surprising result, in which the age of the person featured prominently. More than half of all car owners in generation Y (56 percent) and generation Z (55 percent)—namely, millennials and today's teenagers—stated that they trusted self-driving vehicles. Only 18 and 11 percent, respectively, said that they definitely did not trust self-driving cars. Of the survey participants from generation X (aged 30 to 50 years), 23 percent stated that they trusted self-driving cars (27 percent definitely did not), and in the baby-boomer generation, the distribution was 23 percent did and 39 percent did not trust such cars.[87] There are also differences in the comparison between states and countries.[88] A vast majority of people in growth markets such as Brazil say that they trust autonomous vehicles (95 percent!), whereas in saturated countries such as Germany, the majority does not trust this technology.[89] In India, 86 percent of the population trust self-driving cars, 70 percent in China, 60 percent in the United States, only 45 percent in France and England, and only 37 percent in Germany.

However, the real surprise comes when you finally travel in a self-driving vehicle yourself. As many online videos about Tesla's autopilot show, the drivers remain skeptical for only a short time; then trust the system more than expected. Google had the same experience when it offered its employees the first prototypes for testing, although this kind of trust was in no way justified in that early stage, neither in the autopilot nor in the Google cars.[90]

Tests with pedestrians showed that they were ready to absolutely put themselves in the vehicle's "hands," as it were, crossing the street in front of them without fear or hesitation. This may lead to dangerous situations as long as we are in the transition phase and self-driving and manually controlled vehicles share the roads, the hand gestures used by each differ, or one of the participants does not react as expected.

Humans are not always the most endangered road users. The robots themselves may be the victims, too, in danger from a source we might not be aware of at all: children. Researchers at Osaka University observed that an autonomously moving robot that was put in a mall to answer the shoppers' questions was repeatedly damaged by those little ones.[91] The children—as long as they were unsupervised by adults—stepped in its way, kicked it, hit it, or shook it. The security robots built by Knightscope and used for instance at the Stanford Shopping Center suffered a similar fate. Japanese researchers solved the problem by having the robot approach the nearest adults as soon as it notices that children get close to it. In this case, as in others, the simple presence of an adult helps.

Although children say that they know their behavior is not right, they still do it again and again. It appears that—according to some studies—children learn to feel empathy in this way. The new robot generation that is not yet widely diffused probably experiences a similar situation as the coach drivers did when the first automobiles roamed the streets. They, too, were surrounded by curious children and adults, were touched, and had parts ripped off them. Early motorists had to properly defend their cars against such abuse.

But what should be done if a self-driving vehicle actually causes an accident? When millions of such vehicles are on the roads and someone is injured or even killed? Will the manufacturer be liable? Or the operator? Are the passengers liable? Or the accident victims themselves? The safety driver? And this is not an academic question anymore, because an Uber fatality—the first fatal crash with an autonomous vehicle—occurred in March 2018 in Tempe, Arizona. On that night, one of the then 100 experimental autonomous vehicles from Uber was traveling in autonomous mode at a speed of 43 miles per hour when a woman crossing the street was hit. The victim died later that night. The car's system recognized the pedestrian but could not activate the brakes because they had been disconnected from the system. Also, the lone safety driver in the car was distracted, glancing at her smartphone.

The result was not only a person who tragically died in that crash but also the suspension of all development and test activities by Uber in multiple states. Whereas the company settled with the victim's family and NHTSA and the NTSB seem to have partly exonerated Uber from some fault, for insiders, it came as no surprise that Uber was involved in the first fatal crash. The company's culture and behavior in the past put it at the top of a list among experts of what company could be the first one to be involved in such an incident. Also, the internal pressure to catch up with Waymo and other companies and demonstrate that its cars on average can drive 12 miles without a disengagement may have contributed to the crash. Since then, Uber has tentatively started test activities again, but this time with a new set of measures, including two safety drivers at any time in the vehicle and a much-required more cooperative attitude toward the communities in which the company is operating.

It also had some fallout for the whole industry. Regulators, lawmakers, and the public were putting more scrutiny on the activities of the industry, and any incidents are now being put under a magnifying glass. This also explains the seemingly slow progress, with companies such as Waymo being very careful with announcements of public rollouts and true driverless activities.

Let us take a look at some other (fictitious) examples of what could go wrong and the questions we may have to ask.

Case 1: Wrong Reaction

An empty self-driving vehicle is on its way to pick up a passenger. It arrives at a blind corner, and just as it is driving by, children run out into the road. It is too late for the vehicle to stop, and it runs over the children.

A former colleague at work once told me about a game he played with his friends when he was six. They hid behind parked cars and jumped out at the last moment when a car was driving by. He recounted that game with just that kind of shaking of his head that we would all have when thinking back to such youthful folly. It was more down to luck than anything else that he and his friends never suffered as much as a scratch, and all the drivers were able to react and brake in time.

Our self-driving vehicle now is in a situation that, according to the manufacturer, has never occurred in millions of driving lessons and that is hence unknown to this car. How should a judge handle this? Either the judge points out that parents are responsible for supervising their children and exonerates the vehicle and the manufacturer or owner, or the judge tells the manufacturer to close that gap in its system.

Case 2: Defective Sensor

It is very probable that one or several of the numerous sensors ensuring safe operation of a self-driving vehicle will at some point malfunction. This may not even be due to a technical defect. The low sun may blind a camera, rain may cause the LiDAR to receive a "blurry" image, or baking heat could heat up a sensor so much that its measurements are no longer correct.

In this case, we face the question of whose fault such a malfunction is, just as today the driver is responsible for a vehicle that was not serviced regularly or that was known to have a defective sensor. If a sensor provides wrong data, the first step is to identify the reasons for the wrong measurements. Are they due to the environment, faulty program routines, or incorrect adjustment? Could the car have detected its malfunction itself?

Case 3: Human Instructions

A passenger instructs his or her self-driving vehicle to do something that conflicts with the safety of both the car and its environment. Subsequently, the car is involved in a crash.[92] We have to acknowledge that robots must also be able to refuse human commands. Scientists from the Human-Robot Interaction Lab at Tufts University demonstrated exactly this process.[93] A human instructs a small robot to proceed on a tabletop. When the robot reaches the edge and is instructed to continue, it refuses that command because following it would put it in danger. The robot objects to the commands of its human master. There is a special mode to circumvent this function, the *superuser mode*, which allows the administrator to instruct the robot to forgo the safety measure and endanger itself or a human being. In our case, the robot will continue walking over the edge of the tabletop, where a human will, it is hoped, catch it. The robot must be able to trust its human.

There is a whole further set of possible situations in which a human being circumvents the preprogrammed safety functions of the robot in administrator mode. Humans continually disregard safety measures, sometimes for good reason, sometimes because of stupidity, and sometimes with bad intentions. The responsibility passes from robot to human.

Case 4: External Interference

The matter becomes still more complicated when someone interferes with the vehicle from the outside. Just as recklessness and the search for thrills will always make some people point a laser pointer beam at an aircraft, compromising the pilot's field of vision in the cockpit, people may want to try to interfere with robot vehicles—by suddenly stepping out into the road as the car passes or because someone goes to a lot of trouble to get the car under remote control.

What does a robot do when it cannot perform an action? Don Norman, former professor of cognitive science, proposes that robots should be able to express their frustration.[94] While this sounds strange at first, the proposal does have advantages to it: a self-driving vehicle lost in a traffic circle, unable to find the exit, can ask for human help by expressing its frustration. If it did not, it would continue to apply the same strategy over and over again and fail time and again—and would appear to humans as a really dumb machine. If, however, it indicates that it is frustrated, people tend to try to be helpful.

Frustration also helps us to drop a task and occupy ourselves with another. This same effect helps robots as well. The vehicle in the traffic circle may, for example, because of its frustration algorithm, end the task of finding the correct exit and turn to a new one, say, leaving the rotary at the next exit and trying to reach its destination via a different route.

Robots with frustration mode are thus able to avoid *deadlocks*, basically dead ends in which they cannot complete a task (reach the destination) because another task must be finished first (leaving the rotary at the indicated exit). The questions a robot has to ask itself and decide independently can be limited to five categories:

1. **Knowledge.** Do I know how to complete action X?
2. **Ability.** Am I physically able to execute action X at this time? Am I normally physically able to execute action X at this time?
3. **Prioritizing objectives and scheduling.** Can I complete action X here and now?
4. **Social role and obligation.** Am I obliged to execute action X because of my social role?
5. **Standard acceptability.** Will one of the acceptable standards be violated if I execute action X?

While the first three questions are self-explanatory, the fourth question refers to the person issuing the command. Is this entity authorized to issue instructions to me? Should I react to every person standing at the side of the road and lifting his or her hand, asking me to stop, or should I only do this for my owner or a police officer? And number 5 asks whether the requested action can endanger the robot itself or a human being.

How do we assess the question of responsibility and accountability when an accident happens or a crime is committed? The prerequisite is that a *moral agent*, for example, a person able to distinguish between morally correct and incorrect behaviors, is aware of the consequences of his or her actions and is able to implement an appropriate action.

We can indeed draw on approaches that we have described before, although they may seem repulsive and completely inappropriate at first. Jerry Kaplan, author of the book *Humans Need Not Apply* refers to the (fortunately abolished) slave laws discussing similar questions in the period before the Civil War. Slaves were (tangible) property and had owners. The rules regarding liability and payment for damages caused by slaves and determining who should be punished were set down in the *slave codes* (as were many other regulations, usually to the slaves' disadvantage). Owners were only held responsible in particular cases; in many other cases, the slaves were punished. The determination of who was liable was less concerned with the law than with the slave owner's well-being. Would a slave's punishment possibly entail undue disadvantages for his or her owner? And just for the sake of completeness, even in the seventeenth and eighteenth centuries, those slave codes were not as undisputed as it may seem today.

This leaves us with the unanswered question of how robots and companies should be held responsible in the case of misconduct or any damage caused by them if we are not dealing with an individual person. Should we just punish the persons responsible, the persons implicating, the instructors, or the entire company? Do we concern ourselves with motives, intentions, and the effect on our society as well?

Naturally, a robot cannot be sent to jail. But there are ideas that permit us to make an equivalent. Both a robot and a company have a purpose. Their entire existence is aimed at fulfilling that purpose. If they are sentenced to pay high fines, and if trade licenses or corporate licenses are revoked, the company cannot fulfill its purpose anymore. A judge may even order the closing of a company. All these measures deprive the company of its basis for continued business activity. This may be equal to a death sentence. One example of this is the accident the *Deepwater Horizon* had in the Gulf of Mexico in 2010: the authorities subsequently forced British Petroleum (BP) to assume the immense cost for the cleanup efforts and imposed fines of several billion dollars.

The purpose of a self-driving vehicle is to transport us and our goods. In case of a fine, it can no longer complete the task it was built for. Jerry Kaplan argues that operators of fleets of self-driving cars could be forced to register every car as its own company instead of founding one company for them all together. This would prevent the entire fleet, for example, a robotaxi fleet, from going out of business in case of an accident and insulate the other vehicles against resulting claims for damages because only the one vehicle and the corresponding company would be involved.

Asimov's Robot Laws, or What Is (Still) Legal?

Popular science fiction author Isaac Asimor postulated his laws of robotics as early as 1942. They have a hierarchical order and are as follows:

1. A robot may not (knowingly) injure a human being or, through inaction, (knowingly) allow a human being to come to harm.
2. A robot must obey the orders given it by human beings—except where such orders would conflict with the first law.
3. A robot must protect its own existence as long as such protection does not conflict with the first or second laws.

Asimov later also introduced a zeroth law:[95]

0. A robot may not (knowingly) injure humanity or, by inaction, (knowingly) allow humanity to come to harm.

It does not speak in our favor that these laws of robotics were upheld for half a century without any updated postulation. If you look at the laws more closely, you will soon discover that they fall short, are too unclear, or are too exact. What about animals? Does a robot have to execute commands regardless of who issues them? What if the robot is physically not able to execute a command?[96]

The laws actually were already somewhat modified by Asimov himself for a later science fiction novel that he drafted but could not complete:

1. A robot may not injure any human being.
2. A robot is obliged to collaborate with human beings unless this collaboration leads it into conflict with the first law.

3. A robot must protect its own existence as long as it does not come into any conflict with the first law.
4. A robot is free to do as it wants, unless doing so would lead it to violate the first, second, or third law.

Robots here for the first time are granted free will of their own. Should we therefore allow robots and self-driving vehicles to drive anywhere they want? And if so, under what circumstances?

Bryant Walker Smith, assistant professor in the law school at the University of South Carolina, is of the opinion that self-driving vehicles initially were not illegal in the United States because the United States applies the fundamental legal principles that "something is permitted as long as it is not explicitly prohibited."[97] Whereas the United States thus allows its citizens the greatest possible freedom, Europe often adheres to the contrary principle: "something that is not explicitly permitted is prohibited."

Only recently have several U.S. states issued acts in an attempt to regulate self-driving vehicles. Furthermore, former president Barack Obama instructed the traffic safety authorities to draw up regulations for self-driving vehicles with the explicit requirement of permission, not obstruction. The reason was a first proposal on the part of the California authorities reminiscent of the so-called locomotive act (also known as the *red flag traffic act*) issued in Britain in the late nineteenth century. The act decreed that a person with a red flag or lantern in his or her hand (hence the name for the act) had to run in front of an automobile to warn pedestrians and coaches.[98] The effect was that the speed of cars was naturally limited to walking speed, and British engineers preferred to turn to something else, namely, the development of track-bound locomotives.

The original suggestion from the California DMV from late 2015 only envisioned operation of a self-driving vehicle with a driver in possession of a driver's license on board.[99] However, this passage was deleted after industry lobbyists and representatives from disability associations protested.

One historical anecdote I do not want to keep from you dates back to 1896, when the U.S. state of Pennsylvania was mainly concerned with the emotional balance of cattle and horses. Local legislators adopted a law that obligated motorists accidentally encountering cattle or other farm animals to:

1. Immediately stop the car.
2. Disassemble the automobile immediately and as quickly as possible.
3. Remove the separate parts from view and, for example, hide them behind a bush until the horses or cattle had settled down again.

The bill was never signed because the governor of Pennsylvania vetoed it.

Concessions made by regulatory agencies and a legal principle that initially allows things that are not regulated instead of preventing such matters are conducive to innovation, too. This is another reason why the American manufacturers are so far ahead of the Europeans. If I first have to wait for months or years until I get a concession for a single test drive, I lose any technological edge, or else my technological distance from the competition grows steadily. It took until 2015 for Daimler-Benz to receive permission to test its trucks with autonomous technology on a specific stretch of highway. In December 2016, Stuttgart, the home of Daimler-Benz , no less, granted the corporation a test license for autonomous vehicles.[100] Austria, home to Volkswagen's Porsche and Piëch families, nonetheless enacted a regulation in March 2019 that requires companies to inform the local governor one month ahead of a test drive with an autonomous vehicle. At this point, Google had already covered more than 10 million miles on American roads.

Robot expert Brad Templeton, who was involved in Google's self-driving program, recommends holding off any extremely detailed regulations for the moment. In case of disruptive innovation, the path that is eventually taken is often difficult to foresee. First trends may turn out to be dead ends, and unexpected paths may open up. Overzealous regulators prevent innovation from flourishing, as the example of Uber and other ridesharing companies amply demonstrates. In Germany, Hungary, and France, overzealous regulators under pressure from the threatened taxi lobby prevented this new industry from spreading, which would have benefited consumers.[101]

Uber is also an extreme case, although in the other, namely, the nonregulated, way. Shortly before Christmas 2016, the company began test drives of its autonomous taxi fleet in San Francisco without having previously obtained a respective test approval from the California DMV. California had at the time issued such approvals to three dozen companies and had been working on still relatively loose rules since 2012. After some meetings with agency representatives and under threat of punishment, Uber moved all its vehicles from San Francisco to Arizona, whose governor welcomed them, pointing out that in his state no approvals were required.[102]

Sooner or later, regulations will be inevitable, and several states have already drafted them, issued them, or rejected them.[103] The process for introducing regulations and statutory provisions usually is as follows:

1. This industry/sector deals with items and conducts activities that may have dangerous consequences.
2. This industry/sector does not have any intrinsic motivation for reducing such dangerous activities to a harmless level if it is not forced to do so.
3. The government forces the industry/sector with regulations and statutory provisions to reduce such hazards to a nondangerous level.

The following approach is not common, however:

1. We imagine what the industry/sector could possibly do wrong before it actually realizes its mistake, and we prohibit it in advance.
2. We categorically prohibit anything that is new and allow new elements to be introduced in very small steps and after extensive testing and verification.

Since the technological development for autonomous vehicles has advanced a lot more in the United States, it was not just the immediately involved states that discussed regulations but also NHTSA. In September 2016, this agency published a proposal including 15 safety considerations that was mainly positively received by the corporations.[104] NHTSA has to perform a balancing act between public safety (protection against the technologically not yet mature test vehicles) and the public interest in the development of such technology. The authorities are very aware that all this might in the medium term lead to a much higher degree of safety on highways, which is exactly what it needs to do—and hence its name.

The European Union began to discuss this topic on a transnational level as late as early 2017. An agreement was made that testing of autonomous vehicles may also be done in cross-border collaboration. Until then, each of the 28 member countries had its own regulations.[105]

Again, some ideas for the operation of self-driving vehicles can be found in the regulations for air traffic. The U.S. Federal Aviation Administration (FAA) banned drones, for example, from delivering packages by Amazon or flying over houses to provide photos for real estate websites. Only private drones flying at low heights are allowed, but they, too, have to be registered (since 2016). There are even rules on where they may fly. Areas around airports, air corridors to the airports, or places of special importance such as nuclear power plants or the White House are prohibited zones. And the authorities are not just relying on the collaboration of drone owners. Drone manufacturers have to include microprocessors that contain information about the prohibited zones. A drone with such a device would then refuse any flight into a prohibited zone ordered by the drone pilot.[106]

In this context, there are some interesting questions for self-driving vehicles. Is it acceptable that somebody tells me where I may drive and where I must not go? Can my insurance company order me to avoid certain areas? Even previously, construction sites were one reason to refuse entry, as was closed-off private property. But now there are new possibilities.

In Rio de Janeiro, car owners can already choose to install tracking devices provided by their insurance companies. The devices record the GPS data and know where the car travels. In this way, the insurance companies can see whether the driver regularly moves in a district from which many car thefts and burglaries are reported. If so, the insurance premium is automatically increased. Drivers avoiding such areas pay less. This has potentially toxic political impact, and there is a reason why the insurance companies do not publish which areas they consider dangerous.

How will electric barriers be treated? Probably just like any structural measure allowing entry or exit. However, there is a gateway for abuse included in this, especially for connected cars. What will stop me from setting special signals, for example, from pretending an accident has happened because I want to electronically block my street for everyone and then enjoy an undisturbed night's sleep?

The purchase of self-driving vehicles poses further interesting questions. What happens if the owner does not pay the installments due for the car loan? Can the bank simply order the car to drive back to the manufacturer for repossession purposes? Lawyers interpret this as an unlawful interference with property and believe that contractual clauses asking for advance permission from the buyer would not be legally enforceable.[107]

However, even if cars today already collect a lot of data and entire patterns of movement are recorded, it is clear that regulating authorities will prescribe a kind of black box. The German Department of Transportation is publicly considering such steps.[108]

Little Blemishes and High-Tech with Emotion

Anybody first seeing a Google Koala car will notice two controversial elements:

1. How cute it is
2. How ugly it is

We already discussed the reasons for the car's cuteness in detail. The reasons for its ugliness are far more difficult to grasp because many car owners still indicate that after the price, the car's design and look are important criteria for purchasing a particular model. Is it impressive? Is it racy? What color do I want it in? Which manufacturer? But all this is only relevant as long as I am about to buy a car for myself. A taxi, by contrast, is just a taxi. In a similar vein, most of us will not be able to tell straightaway whether we are traveling in a streetcar by Siemens or one by ABB, or a plane by Boeing or one by Airbus. As soon as the business model of self-driving vehicles used as ridesharing proves to be the dominant model, as is currently predicted, the exterior look will lose much of its importance, in contrast to the design.

An autonomous car is fundamentally different from a manually controlled one not only in terms of looks but also in terms of its entire function. The following quote from *Wer kriegt die Kurve?* (*Who Gets Their Act Together?*) by Professor Ferdinand Dudenhöffer clearly shows how difficult this intellectual leap is for the traditional car manufacturers:[109]

> The third customer value of a robot car is the joy caused by the aesthetics of the digital intelligence. Apple showed how to find design vocabulary for AI, a stylistic idiom that expresses elegant design, high quality, clarity and precision in its iPhone, iPad and MacBook. People do not think in an abstract way; they think in images. Emotions are triggered by associations with images in your head. Therefore the design plays an overwhelming role. The successful new orientation of Mercedes with its CEO Dieter Zetsche is closely linked to the corporation's new design idiom shaped by chief

design officer Gorden Wagener. "Sensual clarity as an expression of modern luxury. The objective is to create clear shapes and smooth surfaces that stage high-tech and yet have an emotional meta-level," Wagener is quoted on the Mercedes homepage. One important design element of all Mercedes models is the so-called dropping line, a line on the car body that first rises and then drops towards the tail, thus creating tension and completing the sculpture. That is the kind of design language the traditional automotive manufacturers also have to try and score when it comes to automated driving. Pure software corporations often have less of an idea of design—possibly one reason why the Google car currently looks like it was borrowed from a Playmobil playpen.

We should remember that it took years for people to intellectually come to terms with the new opportunities offered by the horseless carriage (automobile) in the context of the first railways. In trains that originally looked like a string of squashed coaches, the conductors had to laboriously negotiate dangerous open connection points from one car to the next. It took decades for car designers to shed the image of the coach and build closed units with the respective safe transition points. When it comes to "driverless" cars, the experts find it difficult to depart from the idea that the wheel and the interior design should continue to look like that of a combustion engine vehicle. We already mentioned the possibility of regarding the chassis itself as a battery instead of imagining it as a separate block.[110]

American sociologist Sherry Turkle at the Massachusetts Institute of Technology has been examining the relationship between humans and machines in everyday life for decades. She is still surprised each time to find how quickly people start to trust machines and occasionally develop a relationship with them that may be deeper than that with other people, even close family members.[111] We already heard from Helen Greiner about her experience with the humanization of Roomba vacuum cleaner robots by their owners and learned about hard-boiled soldiers admitting drones and bomb-defusing robots into the robohospital.[112]

The emotional connection Professor Dudenhöffer evokes is less about coolness or power when it comes to robot cars but rather about how humane and trustworthy the user perceives such a system to be. Design, after all, is still important, but in a completely different way from today's automotive designers and experts think.

The Google Koala cars need a very special design so that the sensors have a mostly uninterrupted view of the environment without any blind spots. It emerged for the "smart cars" especially, but for the Google vehicles, too, it is possible and desired to use such cars as wonderful advertising space. Google had several of its vehicles with their side doors decorated with art. Those spaces could also be equipped with neon signs sending messages to the surrounding humans. There are completely undreamed of new fields of applications for such vehicles.

The currently available technology and safety authority regulations affect the design: powerful LiDAR systems are so bulky and expensive that Google has decided to just place one on the roof. Other manufacturers include up to eight LiDAR systems, installed in the front and rear bumpers (two each) and on each corner of the roof (four). As soon as the LiDAR systems become smaller and cheaper, the design can change, too. A 1,000-page NHTSA handbook for automotive manufacturers specifies the control elements the car must have in every detail. Because the current edition of the handbook still presupposes that a human driver will be present, exterior mirrors still have to be attached, although they have no real value for a self-driving vehicle.

Discussing the design of a self-driving vehicle does not stop at the outside. The interior is also going to change. Anybody taking a seat in Google's Koala car at the Computer History Museum in Mountain View, California, will be surprised at how big it actually is. On the road, it looks like a minicar, but as a matter of fact, you can get in almost upright without bending down, as in a London taxi. Because it does not require any steering wheel or pedals or other control elements, and not even a transmission because it is run electrically, the interior is so empty and open that you almost feel lost: there is so much free space in the vehicle.

This offers new opportunities for designers and passengers. BMW presented a concept at the Consumer Electronics Show in Las Vegas in 2017 that even provided space for a small bookshelf—for printed books, mind you.[113] Entertainment systems, workstations, sleeping facilities, and the like are possible. The time not needed for steering the car can now be used in different ways. The interior design will gain importance with autonomous cars and probably fulfill the main function of creating brand loyalty in passengers. We will later also discuss how autonomous vehicles can impact the layout of cities.

Moonshots and Data Business:
Let's Talk About Google's Role

The question of why Google-Waymo of all companies should play such an active role in the development of autonomous vehicles is one that automobile experts still have not answered satisfactorily. Information technology expertise definitely is massively involved, but what is the underlying motive? Surely Google is mainly a search engine? And what are Google's plans for the future?

We have to include two factors into our considerations. First, Google aims for "moonshots"; that is, it aims to solve extremely challenging and difficult problems in order to provide humanity with important improvements. Sending a man to the moon by the end of the decade, and bringing him back again (literally), constituted such a "moonshot," announced by President Kennedy on May 25, 1961—and was accomplished with the moon landing of July 20, 1969. The founders of Google feel that it is their duty to invest the huge amounts of money they have made with the search engine business into such "world-changing projects."

The second reason is more logical and connected to Google's core business. When Google first offered its Google Maps service, I was puzzled by how this fit with other aspects of the business. On the one hand, a search engine, on the other, a map service? When I then saw the concrete application of a real estate agent who showed the location of offered houses on the maps via the open programming interfaces ("Click here to know more"), I finally understood. The virtual internet world the search engine's automated bots combed for information was superimposed on the real world and provided new content. The Google Street View cars taking pictures of the streets were just physical bots complementing the virtual ones. If you can go from a couple of Street View cars to an entire army of self-driving vehicles, you can quickly gather information on all the streets of the world and combine it with virtual information.

Google's motto of "collecting and providing all the information in the world" thus acquires an entirely new dimension. New application purposes were found with the maps and the Street View cars. Thanks to Google Street View, disabled people can plan their routes better. Is a sales room accessible for handicapped people or not? Travelers can look around the neighborhood of their Airbnb accommodation before arriving. For example, this was very useful to me in Paris when the taxi driver did not know the district very well, and I knew from my previous research that there was a chocolate shop directly next to the house entrance. As soon as I saw the store, I had the driver stop.

Do self-driving vehicles mean that Google will start building automobiles as well? I do not think so. It is more probable—and this appears indeed to be the case with a view to its automotive partnerships—that Google is trying to become a supplier and, for instance, provide its self-driving technology to the manufacturers in a box. Today, modern cars already include hundreds of sensors that provide data, and the number is going to increase in the future. Thanks to wireless connection, it is possible to read the precise road map services online, and updates for software and map details can be downloaded. It is also possible that Google provides free services of this kind to the automotive manufacturers—including the black box—and asks for access to the vehicle data in return.

At the same time, it is not at all clear whether a black box is required for self-driving technology. Perhaps it will be sufficient to download an app to your smartphone and then log into a vehicle and control it that way. Admittedly, today's smartphones do not have that kind of computing capacity, but this will not be a problem a few years from now.

Even today, Google does not earn its money by selling road map information or subscription fees for Google docs, but the use of data for additional services. The environmental and behavioral data of millions of vehicles in the real world thus would become immensely valuable. The potential is estimated to be more than $750 billion worldwide.[114] At the end of 2018, some analyst groups estimated the potential value of Google's sister Waymo at a staggering $250 billion.[115]

Energy Efficiency and Accident Prevention: Self-Drivers Have an Edge

Automobile clubs and companies with large car fleets regularly hold seminars to teach fuel-efficient ways of driving: maintaining the correct tire pressure, keeping less baggage in the car and using eco-mode for gear shifting, and especially driving at an even speed and with a lot of coasting, without extreme acceleration and braking, with little idling, and with use of the slipstream. Naturally, not using the air conditioning, heating, and other energy consumers can help save fuel. All this is little known or not done correctly.

The savings seem almost ridiculous if you take the inherent inefficiency of combustion engines into consideration. As we've learned, converting combustion energy into motion wastes 80 percent of the energy generated. At the end of the day, just about 1 percent of the fuel consumed actually serves to transport passengers.

The general public became aware of the high emission share of cars not least in the context of the Volkswagen diesel emissions scandal, and it was noted again how many people die every year from automobile emissions. According to MIT, the number is 53,000 early deaths every year.[116] If the hidden cost was included in the price of gasoline, it would be three times as expensive as it is.

An automated machine, by contrast, can always drive in a fuel-efficient manner. It does not "forget" to do so. As a matter of fact, in one study, it was estimated that self-driving EVs could reduce emissions from today's levels by 87 to 94 percent.[117] The driving style becomes more environmentally friendly, and the weight of the driver can be dispensed with. A self-driving car with a lower accident probability furthermore can be built as a more lightweight design with fewer safety elements. Although much more lightweight materials are used in cars today than could be used formerly, there was also a need for increased installation of safety mechanisms that amply made up for any weight saved by improved materials. NHTSA estimates that every car weighs 125 pounds more just because of "security," and the price for a vehicle increased by a total of $839.13 in 2001.[118] Bigger and heavier cars turn our cars into fatal traps, especially for the smaller of any accident victim. With every 1,000 pounds of additional vehicle weight, the accident probability increases by approximately 47 percent.[119]

Our approach is somehow perverse actually. We equip the cars in such a way that they have a much better chance of surviving an accident unharmed, instead of thinking about preventing accidents in the first place. We should, however, concentrate on the latter: today's driver assistance systems are already well on their way, and autonomous vehicles are even better at tackling this problem—at the root. Safety belts, airbags, protective zones, reinforced passenger cage, lateral cross-beams in the doors—all these elements help to protect the passengers in case of an accident. But actually, it is too late at that point. Damages should not be minimized; they should not be incurred in the first place. The development of self-driving technology serves to catch three birds with one stone: lower cost because of accident prevention, lower environmental cost due to reduced vehicle weight, and lower macroeconomic costs due to less damage and fewer injuries.

Many standard cars today with four and five seats, often occupied by a single person, are equipped for the rare case in which you do not travel to work on your own but instead go on holiday with friends, family, and a lot of luggage and equipment. When ordering a self-driving electric Uber, I can indicate that I just need a one- or two-seater to take me to my destination when I place my order.

We should also not underestimate the waste of energy in traffic jams and while searching for a place to park. A municipal parking lot needs as much gasoline every year as I would need for two-and-a-half journeys from San Francisco to Los Angeles and back. All these factors contribute to an energy-saving potential of between 87 and 94 percent with a transportation system that is completely geared to self-driving vehicles.

No Beating Around the Bush: When Can We Buy a Self-Driving Vehicle?

So just how much progress has been made on the development of self-driving vehicles? When can we use them or actually buy them? The good news first: we can do that already. The prototype tests proved that it is important to remove humans from the control of vehicles as soon as possible. Many events that occurred during the million miles of test drives show that the experimental self-driving vehicles are on the road with at least the same degree of safety as human drivers.[120] The Google project has reached the first internal company milestones and confirms that the schedule will not be easily maintained but that it is feasible. In 2016, Google spun off Waymo, the project formerly known as the Google X Self-Driving Program, now an independent company to commercialize self-driving technology.

There are driverless taxi fleets in test operation in several cities around the globe now, and you just need a little luck to be chauffeured without a driver, such as, for example, by nuTonomy in Singapore. In August 2016, test drives commenced in a district of several blocks. Passengers could travel in the rear while engineers supervised the vehicle from the driver's and front passenger's seats to be able to intervene in an emergency. In September 2016, Uber initiated operation with its test fleet in Pittsburgh. Again, there was an engineer in the driver's seat who could intervene in an emergency. At that time, the journey was free for passengers.[121]

A good place and time to experience self-driving vehicles is always the CES in Las Vegas. Because mobility has been a central part of the show in recent years, many self-driving car companies not only exhibit their technologies but also use them for ferrying conference attendees from their hotels to the venue. In 2019, Aptiv with Lyft and AutoX did just that.

All this pales when we consider that Waymo actually launched the world's first commercial robotaxi fleet in Phoenix in November 2018.[122] Several hundred cars cover an area of 100 square miles for a few hundred so-called early rider families. The program ran for about a year in a test mode, when Waymo officially launched the program and slowly opened it up to more residents. Some of the rides even were completely driverless, without a safety driver in the passenger seat. It is expected that in late 2019 or 2020, Waymo will start a similar service on its home turf in the San Francisco Bay Area.

In fact, Waymo ordered 62,000 Fiat Chrysler Pacifica minivans and an additional 20,000 Jaguar iPaces that it wants to operate in its robotaxi fleets. Waymo also partnered with Renault-Nissan to test driverless vehicles in Japan and France. To put this into perspective, New York City today has around 13,000 cabs, and the whole of the United States has around 240,000. Putting 82,000 robotaxis on the roads over the next few years will have a dramatic effect on this and other businesses.

In addition to taxi fleets, many cities are already using small and large buses or shuttles operated by startup companies such as Local Motors (Olli), EasyMile, and Navya. Lausanne, Amsterdam, Berlin, Salzburg, Perth, Las Vegas, Lyon, Sion, and other regions already provide room for individual test programs, and more are about to start. Cities all over the world are preparing for autonomous vehicles and hope for more cost-efficient and safe means of transport that manage to also create better public transportation connections to previously underserved city districts.[123]

Autonomous vehicles are also the solution for the problem known as the *last mile*. This refers to that part of a journey a person has to travel from a bus or train station to his or her house or workplace on foot. This is the crucial element for public transportation. If a city district is not connected, the inhabitants have no choice but to travel by car. Any population group that cannot afford a car or people who may not be able to drive a car are then excluded from many urban services or options—workplaces, social institutions, schools, hospitals, shopping centers—or else have to make a huge effort to be able to participate.

Over a traveling distance of more than 1.2 million miles on the road in test operation, Google vehicles were involved in only 12 collisions, all at low speed, and only two of which were their fault. Only minor fender benders occurred. This breaks down to about one collision every 125,000 miles. This accident frequency corresponds quite accurately with that of vehicles controlled by humans.[124]

It is therefore encouraging that Google's project "graduated" to Waymo. This means that the company now does everything in its power to commercialize self-driving vehicles and launch them on the market. From its status as a crazy "moonshot," this project made its way and now involves a dedicated company that will make money.[125]

Despite all this good news, the real work has only just begun, and we are far from having solved all the problems. Scientists from Oxford University demonstrated how much still has to be dealt with.[126] In their test program, they concentrated on the same six-mile trip to and from work, all year long, and at any time of day and night. The researchers were eventually surprised at how much conditions on the same roads varied during that time, and not only the light and weather played a role, but construction work did so as well. A small traffic circle was moved to a different location three times in just one year. The researchers divided the changes into short- and long-term modifications. Traffic density, parked cars, sunlight, and streetlights at night are short-term modifications; bushes, hedges, trees with and without leaves, construction work, and new traffic signs are long-term modifications. Some can cause problems for self-driving systems in certain circumstances. If, for example, such a vehicle relies on high-precision road maps, as do the Waymo cars, a sudden new location of a traffic circle can cause confusion. Systems such as Tesla's, by contrast, that rely more on cameras, could handle this kind of problem more easily but would probably encounter other challenging obstacles.

If therefore there are major changes in the same region as a result of the time of day and the season, traffic areas and regulations in a larger city, in a country, or even on a respective continent can differ even more widely. Signal lights or traffic lights, arranged vertically in California, are arranged horizontally in some parts of Texas. In addition, some states allow a right turn on a red light, and others do not. It will certainly be a challenging task to cater to all those different conditions!

Reservations: "But I Like Driving Myself!"

The best self-driving technology is of no use if we reject self-driving vehicles. What are the arguments and the reasons behind them? Should we take them seriously, or are they just irrational objections? Let us take a look at some of the reservations often heard from skeptics in the context of autonomous vehicles.

Objection 1: "But I Like Driving Myself!"

I definitely believe you. It is really great to steer your car through wonderful scenery, enjoy the wind in your hair, and sit in an open convertible. I myself drove through Silicon Valley like that for 12 years and absolutely loved every single outing—when I was able to actually drive, that is. However, the sad reality is that much more often I was trapped in the rush-hour traffic in the morning or evening. Stop-and-go traffic, edging into gaps, missed exits, and the nerve-wracking search for a place to park—of course, just when you are in a hurry anyway. Well, thanks a lot. And I will spare you my adventures when going on vacation. When everyone wants to leave for the holidays, the highway becomes an endless parking lot, and you eventually reach your destination a good deal more stressed than when you left: in situations such as that, "sitting at the wheel yourself" is anything but fun.

In 95 percent of all cases, you just need to get from point A to point B. The enjoyment factor is zero. In the United States, the loss of time due to traffic congestion rose from 700 million hours per year in 1982 to 6.9 billion in 2015, at an estimated cost of $160 billion.[127] In the European Union, the amount is estimated at 1 percent of the European gross national product, 160 billion euros.[128] And all forecasts look bad. For the United Kingdom, experts expect an increase of 63 percent by 2030 and 50 percent for the United States.[129] There are several reasons for this. First of all, there are already more vehicles on the streets than the infrastructure in many countries is designed for.[130] And the current state of the art indicates that the number of vehicles will double again in the next 20 years.[131]

The annual TomTom traffic index lists American cities with the most traffic jams: Los Angeles heads the list, followed by San Francisco, New York, Seattle, and San Jose.[132] In Los Angeles, an average trip takes 44 minutes longer than it would take if traffic flowed well.[133] The worst congestions anywhere in the world are reported in Mexico City (Mexico), Bangkok (Thailand), Jakarta (Indonesia), Chongqing (China), and Bucharest (Romania). In Germany, Cologne holds first place in the TomTom statistics; Hamburg, Munich, Berlin, and Frankfurt am Main follow.

Let us be honest: how often are you really able to experience that frequently invoked "joy of driving"? At this point, a Google employee also started wondering. He was a Porsche enthusiast but volunteered for a one-week test with one of the first prototypes provided by the Google X Self-Driving team when the team called for testers. He could not really imagine that he would like the car, but he soon realized that his daily trip to work and back home actually was much more relaxed. At the end of the one-week test drive, he knew for sure that he only wanted to drive and enjoy his Porsche when he could really take it "on an outing" on the weekends. He hated to part with his Google car.

Objection 2: "I Will Never Get into a Car That I Cannot Control!"

It is a fact, however, that every single one of us today travels with means of transportation that we cannot control. We take the bus, the subway, the streetcar, the train, or an airplane. We cannot even control where we go. A little deviation for the train or the plane, and we end up at a different station or in another city.

We are not comfortable with the idea that our life or death is in the hands of machines. We prefer to have that kind of decisional power to ourselves. But as much as this desire is understandable, it is utterly futile. Many of our means of transportation today are controlled by machines. It just does not always register with us. Driverless subways and autopilots in aircraft or air traffic control allow a much tighter and safer frequency of rail and air traffic than humans would ever be able to organize. And even something as commonplace as an airbag is totally beyond our control: when it is triggered and when it is not triggered, whether we are killed, are severely injured, or get out of the car wreck unharmed. In aircraft, the autopilot has control 99 percent of the time. The percentage on the train is similar. Several subway systems such as the one in Lausanne, Switzerland, and the one in Singapore do not have drivers anymore. As well, travelers at the Frankfurt, New York, and Zurich airports are shuttled between terminals by transportation without drivers. During some flight phases, the autopilot actually has to be in control, for example, when it is necessary to land in a storm or with low visibility. It is only a matter of (a little) time until we will have a similar situation for cars and trucks.

One example from the past shows that we do not even trust human drivers in every single case. When the first female streetcar drivers and pilots appeared, people had similar objections: a male driver or pilot was much more experienced, they said. And today? We find it perfectly normal to be chauffeured by women or by men.

As Professor Dudenhöffer, head of the CAR Institute at the University of Duisburg-Essen, concludes in his above-mentioned book by referring to safety measures such as an airbag:[134]

> If we are honest, we have already answered the question of whether a computer is allowed to make decisions about our lives several million times, and the answer is affirmative.

Many technical systems already include elements that take control without human intervention and make decisions for us. The antilock brake system (ABS) on our cars is one example of this. The ABS decides whether to engage or not when a driver hits the brakes. It is likely to activate itself on a normal dry road but probably not on icy roads.

For the internal Google test program, the participants were asked to always concentrate on the traffic and actively be involved in the driving process so that they could intervene if necessary. After all, they were still dealing with prototype-level technology that was anything but mature. As the footage from the cameras inside the vehicles shows, the employees were very strict to observe this instruction on the first day, remaining attentive at all times. On the second day, they already became more negligent, and on day 3, they started to focus more on the scenery and the objects around the car than on the traffic. The most telling incident was one where an employee wanted to charge his mobile phone and needed to get the cable from his bag that was on the rear seat. He turned around, opened the bag, pulled out his laptop, grabbed the cable, and placed everything on the passenger seat to then fumble around with his phone until it was ready for charging. Only then, after about 15 long seconds, did he look out at the street and the traffic again.

After their initial skepticism, humans start to trust machines with surprising speed, despite the fact that they should remain attentive, as in this example. This means that we have come full circle. Autonomous vehicles in some way lead us back to the time of horse-drawn coaches when we did not have to watch the street and the traffic all the time.

As a matter of fact, emergency-off systems can also have a contrary safety effect. Practical experience shows that human beings are not really good at recognizing emergency situations. We react slowly, often make mistakes, and sometimes are frozen in shock and might switch off the system intentionally or out of negligence, which would put us and the car in even greater danger. The most dangerous period in the transition to autonomous vehicles will come, according to experts, when control is shared between humans and machines. Studies conducted by Ford show just how much of a risk that is. The time needed for a human to "understand" the traffic situation again when the computer passes on responsibility to him or her is more than 20 seconds.[135] We will only be able to use the increased safety predicted for autonomous systems if we completely relieve humans from control while driving.

Objection 3: "Most Collisions with Self-Driving Vehicles Are Caused by Cars Driven by Humans"

This objection goes on to say, "but the only reason for this is that the former—like senior citizens—drive overcarefully. Small wonder if other drivers get nervous." One aspect we can detect from this objection indicates a certain ambivalence with regard to the requirements for a self-driving vehicle. One the one hand, we accuse these vehicles of being too hesitant; on the other hand, we view them as uncontrolled and crazed machines. One moment we wish that they would show a little more spirit at a green light or when turning a corner; the next moment we never want to encounter them on the road again or we want to curb their maximum speed drastically. People complain about other drivers being aggressive but like to drive in a "sporty" style themselves. People apprise slow occasional drivers of their stupid maneuvers by flashing their lights at them. But self-driving cars are programmed for politeness and leave the right of way especially to more vulnerable traffic participants—precisely those situations in which human drivers quickly lose their patience, take the right of way, squeeze in, do not let a pedestrian pass, or overtake a cyclist with very little distance between them. All the time, mind you, not only when they feel like it.

What does it say about us human beings when we call that kind of driving "stupid" or "dangerous"? If, however, something were to happen, if, say, self-driving vehicles were traveling just as fast and aggressively as many "normal" drivers today and then caused an accident, the same people would pipe up and say, "Well, I told you so! That technology is not safe!" As a matter of fact, Google introduced all those safety features into the algorithms because self-driving vehicles mainly react to human errors and have to correct them.

Objection 4: "Human Drivers Will Always Steal the Right of Way from Self-Driving Vehicles, Without Any Consideration"

This objection goes on to say, "Then self-driving vehicles will simply stay where they are because nobody will ever let them merge with the traffic flow." The robot vehicles' friendly, helpful behavior can be compared with that of novice drivers, and it can be mistaken for insecurity. And if some drivers then think that they need to be aggressive toward them, that is their problem. However, it should never be tolerated, not with human drivers, not with robot vehicles.

Objection 5: "Terrorists Will Load Autonomous Vehicles with Explosives and Use Them as Moving Bombs"

As a matter of fact, a survey among 700 people in 2016 showed that 43 percent of Germans and Austrians, 40 percent of Americans, and 41 percent of South Koreans are worried that autonomous vehicles could be abused by terrorists.[136] The solution for this problem might come from a completely different industry, namely, blockchain, a decentralized data structure. This idea was originally intended for the financial sector only but has already proved to be incredibly versatile.

Without going into too much detail, the *blockchain* simply is meant to secure a transaction, for example, withdrawing money from one account and crediting it to another. This also works with property registers. The title is transferred from one person to another. The main feature is that the blockchain is not centralized but organized decentrally. The property records are not kept in a bank or the land registry but can be publicly viewed by everybody and are updated in just a few minutes. Because the entire transaction history of an object is maintained in the blockchain—transaction blocks are put in sequence, that is, linked—it can be seen immediately who was involved. An autonomous car managed via the blockchain has its entire transaction history present and viewable at all times; any abuse becomes instantly visible.[137]

Some unique features of the blockchain technology might still obstruct their application in the automotive sector, but it is definitely one possible solution for cybersecurity issues.

Sickening Driving Styles

When my family was riding in the car with my little sister when she was only a child, we always had bets going: would she make it without being sick? Her sensitive stomach was not reserved for the car only. She also felt sick riding in a tram, not always, but regularly, so my parents were drastically limited in their mobility. All that has passed since she got behind the wheel herself. It soothes her stomach that she has the vehicle under control and can decide on its course.

But what is the reason for that nausea in moving vehicles? It occurs when we feel a discrepancy between the movement we see and the movement we experience with our body and when we cannot predict or control the direction of travel. Nausea usually affects passengers, almost never the driver. To put it plainly, does this mean that self-driving vehicles will simply be sickening?

The answer is: only if their style of driving is once more like that of a moving horse. I took the company bus or public transportation for many years on my way to work. In the first weeks, I simply could not read anything during travel because I felt woozy almost immediately. I got used to it over time and had no problem reading while riding after a while. However, there was one exception: when one particular bus driver was on duty on our route. He constantly switched between acceleration and braking, and it made me nauseous almost instantly. As my colleagues on that bus said, "Only riding a horse is nicer!"

We know this from our own experience. If we are in the car by ourselves, we drive differently from when we have the entire family with us. When we have passengers, we are more careful, drive more slowly, to ensure that nobody feels ill. If another person is driving, our well-being depends on whether we are "compatible" with the driver's style of driving. A self-driving vehicle must move in a way that agrees with everybody.

Scientists at the University of Michigan expect that more passengers will feel nauseous as self-driving cars become more common because nobody will actually be at the wheel anymore.[138] Activities such as reading or playing video games can also contribute to this. Between 6 and 10 percent of all adults are estimated to suffer from car sickness frequently; another 12 percent occasionally.

How can this be prevented? By manufacturers that extend the field of vision with large windows and install screens in such a way that passengers look forward. This would also mean avoiding pivotable seats and actions requiring a lot of head movement and instead installing seats with adjustable backrests.

Driving Around in Circles Gets Boring Somehow

Will self-driving cars be entirely limited to public roads? Absolutely not! Land Rover is experimenting with autonomous vehicles that are able to go off-road, not requiring any asphalt or marked roads.[139]

Anybody who believes that self-driving cars can only drive slowly has never seen a self-driving car race. The cars are definitely able to drive fast, even very fast, as Audi proved with its RS7 prototype vehicle. It made an incredible 160 miles per hour.[140] There is even a special race series for autonomous vehicles, initiated by Silicon Valley entrepreneur Joshua Schachter.[141] Three hours' worth of driving north of San Francisco brings you to the Thunderhill race course in Central Valley, where the first race took place at the end of May 2016.[142] A dozen teams participated. Although the event was less like a Formula 1 racing event and more like a presentation and advertising event, it still was reminiscent of the origins of automobile and aviation sports. Many vehicles at that time were built by tinkerers, some of them not even making it to the start or failing in the first turn. There were no official manufacturers of cars or airplanes then, and those daring pilots and drivers went to the start with their home-built constructions. In 2017, the world also saw the first test races of autonomous racing cars in the Roborace Series in Paris and Buenos Aires.[143]

Just how car racing as we know it today will evolve is difficult to assess. Currently, the drivers are still the ones who get the affection and carry the hopes of the audience, the fans, and entire nations. They may lose their popularity with autonomous racing cars, or they may not. However, if I look at the enthusiasm in the fans' faces at the ComBot Cup and the RoboGames, I am not overly worried.[144] I went to two such events with my son when he was six years old. It was hellishly loud; the remote-controlled robots attacked each other with flame throwers, circular saws, hammers, and all sorts of other tools and captured the audience, consisting of at least 50 percent children. If I may be allowed to predict something: race courses of the future will look more like a Mad Max kind of race without drivers, and anything goes. And children growing up will be unable to comprehend why Formula 1 or—even more boring—NASCAR once was so popular. As the late Niki Lauda, triple Formula 1 world cup winner, said many years ago, "I'm sick and tired of driving around in circles."

Clever Saving: For the Environment's Sake

In order to determine the level of environmental friendliness of cars, we usually look at production, usage, and eventually recycling. Because of the specific characteristics of autonomous automobiles, the hopes are high that there will already be advantages from production. Experts agree that autonomous vehicles will help to significantly reduce the number of accidents once they have been introduced, and many safety features can be discontinued. From a certain degree of safety of autonomous systems and once manually controlled vehicles are banned from the roads, fewer and lighter materials can be used for the car design, and the vehicles will thus be more affordable. Lighter vehicles are better for the consumption of resources and need less battery capacity and weight while maintaining similar reach, as long as we are talking about EVs.

A car that is on average parked for more than 23 hours every day is a huge waste of resources, which is anything but sustainable. Electric robotaxis transport passengers and goods around the clock, and they are available as one-seaters, two-seaters, or multiseaters. A fleet of robotaxis fueled via sustainable energy sources such as solar and wind power might reduce greenhouse gas emissions, respectively, by up to 87 and 94 percent.[145]

We could also save up to 15 percent of fuel cost if we select the best route and drive in a way that is not based on a "human" alternation of stop and go.[146] Stopping at intersections and then starting and accelerating again consumes kinetic energy. If the vehicles communicated with one another and synchronized at intersections so that they would not have to stop at all, they would even save as much as 30 percent of fuel cost.[147]

Vehicles could also be joined to *road trains* as a convoy. Cars driving together in a coordinated fashion also do not cause any congestion. This would eliminate hundreds of billions of dollars of lost production cost and fuel burned in traffic jams. The method of *linking* several vehicles from different transportation providers and driving at a distance of 40 to 50 feet one behind the other is particularly interesting for trucks. Swedish manufacturer Scania is testing this principle in Singapore.[148] The first car saves 4.5 percent in fuel cost; any of the others, 10 percent.[149] If you consider that the cost of fuel amounts to up to 40 percent of the operating cost of a transportation company, this is a significant amount of savings. Although trucks are only 4 percent of all vehicle types, they cause 25 percent of the emissions; 42.7 percent are caused by sedans and 17 percent by pickups, small vans, and SUVs.[150] In total, the transportation sector was responsible for 28.9 percent of all 2017 greenhouse gas emissions.[151]

Intelligent systems helping a vehicle to find parking or the best route can save another 5 percent. If we consider the number of cars that are looking for a place to park today at certain times of day, we come to value autonomous vehicles without their need for parking lots or electronic parking guiding systems even more.

Not everyone thinks that autonomous vehicles are the only solution. One study concludes that the realistically possible savings due to less traffic congestion come to a maximum of 5 percent. The efficiency of road trains, for example, depends on the distance between the individual vehicles and the total length of the entire convoy.[152] Existing infrastructure and legal regulations could also be limiting. Perhaps we would even be tempted to drive more and more miles than before, once we can simply order a car via app and do not have to worry about anything else anymore.

Cars are the consumer product with the highest degree of recycling. Almost 100 percent of the iron and steel used, a total of 60 percent of the vehicle weight, is recycled.[153] Even if aluminum car bodies became the norm, the amount of recycling would not change because that material has an even higher recycling ratio since it can be reused without quality deterioration. It is cheaper than producing new aluminum as a matter of fact.

Self-Driving Vehicles "in the Wild" and on Test Tracks

At the children's traffic education course, every kid waited impatiently for his or her turn to sit in the car instead of playing the pedestrian. There were some tricycles and even bicycles, but everyone wanted to sit in the pedal car and ride through this miniature version of a city. Everything was available: traffic signs, crosswalks, traffic circles, and miniature traffic lights. Before we children could be trusted to move in the real world of traffic, we had to practice in a safer version. We felt like adults but sadly were only going there once.

Just as the children could not immediately go out into traffic, prototypes of self-driving vehicles cannot immediately be left to their own devices on public roads. After all, those cars are robots with a weight of 2 tons that may cause some serious damage if something goes wrong.

Manufacturers previously tested their cars on their own separate courses, such as Daimler on the Hockenheim race course or at the test center in Immendingen, which will be available in an extended version from 2018 onward.[154] Often race tracks or former air fields are used for this. Still, at some point, you have to go out onto the street with your new models. There is a specialized group of photographers and journalists trying to spot such camouflaged new car models to present them to automobile enthusiasts in various specialist journals.

The test tracks available today, however, are no longer appropriate for the requirements of driverless traveling. It is not sufficient to simply do some tests on a simple stretch of road. Manufacturers need to be able to find real conditions with the usual traffic volume, varying road and weather conditions, buildings, subjects, and objects on the street and the sidewalk and with surprising situations. In the United States, there are several test areas such as the GoMentum near San Francisco. They offer diverse conditions already, and many more such areas are currently being designed. The streets of GoMentum are a representative mix of conditions found on many California roads, with potholes, faded lane markers, railroad crossings, highway ramps, bridges, and underpasses but also with buildings, intersections, worn road signs, and—if desired—other vehicles and pedestrians. This test area offers everything a designer of autonomous vehicles could wish for. Because the area is still considered a military area, it is guarded and not open to the public. For manufacturers that wish to continue their research without being watched, this is an invaluable advantage.[155] Acura, Honda, and EasyMile are or were testing at the GoMentum Station, and a dozen other manufacturers are negotiating with the operators. The NASA Ames Research Center is also currently being used for testing by companies such as Google, Nissan, and Peloton.

Since former California governor Jerry Brown approved the legislation, there are also two clearly delineated road segments where autonomous vehicles without a human driver on board may move around, namely, at the GoMentum Station and at the Bishop Ranch business center near San Ramon on the East Bay.[156] The California DMV even took things a step further. From April 2018 onward, autonomous vehicles may travel without a driver on public streets in California. In October 2018, Waymo received such a license as the first company in California. The license allows the company to operate truly driverless cars in Mountain View, Sunnyvale, Los Altos, Los Altos Hills, and Palo Alto, an area of about 72 square miles with about 340,000 inhabitants. Up to 40 such cars are allowed to drive during day and night and even in light rain. But first the company also had to train emergency responders how to deal with the cars in case of an emergency. In March 2019, the company trained firefighters, police officers, and city officials in how to contact remote operators, turn vehicles off, or signal them.

The U.S. Congress has a particularly extensive proposal to offer: if it is approved, up to 100,000 autonomous vehicles could travel on all U.S. streets without the need to maintain existing safety standards. Why would Congress do this? In order to support development instead of impeding it. The advantages offered by autonomous vehicles seem so promising to the members of Congress that they do not want to create any obstacles for them.[157]

M City in Ann Arbor, Michigan, is much smaller and has a different focus. It is operated by the University of Michigan and offers artificial street scenarios with ideal conditions. All road signs and markings are new, and the street is optimized and adaptive. The weather in Michigan is better suited than the climate in California for testing vehicles under bad weather conditions such as rain and snow.[158] Since December 2016, Michigan has been permitting driverless cars without a driver or a steering wheel on board.[159]

And there is yet another test area in planning for "connected cars" and autonomous vehicles in Michigan, under construction in the former bomber aircraft plant Willow Run in Ypsilanti. The groundbreaking ceremony was in November 2016.[160] Compared with M City, the planned capacity of traffic scenarios will increase tenfold. This test area is also operated by the University of Michigan under the name of American Center for Mobility.[161] In Flint, also in Michigan, Kettering University operates a test track in cooperation with General Motors, and Virginia Tech inaugurated a test area in Blacksburg, Virginia.[162] Florida did not want to be left behind and started building a test area in Polk County in the spring of 2017.[163]

Google uses as its test area the former Castle Airforce Base, 2½ hours east of San Francisco near Merced. It is also a training center for drivers, or rather, nondrivers, who have to observe the vehicle and its surroundings to be able to intervene in an emergency.[164] The driving instructors provide instructions for the respective test drive in order to collect data and adjust the algorithms.

The test center includes a storage room with lots of items that can be placed onto the test track or be used by other employees simulating traffic participants. The test drivers help the car to recognize and categorize the items as well as to correctly interpret traffic situations. This allows them to pinpoint certain types of behaviors and inform the designers about them. This task is actually not as easy as it sounds because test drivers and engineers sort of doze off occasionally.[165]

In January 2017, the U.S. Department of Transportation published a list of 10 official test sites and public test areas that are to support the further development of autonomous vehicles, the drafting of standards and legislation, and the exchange of information and experience.[166]

1. City of Pittsburgh and Thomas D. Larson Pennsylvania Transportation Institute

2. Texas AV Proving Grounds Partnership
3. U.S. Army Aberdeen Test Center
4. American Center for Mobility (ACM) at Willow Run
5. Contra Costa Transportation Authority (CCTA) and GoMentum Station
6. San Diego Association of Governments
7. Iowa City Area Development Group
8. University of Wisconsin–Madison
9. Central Florida Automated Vehicle Partners
10. North Carolina Turnpike Authority

Every municipality, every city, and every state in the United States has the strong desire to be a part of this race for the favor of the automotive industry. By offering flexible regulations and public infrastructures, local administrations hope to become the site of a test area and, eventually, a production and development site.

So what about Germany in this regard? Although German manufacturers and research facilities started work on autonomous vehicles as early as the 1980s, and 58 percent of all patents in this field since 2010 are held by German institutions, they are falling behind.[167] Applying for patents and driving the technology forward are not necessarily a hand-in-glove process. While many test sites are operating in the United States and many more are in the planning stage, while some U.S. states conditionally allow autonomous driving on public streets and are preparing a driving permit for autonomous vehicles with no humans on board, and while NHTSA is already working on a comprehensive set of rules, Germany is not even ready to start yet. In 2016, there was an announcement that a part of highway A9 would be approved as a test track, but implementation was delayed. Anybody driving along A9 and A93 today can, however, see some signs that are about 28 inches (70 cm) wide installed on the side of the road (see Figure 7.1): they are intended as points of orientation for driverless vehicles and should allow data matching with the GPS signal down to less than a half inch (a centimeter).

In 2017, a test with autonomous BMWs was announced, to be conducted in Munich city traffic to gather experience.[168] Mercedes Benz also received a license from the city of Stuttgart in late 2016 to test autonomous cars within the city limits,[169] but not much has been done. A few cars here and there, but in a very timid manner.

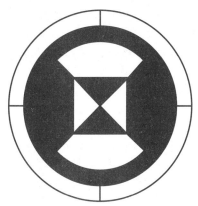

FIGURE 7.1   German road sign for autonomous vehicles

At the same time, the German Department of Transportation announced the test operation of a driverless van by U.S. startup company Local Motors on the EUREF business center campus in Berlin.[170] A similar test operation has begun in Leipzig in collaboration with French manufacturer Easymile. Both trial operations are run by Deutsche Bahn. In Hamburg, traffic planners are also preparing for a driverless bus route, although this will not occur until 2021. Braunschweig and Kassel have also been designated as test sites. Friedrichshafen, home to transmission manufacturer ZF, also has a public test track.

At present, German experts and legislators do not completely agree about what such public test tracks should be able to handle before they are released for general use. What about sensors enabling communication with the vehicles? In Berlin, one test track is equipped with sensors for the enormous amount of $4 million in subsidies. Partners of this DIGINET-PS project are the Technical University Berlin, the FOKUS Fraunhofer Institute, Daimler, Cisco, and the Berliner Verkehrsbetriebe (BVG, Berlin public transportation services).[171]

Just note the choice of words. While Germans talk about *test tracks*, the United States is dedicating entire states as *test fields*. In addition, Germans would talk about *automated* or *highly automated* driving, whereas in the United States we talk about *autonomous* driving. These are small differences in the words of choice that speak volumes about the ambition levels. Here just the next level of a driver assist system, there a fully self-driving car.

U.S. manufacturers are not bogged down by such doubts and reservations that Germans may have. A $4 billion initiative from the government running parallel to any already existing trials aims at preparing an entire highway as a transit route for driverless vehicles. Highway 83, running 2,000 miles through North Dakota, South Dakota, Nebraska, Kansas, Oklahoma, and Texas, might be that highway.[172] American companies have even set up a transindustry platform to help develop the framework conditions and technology for autonomous driving. The Self-Driving Coalition for Safer Streets unites Uber, Lyft, and Google with manufacturers such as Ford and Volvo. Similarly, organizations such as Mothers Against Drunk Driving, the National Federation of the Blind, the United Spinal Association, the R Street Institute, and Mobility 4 all have joined this coalition, headed by former NHTSA manager David Strickland.[173]

And what is Europe doing in the meantime? Hungary announced the building of a test site in Zalaegerszeg. Austria will follow suit but has not yet implemented anything.[174] Legal issues, skepticism, and a total lack of any sense of urgency have make the efforts grind to a stop. In late 2016, only the automotive supplier AVL List completed one drive on Highway A2 near Graz, and there was a test run of an autonomous minibus by French startup company Navya in the historical city center of Salzburg.[175] In Switzerland, on the campus of the Technical University of Lausanne (EPFL), first tests with such an autonomous small van already took place in 2015, together with the fleet management startup company BestMile and the Swiss Postbus.[176] In Sion/Sitten, an autonomous bus in test operation has been taking passengers on board for several months.

German manufacturers so far can only test their vehicles in California and Nevada. While they are still struggling to get a testing license in their own country, American and Japanese manufacturers are already moving into Europe with their vehicles. From 2017 onward, Ford began test operations in Essex, United Kingdom, followed by tests in Aachen and Cologne, Germany.[177] Nissan, in turn, started trial runs in London in the spring of 2017.[178]

Similarly, Canada is speeding up, despite the fact that it has hardly any automotive industry worth mentioning. For example, Ontario adopted laws permitting the University of Waterloo, the corporations Blackberry and General Motors Canada, and German manufacturer Erwin Hymer Group to test autonomous vehicles on public roads.[179]

Even a country such as India with its chaotic traffic conditions is active; there are already two Indian organizations that have applied for a test license for a "mission impossible," namely, Tata Elxsi and Automotive Research Association of India.[180] Russia, too, desperately needs self-driving cars, if you bear in mind that two-thirds of all traffic fatalities in Europe happen in Russia.

Cognitive Technologies and Russian internet giant Yandex in Moscow are working on the problem.[181]

Traffic Activities and Other Side Effects

There is a somewhat obscure but very funny Czech movie titled *Knoflikari* (in English, *The Buttoners*). It was made in 1997 and tells the story of a couple taking a taxi. They ask the driver to drive along a deserted street in Prague. While the driver—who has seen a lot in his life—concentrates on the nighttime traffic interactions, the two passengers on the rear seat enjoy a different kind of interaction. They try out various positions; once their legs appear between the front seats, then their heads. After a while, they both just sit there frustrated. Finally, the woman speaks up and tells the driver, "He can't do it when you drive so slowly!" The taxi driver knows a different street where he is permitted to drive faster than the official 35 miles per hour, so the two passengers finally can enjoy the ride.

Is this the kind of use we will have for autonomous vehicles in the future? Will we use the time on our way to work for a lovers' tryst instead of booking a hotel? Will self-driving cars end up abused as moving brothels? Will the red-light sector become the main driving force in expanding the new technology and continue its development, as has happened before for video streaming or the improvements made in image quality? It seems that Tim Kentley-Klay, founder of the startup Zoox, has at least toyed with the idea already. At a conference, he made an amused comment about a possible "lights-off mode" that would switch off all recording devices in the car so that people could "have some fun."[182]

The question is whether self-driving vehicles will eventually lighten traffic or cause even more stress and increase other kinds of "intercourse." On the one hand, a more intensive use of cars means that, in total, fewer cars will be needed, which would result in less congestion. On the other hand, people realistically fear that it will be easier to cover longer distances, which could, in turn, fuel urban sprawl. Both effects were demonstrated in a simulation.[183]

The limit for commuting today is the time needed for the distance. If you have to spend more than one hour every morning and every night on your way to and from work, the perceived cost for the time lost is too high. If, however, you can use that time sensibly for work, reading, or something similar, then perhaps two hours spent commuting will seem quite acceptable, and it lets you have a home with a garden somewhere far away from work.

Ford conducted a survey to find out what people would do if they had free time in the car. Despite the fact that very few people have, in fact, traveled in such a car, most people had very clear ideas about what they would do: 80 percent would simply relax and enjoy the view, 72 percent would make phone calls, and 64 percent would enjoy a relaxed meal.[184]

Another interesting aspect is what my car (as long as I still own one myself) would do in the meantime, after it took me to work or back home. After all, it could make some money, picking up passengers and transporting them somewhere. And that was exactly what Elon Musk elaborated on in the Tesla Autonomy Days in 2019. Tesla's plan is to operate its own robotaxi fleet made up entirely of Tesla vehicles. Tesla owners can send their car to be part of the Tesla robotaxi fleet during the day, making money while their owners are at work. At the end of the day, the vehicle comes and picks up its owner from work.

It gets exciting when you start looking at the cost of accidents, although we should bear in mind that autonomous vehicles can mostly avoid accidents in the first place. A study in several countries contrasted the cost incurred today with a 90 percent drop in accidents with driverless cars. For Germany, expenses would decrease from $40 billion to $4 billion annually, Austria would enjoy a decrease from $12 billion to $1.2 billion, and Switzerland would need to pay only $0.7 billion instead of the current $6.6 billion. The biggest expected savings are in the United States, where the drop of $307 billion to $34 billion is the most dramatic in absolute numbers.[185]

By the time the self-driving technology is ready to be launched in the market and EVs with this technology are available in a sharing model, the rules will undergo such a fundamental change that no stone will remain unturned. A study commissioned by Nissan estimates the economic importance of autonomous cars in Europe to be $20 trillion by 2050.[186] Morgan Stanley expects autonomous cars to be able to provide cost savings of $1.3 trillion in the United States alone and mentions $5.6 trillion as a global figure.[187] The amounts are a combination of factors such as traffic accidents, loss of productivity, and fuel consumption as a result of driving wastefully, looking for parking spaces, and waiting in traffic jams.[188]

Design Quirks and International Differences

Is it enough if cars are just self-driving? Google, at any rate, believes that a driverless car needs to have a little more. The designers are convinced that human expectations will make other functions necessary, such as electric doors on a taxibot that open and shut automatically. This is the reason why Google-Waymo selected the Fiat Chrysler Pacifica minivan as a rolling platform for future tests: its sliding doors for the passenger compartment open electrically.[189] This is a sensible approach. A passenger who, in a hurry, forgets to close the door of a robotaxi properly would create a problem for the vehicle if it could not close its doors itself and would have to wait for a human to come along before it could drive on.

Doors that close automatically are just one of many design problems in the interaction between robots and humans. It is more critical if we are not careful enough and somebody's life depends on it. For example, many safety features in the automotive industry were developed on the basis of tests with crash test dummies that were modeled using the body measurements of adult male drivers. This led to a 47 percent higher injury risk for women because the seatbelts simply were not made for them. Their body measures have been taken into consideration during testing only since 2011.[190]

Carol Reiley, cofounder of the startup company drive.ai, also illustrated some gender-specific problems in her doctoral thesis. She had designed a voice-activated surgical robot. The voice recognition software she used had been developed and used by men, and it did not understand her voice because her pitch was not low enough. As a consequence, whenever she did a presentation, she had to ask a male student to say the voice commands for her.[191]

If you have watched YouTube videos of traffic scenes in other countries, you have probably asked yourself why the number of fatal accidents is not even higher. Chaotic intersections in India, Pakistan, and Brazil have very few similarities with the comparatively disciplined procedures in Europe. It is a small wonder, then, that self-driving vehicles first have to adapt to the country's specific rules before they can be sent into mixed traffic, interacting with human drivers.[192]

Here in Silicon Valley, cars often stop far away from a pedestrian crossing to tell the people waiting there that it is possible for them to cross the road unharmed. In San Francisco, however, life is like it is in every other metropolis, with pedestrians crossing the street between vehicles. Furthermore, if you look at conditions such as those in the big traffic circle at the Arc de Triomphe in Paris or at intersections in Bangalore, you will immediately recognize that they pose special challenges. Tourists have to first learn and observe how to best deal with such situations.

A gradual expansion of the areas in which autonomous vehicles may be used will probably also have to include construction sites, which at the moment are still very complex and difficult to program for. Every road marking there is invalid, after all, or possibly made out in a different temporary color; traffic cones indicate the lanes; workers hold signs or signal with their hands to guide traffic; heavy vehicles maneuver forward and backward; the street is dirty; and any of those conditions can vary from country to country or, as in the case of the United States, from state to state. Hence we should not be amazed that today's prototype vehicles (for now) have problems with all that.[193] Furthermore, construction sites can be set up at any time, as soon as the construction company receives permission for commencing work. Data on this are not always available centrally, so a vehicle cannot obtain this information beforehand. One temporary solution could be that driverless cars first avoid construction sites and choose an alternative route. Or as Nissan suggests, a service representative remotely controls the vehicle and safely navigates it through a construction site.

There are ample opportunities for anyone who would like to join in the development of autonomous cars today and has some spare cash. In mid-2016, there were already several hundred companies working on individual technology components or complete vehicles. The industries range from sensor technology (radar and ultrasound), LiDAR systems, and cameras to processors and high-precision GPS systems. Software is another important factor in the development: map solutions, algorithms, AI and machine learning, cybersecurity, solutions for fleet management, and ride sharing.[194] Investment platforms such as Angel List and Venture Radar list dozens of startups in the various technology areas.

Become a Tinkerer with Open Source

If you have a desire to delve into the details of the design of autonomous vehicles, it is not absolutely necessary to be able to possess the massive resources of automobile companies and internet giants. Thanks to many open-source projects, interested parties can obtain software and data sets and even hardware kits for a reasonable price.

The most prominent projects are the ones managed by Udacity, providing not just the code students developed and tested on Udacity's vehicle but also the drive simulator and respective data needed to test real-life traffic situations.[195] George Hotz, founder of comma.ai, provided Open Pilot, whose source code and 3D print files are available for several types of vehicles.

Using it, you can turn your smartphone into a drive assistant.[196] The Open Source Car Control Kit, in turn, supports those interested in modifying their Kia Soul into an autonomous test vehicle.[197] The Technical University of Braunschweig also plans a release of its data and simulators as open source. And then there is the latest newcomer: an open-source project based on the Android system.

The most ambitious open-source project comes from Chinese internet search giant Baidu. Its Apollo initiative tries to tackle every part of a full-fledged autonomous driving operating system. And since its launch in mid-2017, the project has progressed rapidly, with a current heavy focus on China. Baidu's strategy behind the Apollo program seems to copy Google's playbook when it took on Apple's closed iOS with the open-source Android. But this time, for its autonomous drive operation system, Google (Waymo) is the one with a closed system, and Baidu has an open-source system.[198]

In addition to the software, some organizations also provide images of street traffic scenes, as the Max Planck Society did with the so-called Kitti data sets. The images show cars, trucks, pedestrians, cyclists, trams, and other objects with annotations—a marking on those objects.[199] The website Common Objects in Context (COCO) provides similar material as a download, as does the California Institute of Technology in Pasadena.[200] However, not everybody is convinced that the quality of the data available for free is sufficient. A group of researchers from the Max Planck Society criticized the Pasadena data in a study.[201] Oxford University, by contrast, makes standardized images for classification available in its PASCAL Visual Object Class project. This data set provides annotations and images of aircraft, bicycles, birds, boats, buses, cars, horses, dogs, sheep, and even potted plants.[202] After all, it is better to be prepared if a potted plant should make a driving mistake! And last but not least, I would like to mention the data sets provided by Cityscapes, with 5,000 annotated images of high quality and 20,000 images of medium quality from 50 cities.[203]

Although all these data sets are available for free, not all of them may be used for commercial purposes. For once, it is a good idea to actually read the general terms and conditions.

Safety First: Also for AI

Both the American NHTSA and the European New Car Assessment Program (NCAP) conduct safety tests with cars. There are, however, significant differences. While NHTSA mainly focuses on structure and retaining tests and its main question is how any collision affects (adult) passengers, NCAP considers a larger range of scenarios, including the safety of children in the vehicle and any pedestrians potentially colliding with the vehicle.

Self-driving cars pose new safety-relevant challenges for authorities and manufacturers. How can you test an algorithm? How can you test the neural network of an AI system? The programmer is no longer the reference point giving instructions for safe behavior; it is the system that gains experience and learns, although supported by humans and under supervision. Philipp Koopman, researcher at Carnegie Mellon University, currently the leading research center for robotics and AI, says:

> This is an inherent risk and error mode of machine learning. If you look inside the model to understand what it's doing, all you get is statistical data. It's like a black box. You don't know what exactly it's learning.[204]

It is thus very difficult to understand whether the system reacts as it should and how it makes decisions. In Gothenburg, where Volvo is testing its autonomous vehicles in a field experiment, researchers from Chalmers Technical University now work on procedures for testing autonomous vehicles in terms of safety.

CHAPTER 8

Hey There: Connected Cars in Conversation

*True love is a lack of desire to check one's smartphone in
another's presence.*

—ALAIN DE BOTTON ON TWITTER

When you are driving, you are constantly checking your surroundings
and communicating with other road users—a look into the rear-view
mirror, a nod to another driver, a gesture to a pedestrian, flipping off the
guy who cut you off. People are very versatile and inventive communicators.
Language and gestures have evolved and changed over the millennia, but they
regulate our lives. Although the intentions are not always clear and unam-
biguous, we usually get the drift. Mostly.

But how do vehicles with no people at the controls communicate with
their environment? And what about the passengers in the car, if the car is the
decider and controlling force? We already learned about some of the commu-
nicative options; defensive driving and digital displays outside as well as voice
messages to other road users are just some of them.

It becomes really interesting when self-driving cars can communicate perfectly naturally with other vehicles, road users, and objects in the vicinity.
Imagine getting to an intersection with no traffic lights, and there is a lot of
uninterrupted traffic from left and right. When will your car be able to make
a right turn? Will it have to wait until someone takes pity and lets it in? Or

should it just press in and hope that the other traffic participants are attentive? Depending on the country you are traveling in, one way can put you into grave danger, and the other one can leave you there at the intersection until nightfall.

A connected car, by contrast, can communicate with other vehicles, that is, exchange information and intentions so that traffic continues to flow. This is called *vehicle-to-vehicle (V2V) communication*. At an intersection, the situation therefore could evolve like this: my car wants to turn right and consequently plans to arrive at a certain speed, while an approaching car announces its own speed and notifies yours that it will adjust its speed and allow your right turn. Another approaching car in the middle lane could also communicate that it is not planning to change lanes and that it is safe for the turning car to complete its maneuver. This communication with one another potentially allows for a larger range of situations than the one offered by sensors that often cannot capture the entire relevant section of traffic events.

Toyota is already using V2V communication for the Prius, Lexus RX, and Crown. The vehicles communicate their location, speed, and destination in a range of 1,000 feet around them.[1] The models can also communicate via *vehicle-to-infrastructure (V2I) communication*, for example, with traffic lights that announce when they will switch from red to green. This allows the cars to adjust their speed accordingly. Audi already demonstrated this procedure in Las Vegas, and Mercedes is planning to equip its S class cars with V2I communication from 2019 onward.[2]

Emergency personnel, streets, lights, parking lots, or buildings might exchange information with the vehicle and keep it updated on the traffic situation or give instructions. It would be possible to reroute traffic electronically via a signal in case of an accident and subsequently blocked roads, a process that would also be possible for occasional traffic calming for a good night's sleep or during events. Ground sensors can be used for alerts if the road gets icy or slippery. Solutions offered by companies such as Inrix and UPark help to establish such approaches.

Of course, appropriate safety measures have to be adopted for seamless operation, making sure that instructions and information can only be sent and received by an authorized subject. We can easily imagine that someone electronically communicates that the street where he or she lives is blocked due to an accident, diverting traffic away from the street just to have a quiet night. False information might indicate safer road conditions than are actually the case. Electronic signals may force connected cars to stop so that they can be robbed.

It is obvious why researchers looked into electronic security protocols from the start, drawing inspiration from online shopping portals. Safe transactions involving the exchange of sensitive data such as credit cards and bank account details use the Secure Sockets Layer (SSL) protocol. This protocol encrypts information during transmission: only someone in possession of the electronic key gains access. A central admissions site issues a certificate with a preset expiration date, much as if your house key were only valid for a year and then had to be renewed. In this way, everybody knows that the subject about to read an SSL protocol is definitely a certified site.

The difference is that every involved device—whether it is the car or a ground sensor—must be able to read such certificates because it is both sender and recipient of important information affecting the traffic behavior of other road users. It must be ensured that the certificates are valid and that the sender has the right to issue them. The biggest challenge will be to get the verification process to take place rapidly enough. Waiting several seconds for data transfer is unacceptable in a dynamic environment such as street traffic. Wobbly internet connections are not helpful, and data volume as well as authorized objects must be predictable. If there are not only 20 vehicles communicating at an intersection but also hundreds of sensors, you will need a very powerful network. The authorities keep a special radio wavelength range for this application, the so-called dedicated short range communication (DSRC) channel, which is, for example, used in Europe for electronic toll collection.

We should also expect hacker attacks. Distributed denial-of-service (DDoS) attacks are already the most frequently used method to crash websites. Thousands of so-called zombie computers—PCs infected with malware whose owners have no knowledge of this and certainly did not agree—send so many queries to the server of the website under attack in a short time that the website crashes. This kind of attack can also be imagined for connected cars: they could be bombarded with information from too many services at the same time and take too long to identify the relevant senders and data. And last but not least, one's own data must be protected properly. Not everybody (and certainly not a person of public interest) would want his or her usual travel routes and preferences known to everybody else.[3]

The following example shows how dangerous attacks can be.[4] A journalist from the technology journal *Wired* tested a Jeep traveling at 65 miles per hour on the highway. The car was, as had been agreed beforehand, remotely controlled by two friendly hackers. They first turned up the radio and then the ventilation, and then slowed the car down and sped it up, eventually guiding it onto a patch of grass—and the driver was unable to do anything about it.

IT security expert Craig Smith published a manual on this topic bearing the catchy title, *The Car Hacker's Handbook: A Guide for the Penetration Tester*.[5] It was originally intended for automotive manufacturers to provide some inspiration and information about security gaps, but today the main readers are mechanics and car owners who use the book and the respective Wiki because they feel patronized by the electronic systems and restrictive interpretations of warranty regulations that the manufacturers favor. Vehicle maintenance and tuning are increasingly a question of software, and manufacturers increasingly resent interference from third parties.

Once the connection with the network fails, the use of the vehicle may be limited, and safety problems could ensue. In August 2016, Tesla's data network was down.[6] Although it was still possible to manually control the cars, it was no longer possible to download the very detailed road maps for the respective areas in which the cars were moving. This is especially critical if the cars are traveling in autopilot mode (or, in the future, in fully autonomous mode).

However, connected cars do not limit their communication to the outside. We already connect our smartphones with the vehicle, using them as navigation and entertainment systems. After all, people don't want to copy all their music and contact information into the car's memory if they already carry all that information with them. Such systems will be more extensive and relevant in the future. Self-driving vehicles relieve humans from the task of driving, but they will also give them more time for relaxing, for entertainment, or for work. The ridesharing service Uber already lets passengers play their own music on the car stereo system. In the same way, other settings could also be controlled via smartphone—seat position, preferred driving style, destinations, and much more. Still, we cannot yet predict the extent of *digital customization* in the future.

Automotive manufacturers and digital companies fight for first position for every aspect. The question of dominant operating systems and standards for connected cars remains completely open. Google founded the Open Automotive Alliance, which aims at integrating Android into cars.[7] At present, most of the major manufacturers, such as Volkswagen Group, Jeep, Mazda, and Ford, are part of this alliance. In addition, the Automotive Open System Architecture group, abbreviated AUTOSAR, is working on an open standard to define electronic control units. This includes electronic components for entertainment systems, software test routines, interfaces to read data and connect devices, and the computer capacity autonomous vehicles require.[8]

The authorities may be interested in equipping cars with electronic identifiers to "track" them.[9] The Chinese electronics hub metropolis Shenzhen issued 200,000 electronic licenses for cars, including trucks and buses, as part of a pilot project. The aim of this project is to determine, for example, the routes taken by hazardous goods transporters and school buses. Such licenses would, for instance, allow the assignment of special routes to particular types of vehicles, depending on the time of day, the weather, and other conditions.

Volvo, in turn, uses connected car technology for a delivery service. Volvo In-car Delivery allows suppliers to deposit ordered goods in the trunk of a parked vehicle.[10] The application shows where the vehicle is located and permits opening the trunk via a smartphone app.

The infrastructure required for vehicle-to-X communication (i.e., communication between vehicles and objects) also lures traditional software providers such as Microsoft into the business. Data generated in and through traffic have to be shared with others and stored both in the car or object and externally. Microsoft used the opportunity to seize a piece of this ever-growing cake with its Cloud Azur solution.[11] The Gartner Analyst Group is expecting to see up to 250 million connected cars on the roads by 2020.[12] All the ensuing data will create a new billion-dollar industry providing additional services for drivers, passengers, taxi fleet operators, and others, similar to the development when smartphones came on the market.[13]

During the transition period in which autonomous and connected cars share the roads with vehicles controlled by humans, the information exchange between connected cars might seem a little like a whisper through a grapevine. The cars will let each other know which of the human road users drive badly or aggressively. Anca Dragan, human–machine researcher at the University of California Berkeley, finds this scenario quite interesting but believes that it has reverberations relevant to data-protection legislation.

How, then, should we send large data packages? Today's networks are limited when road maps have to be updated, new traffic information is received, or communication with other vehicles and objects takes place. Just as the CEO of AT&T once planned to limit the use of certain data services such as YouTube videos to 20 seconds before the first iPhone was launched because he was worried the networks would crash (and was scorned by Apple's CEO Steve Jobs), the capacity for those new, additional participants—namely, the cars—must be extended, although the providers of data services in automobiles also have to ensure that they do not bog down the network with useless data volume.

However that may be, companies providing such infrastructure are already starting to expand capacity. Qualcomm, for example, is testing its first 5G networks that can handle data volumes of up to 45 gigabytes per second. The company plans to install them widely after 2020.[14] By comparison, current downstream speeds in Germany are just 13.7 megabytes per second, 26,000 times slower (and just for your information, half the speed found in South Korea).[15]

Europe's aversion to digital solutions and its data-protection paranoia will be the greatest disadvantage imaginable. Even today, Germany lags behind in terms of distribution of the latest models of smartphones and access to private/public WiFi nodes and thus has less experience with the digital economy. This lack of experience will become much more poignant when 5G networks become an essential part of products such as, you guessed it, cars.

The number of sensors and chips required increases massively once we have autonomous connected cars. At present, up to 170 sensors and 100 chips measure anything you can imagine, from tire pressure to exterior temperature in cars. Self-driving vehicles add an entire new range to this. The LiDAR system presented earlier is probably the most demanding element in all this. Tesla hinted at new uses of sensors when presenting the Model X. Some bloggers saw this as nothing less than preparation for the link of self-driving electric vehicles with drive service functions.[16] A sensor in the driver's door recognizes the passenger, automatically opens the door, and arranges the seats to the driver's preferences. Under the seats, there is room for a purse or laptop bag.

Electric and electronic architectures will have to be modified for the new vehicles. Instead of assigning a chip to every sensor and component, they will be concentrated onto a few clusters. These systems must be secure and therefore extensively tested beforehand, especially the parts responsible for drive behavior.[17] And who could have more experience with this than companies with a strong background in software and hardware, such as Google, Apple, and Tesla! Once again, the expertise for this is something the traditional automobile companies still have to acquire and make part of their own "systems."

A Literally New Zeitgeist?
The Sharing Economy in Progress

We are standing on Market Street in San Francisco. The executives from a German pharmaceutical company are planning to return to their hotel after their visit to a startup accelerator. It is rush hour. There is no free taxi anywhere in sight. Eventually, somewhat reluctantly, the executives give in to my request to simply order an Uber. After downloading the app, they press the order button and immediately see information about their Uber car light up. On the map, they can see it coming round the corner, and a moment later, it stops in front of them. They are flabbergasted. It took less than a minute for the car to arrive and 10 minutes to get to their hotel.

It could have been any of the other car-sharing companies that sprang up like mushrooms in recent years: first, the inner-city car-sharing startups such as Uber, Lyft, Gett, and Via in the United States, Didi Chuxing in China, and Haxi in Norway and then the providers offering long-distance services, such as BlablaCar in France. More than two dozen startup companies, some of them with excellent capital situations, compete for pole position in the car-sharing industry. Although in this book I decided to subsume them under the same heading, there are actually at least three variations. First, there is true car sharing between people (peer to peer), as offered by GetAround, JustShareIt, and Turo. Second are membership-based rental services such as Zipcar, Car2Go, and WeCar. And finally, there are providers of on-demand taxi services—Uber and Lyft.

Despite the fact that in Western countries we usually mention Uber when speaking of ridesharing, Uber's Chinese rival Didi Chuxing is far more impressive. The company states that it had up to 300 million users worldwide in May 2015, among them 14 million registered drivers; the company is active in 400 cities—and that is just in China.[18] For its global impact, the Chinese sharing service has closed agreements with other companies. For example, it invested $350 million in GrabTaxi from Southeast Asia and $100 million in Lyft and $500 million in Ola in India.[19] Didi was so dominant in China that Uber eventually sold its Chinese subsidiary to Didi in August 2016.[20]

Wherever ridesharing cars appear, conflict is in the air: lawsuits are filed, operations are prohibited, penalties are threatened, and local taxi services (which feel cheated out of their work) attack viciously. And taxis are fully justified in feeling threatened. If you have ever had to find a taxi in an American city (or rather ever had the luck of getting one), you could not possibly overlook the vehicle's low quality, not to mention the feeling of being in a cage in which the driver and the passenger only communicate with each other through a tiny sliding window. And then come the inevitable laments if you want to pay with a credit card instead of cash. I experienced a wait time of 50 minutes for a car in front of the *New York Times* building during afternoon rush hour. Mind you, this is not because there are no taxis. At least every second taxi that stopped drove off without picking up a passenger, perhaps because the passenger wanted to go in a different direction than the driver wanted to go or because the taxi drivers were about to finish their shift and were on their way to a handover station somewhere in town. Can you believe it? During the busiest time, a large number of taxis are traveling without passengers because they need to hand the vehicle over to another driver somewhere. And the drivers did not even think this strange; thus is the result of a survey done by the newspaper whose journalists started to wonder about this phenomenon.[21]

The cost for the taxi licenses indicates how much ridesharing companies disrupt the old power balance. In the United States, *cab medallions* are required to run a taxi service. Because the number of those medallions is limited by the respective cities, the price per license has risen to more than $1 million in recent years. While the metropolises grew with breathtaking speed in the last decades, the number of licenses did not—until Uber arrived and caused prices to drop drastically. Cab medallions suffered a 50 percent loss in value and are now the subject of legal action initiated by taxi companies and loan providers in New York.[22]

In the minds of investors, however, the number of legal actions and the degree of resistance against a startup company are an indication not only of the inherent risk but also of the degree of disruption it introduces into an existing market. The equation is simple: the more legal claims there are, the bigger is the disruption, and the bigger is the chance of generating high returns on investment.

Whereas European startup companies bow to existing regulations, the newcomers in Silicon Valley question regulations and modify them. Regulations in the taxi industry are not without history. There used to be an imbalance in the information available for taxi service providers and passengers. A visitor often does not know a city or region very well or does not know it at all. And many drivers take advantage of this. The experience of one of my Spanish friends may serve as proof: after a conference in Las Vegas, the woman was traveling to the airport in a taxi early in the morning when a motorcycle cop knocked on the window at a red light and asked the driver about his destination. He claimed to be on his way to the airport. The policeman asked why he then was going in the opposite direction. The driver's reply that he had to take a detour because of traffic did not satisfy the policeman because there was absolutely no sign of increased traffic at 6 a.m. The policeman noted down the taxi's license plate number and driver's license number and told the driver to turn around immediately and take the direct route. He then handed my surprised friend a business card with his phone number and the offer to call him if she had any additional problems or if the cost of the trip came to more than $20.

Are taxi drivers in Las Vegas any different? No more than in Vienna, or Munich, or Berlin. When I once was telling this story, one of my friends from Vienna contributed his own experience: after a dinner with friends, he and another acquaintance from Switzerland took a taxi together and started an animated discussion in the rear of the car, speaking English together—until my Viennese friend noticed that they had passed the town hall twice on the same road in the course of 10 minutes. The taxi driver had thought his passengers did not know the city and wanted to exploit the situation.

In order to stop such business practice, many towns and districts adopted regulations and statutory provisions, allowing them to issue taxi licenses, demand driving tests, and regularly check compliance. With an Uber offer, however, the passenger knows beforehand just how much the trip will cost and how the driver was assessed by other riders, and the driver can also access information on the passenger entered by the driver's colleagues. There is more transparency—who is this and what is the expected price-performance ratio?

In the wake of Uber's popularity, the taxi services tried to use the regulations to which they are subjected as a weapon because their competition claims that it is not subject to those regulations. Suddenly, the authorities found themselves confronted by the fact that regulations intended to protect passengers against taxi services now were being used by those taxi service providers to protect them against Uber. Some interesting situations arose from this paradox: Uber users loudly protested during community meetings, and official representatives had to remember why those regulations were introduced in the first place. The new situation made the regulations partially obsolete and required new arrangements such as insurance coverage for Uber passengers.

Things can get somewhat absurd at times, for example, in the case of the new taximeter calibration directive. If you visit Amsterdam, you cannot help but notice the huge number of Tesla Model Ss being used as taxis. As a matter of fact, there are two taxi service providers running fleets with 167 Teslas.[23]

Environmentally friendly and comfortable. This is not possible in Germany because of a calibration directive. In order to prevent manipulation and incorrect calculations, only vehicles the manufacturers offer as taxis may be used as such in Germany. At the time, the EVs offered by Tesla and Renault did not meet this requirement.[24] However, the German government noticed this flaw and modified the directive.

Small pinpricks—a directive here, a regulation there—with the intention of slowly but surely choking everything that might blossom. As a matter of fact, taximeters are gauges from prehistoric times. They calculate the duration of the trip and determine the price for the trip on the basis of the distance covered and the time needed plus some confusing lump sums, tariff categories, and surcharges. Once a passenger has arrived at his or her destination, he or she is often surprised that the price is higher than he or she thought. This would be very different with Uber! But then, Germany banned Uber. Quoting Christoph Keese, author of *Silicon Germany*:

> Germany is the technology museum of the 20th century. Neophobism rules supreme, namely, hostility towards anything new.

So why are car-sharing and ridesharing companies so popular? The reasons are simple: motivational structures change, and the new generation also has new priorities. The young are less concerned about owning a house and a car and more about doing something sensible with their lives. In contrast to their parents, younger people are more willing to forgo higher wages as long as the work itself is important to them. Basically, the video game generation is looking for an "epic mission" in their life's game.

This approach is a problem for the automobile industry. A generation that does not want to have a driver's license and sees no point in owning a car must want to have a different kind of service offering.[25] Car-sharing companies fulfill this need by making digital technologies the core element of their mobility services. The previous dominating model, the *owner model*, is now regarded as being too extravagant and costly. The economy consequently is moving from a concept of owning resources to one of having access to resources, which affects the sales figures. A survey in 10 American cities indicates that every vehicle added to a ridesharing fleet decreases the number of vehicles sold by 32.[26] Eighty percent of Zipcar members sell their own vehicles once they have become members. This means that one Zipcar takes 15 private cars off the road.[27]

Boston Consulting Group estimates that car sharing will have a particularly strong effect on car sales once driverless cars are fully available.[28] The degree of comfort will be unparalleled. Until then, there will be three cars less sold for each car accepted into a car-sharing program. In the United States, the decrease is 1.2 cars, and in Asia, it is 4.6.

The different models offered by Uber are also a remarkable development. Uber Pool and the model adopted by the now-defunct startup company Chariot, with several passengers sharing a vehicle and choosing an approximate destination, are somewhat similar to the fixed-route taxis popular in, for example, Russia. As a matter of fact, Uber is surprisingly successful with exactly that business model. More than 50 percent of all fares in San Francisco are booked as Uber Pool rides.[29] Some friends of mine also use Uber to have their children picked up from school if they have problems getting there on time. Daimler was testing a similar model with a pilot program called Boost.[30]

While Uber & Co. is disrupting the taxi business, others are trying their luck with rental-style companies. Zipcar and Car2Go are nothing new as such. I already was a member of such a club during the 1990s. The club provided several vehicles at train stations and other key locations in town. There were three types available: C-class vehicles, sedans, and vans. You can rent the desired car type by the hour.

Flightcar followed a slightly different concept.[31] Travelers who often have to leave their vehicles in expensive airport parking areas can rent them out while they are traveling. The advantages for both parties are obvious. Instead of paying expensive parking fees, your car can actually make you some money. And once you have arrived at your destination airport, you could also book a private vehicle via Flightcar instead of getting a normal rental car. Alas, this idea didn't pan out, and the company closed in 2016.

Automotive manufacturers have no intention of leaving the sharing economy to its own devices and thus worked on their own programs—BMW offers DriveNow and ReachNow, and Daimler has set up Car2Go and now Croove.[32] The latest addition is from Renault-Nissan, which announced its development of a fleet service for self-driving vehicles that is supposed to be a substitute for public transport and will be tested in Paris.[33] Daimler predicted that Car2Go will become a Car2Come as soon as self-driving technology is marketable.[34] The latest development, to the surprise of industry experts, was the announcement in 2019 by Daimler and BMW of their combined car-sharing effort.

Audi restyles the company as a premium provider of ridesharing services, too. Ulrich Quay of BMW Ventures expects that more people will have access to premium vehicles via sharing models. If, thanks to sharing, the cost of mobility of $6,500 in Europe and almost $12,000 in the United States per year for the private ownership of such a car decreases to just one-tenth that amount, a couple of hundred extra dollars every year for access to premium brands will not make any difference. We therefore might have good reason to believe that Mercedes, Porsche, BMW, and Audi may actually gain importance in the sharing economy.

What is not so certain is whether users will indeed be interested in premium vehicles after all. Emily Castor, responsible for transportation regulations with Lyft, reported at a panel discussion on the topic of mobility at Stanford that the users of ridesharing services are not too concerned about the quality of their vehicles but are very much interested in the quality of the point-to-point connection.[35] She claimed it was crucial to be able to recharge your phone during the 10-minute drive but not so crucial whether your tender behind was firmly placed on heated leather seats.

However, people who believe that the disruption that ridesharing and car-sharing services cause would be limited to driving itself as well as the taxi and rental companies are missing the bigger picture behind all this. The companies' digital nature generates an extremely detailed image of the mobility requests and the transportation patterns in a region. This lets companies make predictions that allow for improved traffic planning as well as the solution of potential problems. Anybody who owns such data has struck gold. The potential applications range from expenses for urban planning to increases in real estate values and the best possible use of advertising budgets along transit routes. Small wonder that digital giants such as Google, Baidu, and Apple have turned into gold miners and are investing in this kind of technology. If you follow this train of thought, Uber is already transforming from a pure mobility service provider to a company whose AI systems are dedicated to predicting human behavior, whereas traditional companies are still trying to design better or worse copies of the Uber model. This is a process we already know from the example of Netflix and Blockbuster. Anyone who is not innovative will disappear from the market.

Because the Uber system is mainly based on a digital application, business models and services can quickly be expanded. One possible expansion is the so-called surge pricing, which allows a short-term price increase when demand increases. Whereas this approach simply follows the market economy of supply and demand and seems like a general disadvantage for customers, this may actually hold some benefits for passengers in some cities. It is often difficult to find a taxi in New York when it is raining.[36] In a study from 1997, researchers discovered that taxi drivers actually adjusted their working hours to their daily revenue. If the day was good, they finished their shifts early. On other days, they worked extra time to get their required daily wages. This market behavior ensured that the drivers reduced the supply on days when there were many passengers and thus an increased demand for taxis by going home earlier simply because they had reached their self-imposed daily target earlier.[37] Uber's surge pricing would let the drivers earn their normal fares several times over and encourage them to work longer shifts and massively increase their profit margin.

Today, 50 percent of the cost of a taxi is the driver's wages. The rest is spent on ridesharing fees, amortization, gas, insurance, servicing, and other expenses.[38] Despite all the sharing model enthusiasm, the automobile manufacturers are still skeptical. Managers do not believe that private ownership of vehicles will entirely become a thing of the past despite the fact that some already had to admit that they personally know younger people who do not own a vehicle anymore or hold a driver's license and have no intention of getting either. Today's sales figures, which are rising thanks to the economic boom in countries such as China and India, should not simply be extrapolated. The peak number of horses on the streets in the "olden days" was only reached long after the automobile had been invented.[39]

The population growth and continuous expansion of the metropolis nevertheless necessitate a thorough investigation into the private ownership of vehicles. As of today, the majority of people live in urban areas.[40] Whereas their share was 34 percent in 1960, the urban part of the population in 2015 was already 54 percent on a global level and almost 80 percent in Germany and the United States. Today, there are 28 megacities in the world in which more than 10 million people live. By 2030, 41 cities will have grown to such numbers, half of them in Asia.[41] In China, there are more than 130 cities today with more than 1 million inhabitants,[42] and the number of cities is expected to rise to over 200 by 2030. Just as a comparison, in the European Union, there are just 35 cities with a million inhabitants.[43]

As Twitter and other social media platforms have shown in the past, the approaches from Silicon Valley are potentially explosive beyond the economic and technological range. Social media had their share in the "Arab Spring" that swept away the leaders and despots in several countries almost at the same time. There are reasons over and above those of a good investment that Saudi Arabia invested $3.5 billion in Uber. Until recently, Saudi women were not allowed to drive cars, a law that drastically reduced their mobility. The

Saudi Arabian government hoped to offer a transportation solution with Uber instead of allowing women to drive, which would be easier and better, provided that women are permitted to get into a vehicle owned by a male stranger (every woman in Saudi Arabia has to have a legal guardian). Saudi women themselves are less than enthusiastic about this; their argument is that not only does their government keep them from driving, but it is also now making money with the antiquated drive ban.[44] Dubai's "pink taxis" show that there is another solution. Those taxis are exclusively driven by women and cater only to female clients.

The economic and convenient ridesharing services might generate some welcome side effects, for example, a reduction in the number of drunk drivers. According to company information, the accident rate among people under age 30 who had drunk alcohol before driving decreased by 6.5 percent in 17 California cities as soon as Uber started offering its transportation services.[45] However, other studies contradict these findings.

A survey from the University of California that analyzed the mobility behavior of Car2Go members in Seattle, Calgary, San Diego, Vancouver, and Washington, D.C., found that exhaust emissions also may decrease because the members drive less overall than if they had their own cars.[46]

Sharing models replace not only taxis but also public transportation, and for the first time, undersupplied regions can be connected to general traffic. In a test program, the city council of Altamonte Springs, Florida, uses Ubers to replace missing bus services.[47] The city previously had experimented with a flexible bus line, but Uber turned out to be a cheaper and more adaptive service. The city subsidizes the Uber trips, and everybody benefits. The citizens just pay the same price as for a normal bus ticket and are even dropped off in front of their own house, so the service is faster and more convenient than a bus ever could be, and the city pays less than it would if it ran its own bus line. Although not all the problems have been solved yet—for example, customers have to have a credit card and a smartphone, and Uber is not required to have handicapped-accessible vehicles—the potential is obvious.

But what are we going to do when private companies are replacing public local transportation? Many transportation corporations today are either fully or partially state owned. Cities and regions are trying to provide some social compensation, allowing all population groups to participate in public life. As soon as local transportation is completely removed from public responsibility and instead entrusted to private corporations, possible discrimination of certain groups in the population must be avoided. The simple fact of not having a credit card may become an insurmountable obstacle if a ridesharing provider demands one. Similarly, ratings for passengers from drivers, as Uber provides them, may cause a nontransparent refusal to mobility.

Skepticism with regard to sharing service providers also comes from another source. Do we create even more low-income workers who have no claims to social benefits because Uber, Lyft, and others regard them as freelancers? Despite the positive experience with Uber in San Francisco, the German pharmaceutical managers mentioned earlier also pointed out this aspect, and I have to say that I agree with them. This problem has to be solved. As a matter of fact, it is not just an issue that is limited to new transportation service providers but one that also concerns traditional taxi services. A taxi company in Vienna became the target of prosecutors when evidence of tax evasion surfaced: revenue had not been declared to the tax authorities, drivers had been falsely registered as marginally employed, and consequently, social benefits had been embezzled.[48]

The Chinese government is currently considering a ban on any illegal city inhabitants acting as drivers for ridesharing services.[49] In China, people cannot just choose where they want to live but have to get a permit if they want to move to another town. However, economic changes led many country people to (illegally) move to the big cities. Work as a driver is often their only way of earning some money.

And there is another thought to consider: the companies involved and many of us preferably call this industry the *sharing economy*. The *economy of sharing* is actually meant to refer to a noncommercial personal exchange without the need for exchanging money, basically neighbors helping each other and people getting access to previously unused resources.[50] While some of the sharing platforms may actually have started out like this, their development led to something quite different from what the now established term for it suggests. Initially, there were indeed mainly individuals offering an exchange of rooms or vehicles, but the relationship shifted heavily in favor of professional and profit-oriented providers of apartments and fleets. Uber and Airbnb are quite clearly examples of the aggregation economy—of ride services and free rooms. Harvard Professor Yochai Benkler called it nonsense to regard Uber as part of the sharing economy. In his opinion, Uber has used the availability of mobile technology to found a company that lowers travel costs for consumers, no more, no less.[51]

People also tend to overlook the fact that platform operators such as Uber and Airbnb increasingly gain control of those aggregated resources. Thanks to the network effect, they have the power not only to recommend prices but actually to dictate them and determine the standards and requirements to be met by those who wish to rent. They also have the power to decide which providers they want to accept on their platform and which ones they do not. Uber drivers who decline rides too frequently or whose ratings are too low are excluded from the platform without any explanation and have no legal lever to dispute that. The rather absurd effect is that Uber drivers are almost uniformly rated with five stars because passengers want to protect them, even though they may not have been satisfied with the service. Consequently, the rating is pointless and leaves a bad aftertaste for many of those involved.

Because Uber does not define itself as a taxi service but rather as a brokerage platform between offers of ride services and demand for such, the company has so far avoided expenses for insurance, value-added taxes (VATs), vehicle inspections, and accessibility requirements. Taxi companies are subject to such regulations, however, and have to provide a certain percentage of vehicles for disabled people. Uber cites Section 230 of the U.S. Communications Decency Act, which was initially intended to exonerate website providers from liability for the content of linked websites and/or comments and posts of users.[52] Just as a telecommunications corporation cannot be held responsible for the content of telephone conversations, Uber uses this act as a lever to relieve itself of responsibility for the potential misconduct on the part of a driver. After all, all it does is run a website and an app, and what drivers and passengers then do there is nobody else's business.

Small wonder that Tom Slee, author of *What's Yours Is Mine*, calls Uber's conduct in the towns where the company operates as "parasitic" behavior. The company is under pressure from all sides, as the 2017 disclosures about sexism and sexual harassment of female employees show, as well as the misconduct of Uber founder and now former CEO Travis Kalanick toward a driver, the criticism of Kalanick resulting from his short-lived consultancy for President Trump as a member of the latter's Strategic and Policy Forum, and the legal case of Waymo accusing Uber of theft of intellectual property all show. The corporation made many enemies and hardly any friends.

However, even Uber is not immune to disruption. Former Google employee Mike Hearn presented the concept of a "TradeNet," an attempt to return control to the vehicle owners by means of blockchain technology on a kind of auction platform for services—such as access to an autonomous vehicle. When a user asks for a ride, the vehicle owners can make this user an automated offer based on the user's profile. Instead of Uber determining price and service provided, the platform users agree on both among themselves.[53]

"While most technologies are laying off employees on the periphery doing simple tasks, blockchain makes the head office redundant," comments Vitalik Buterin, founder of the blockchain startup company Etherum. "The blockchain doesn't take work away from the drivers, but from Uber. Taxi drivers now can work with their clients directly."[54] Similarly, the Israeli startup company La'Zooz tries to take up the criticism encountered by the central transportation network providers. Just as Bitcoin is a decentralized currency, La'Zooz is a decentralized transportation network based on the blockchain and not owned by anyone. This should cause a power shift from the provider to the drivers.[55]

Today, vehicle owners still mainly provide their cars to the service providers; most drivers in San Francisco use Uber and Lyft. However, as soon as the manufacturers offer autonomous vehicles and fewer people are likely to own cars, using fleet services instead, car manufacturers may as well run those services themselves. Who would need the intermediaries anymore?

Their own business model is the greatest danger for Uber, Lyft, and others. They do not own vehicles and are in the market solely as intermediaries. As soon as a situation occurs in which users and drivers lose their trust in the intermediaries, they are out. Uber had to find this out the hard way during the social media campaigns such as #DeleteUber and the revelations around sexual harassment. Hundreds of thousands of users deleted the app from their smartphones.

Despite all the euphoria regarding the potential of ridesharing companies, there are still some advocates of private car ownership. For many people of the middle and older generations, the car still represents freedom—an option for simply heading out to where they want to without preparation, without waiting, and without the last mile to public transportation. They accept that this freedom comes at a high cost and actually limits many other options. First, there is the cost for service and upkeep, the necessity of finding and paying for a place to park, hiring fees for any transportation vehicle required in some situations, and the many traffic jams and accidents. The idea of freedom still prevails because there still are no alternatives, or at least none that have been tried yet, and it is not so easy to give up old habits. Funny enough, several people independently of one another gave me the same argument about why they absolutely have to have their own car—namely, the possibility of always keeping their golf bags in the trunk of their cars and being able to spontaneously drive out to a golf course. Most people probably will just laugh about that and shake their heads. They really have quite different problems.

Meanwhile, the parking situation has become so dramatic in many cities that I personally try to avoid driving into town whenever possible or planning my in-town ventures in such a way that I have the least amount of driving in the city itself. I heard from some friends that they drive to the city limits in their own car and then get an Uber. As soon as autonomous vehicles are available everywhere in ridesharing mode, the advantage will be so self-evident that even many of today's skeptics will be convinced.

CHAPTER 9

Research, Innovation, Disruption

More Money, More Features

*If everything seems under control, you're just not going
fast enough.*

—Mario Andretti

Imagine that you are building a car containing components that are not used
at all. Not only that, but those superfluous components increase the price
of the vehicle for the end customer by several percentage points. You tell
your sales staff not to talk about them or list them in any specifications. The
reaction is predictable: you would lose your job very quickly.

This is exactly what Tesla did. There are some components already included
in the software package that will only be supplemented in later updates or for
an additional license fee. The Tesla Model S contains such components; after
a software update, the vehicle suddenly sported a driving assist system named
"Autopilot" and parking assistance function, much to the owners' delight.
Tesla's Model S was furthermore offered in variations that already included
the more powerful and therefore actually more expensive battery, which was
activated once a surcharge was paid. This allows Tesla to provide features with
"one click" that would require a shop refit for traditional vehicles first. Tesla's
fleet of Model S, X, and 3 includes the autopilot hardware that at the time of
this writing was ready but could not yet be used for autonomous driving. The

software update is still in development and at least two years out before it is ready for download onto the cars.

This kind of procedure is a problem for traditional manufacturers. They have a hard time justifying the installation of components a priori without being sure that they will ever be used. Admittedly, Tesla uses this procedure in the higher-priced premium sector where higher margins make it economically feasible. However, there is a marked trend in this direction, especially when digitalization in the vehicle becomes more important. Because sharing models are among the first to benefit from more expensive components owing to their economic model of operation, the vehicle can be adapted according to the individual user. If you pay the lower tariff, some functions will be disabled. Paying more means more features, similar to air travel today, where different seating categories determine the choices of entertainment and food available to you.

Are the Companies with the Biggest Research and Development Budgets the Most Innovative?

> Progress is brought by those lazybones who are looking for easy ways of getting things done.
>
> —ROBERT HEINLEIN

If you look at the expenses for research and development (R&D), automotive companies are quite ahead. There are no less than seven automobile manufacturers among the top 25 corporations. Volkswagen was at the top of R&D expenses in 2015 and 2016 on the list of publicly traded corporations, listing $13 billion.[1]

If you look at the expenses and the turnover, however, the efforts Tesla makes dwarf those of all the other car manufacturers. Tesla is far ahead with 17.7 percent, VW comes in second but with only 6.4 percent, and BMW is third with 6 percent. GM spends 4.9 percent, Daimler 4.4 percent, and Toyota 3.7 percent[2] (see Figure 9.1).

European innovation successes fade compared with those of Silicon Valley or Asia. An article in the *Washington Post* showed that the innovation gap in Europe cannot be closed soon.[3] A report on internet trends by Kleiner Perkins Caufield & Byers, a venture capital company from Silicon Valley, showed as early as 2015 that the 15 most valuable internet companies have a market

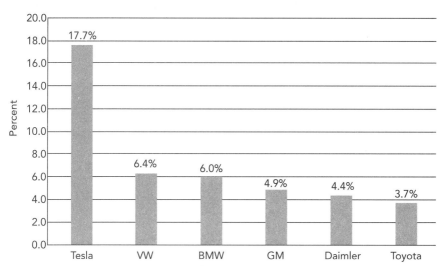

FIGURE 9.1   Research activities of car manufacturers

value of almost $2.5 trillion. Not a single one of them is from Europe, compared with 11 from the United States and another 4 from China.[4] If anything, this gap has grown, as mentioned earlier.

In 2018, the combined market value of the five most valuable U.S. companies—Alphabet, Amazon, Apple, Facebook, and Microsoft—was $4.2 trillion, whereas the 30 companies on the German Stock Index (DAX) came to about $1.4 trillion. While the five U.S. companies were all younger than 45 years, three of them even younger than 25 years, 24 of the 30 companies on the DAX were older than 100 years. The only digital company from Germany of global significance, SAP, was the most valuable, with about $140 billion, but still with a market value of about a fourth of the smallest one among the top five U.S. companies, Facebook.

European governments and the economy have a desire to change that, but it seems as if the answers they have are always the same well-known and little effective procedures: new subsidy programs, increased spending on R&D, money for universities. The latest German initiative is a program the government proposed called Industrie 4.0.

Would there be more innovation if the money were instead channeled to companies or more incentives were offered? This is a question the management consultancy firm PricewaterhouseCoopers posed, and the company proceeded to analyze the R&D spending of companies listed on the stock exchange.[5] The twenty-five companies with the biggest R&D budgets included eight companies from Europe, four of them from the clinical pharmaceutical sector, which was no surprise, one from capital goods, and the other three from the automotive industry (see Table 9.1).

PricewaterhouseCoopers wanted to discover whether there was a correlation between innovative power and R&D budgets and therefore asked managers about which companies they regarded as being most innovative. The results were sobering news. Apple, the company rated as being the most innovative, only came in sixth in 2018 in terms of R&D spending (see Table 9.2). Consequently, more spending on R&D does not necessarily mean more innovation. There was not a single European company named in the survey in the top 10. Let me repeat that: not a single European company.

How is it possible that higher expenses for R&D do not automatically lead to more innovation in a company? Well, there are many factors to consider.

A dedicated R&D division conveys the message that innovation is mainly expected from the people employed in that division, not from everyone else in the company. Furthermore, isolated R&D divisions often are too far removed from customer demands and do not sufficiently interact with the rest of the company. This deprives them of the necessary direct feedback and subsequently makes it more difficult to make changes quickly or even pivot—possibly involving a change of application or business model. R&D staff, often engineers with profound expertise in their fields, are often evaluated on their patent applications or scientific publications—just as they are in university research—but not necessarily on the number of product launches. As I indicated earlier, disruptive innovation often comes from nonexperts who get their new ideas in unusual ways and from the intersectionality with other disciplines. Engineers often regard entrepreneurial thinking—as businesspeople have it—not as their main task, delegating it to staff from other departments in the company. Furthermore, availability and use of too many resources may hamper innovation because too many problems are tackled simultaneously.

Researchers analyzed how innovation actually happens. They looked at the winners of the R&D 100 Awards and found that only 6 of 100 award winners were listed in the Fortune Global 500 in 2006. The authors of the study surmised that larger corporations are more dedicated to incremental innovation (i.e., innovation of existing products) than throwing themselves into radically new ideas.[6] As a consequence, many bright people leave the R&D divisions of such corporations and prefer to work in state research centers, universities, and smaller laboratories.[7]

TABLE 9.1 Top 25 Companies with Highest R&D Expenditures

| 2018 RANK | COMPANY | COUNTRY | INDUSTRY | R&D EXPENDITURES (BILLIONS OF U.S. $) | | | | R&D INTENSITY | | | |
				2015	2016	2017	2018	2015	2016	2017	2018
1	Amazon.com Inc.	United States	Retailing	9.3	12.5	16.1	22.6	10.4%	11.7%	11.8%	12.7%
2	Alphabet Inc.	United States	Software and services	9.8	12.3	13.9	16.2	14.9%	16.4%	15.5%	14.6%
3	Volkswagen AG	Germany	Automobile and components	13.9	14.2	13.8	15.8	5.7%	5.6%	5.3%	5.7%
4	Samsung Electronics Co., Ltd.	South Korea	Technology hardware	13.9	13.5	14.3	15.3	7.2%	7.2%	7.6%	6.8%
5	Intel Corporation	United States	Semiconductors	11.5	12.1	12.7	13.1	20.6%	21.9%	21.5%	20.9%
6	Microsoft Corporatiom	United States	Software and services	12.0	12.0	13.0	12.3	13.9%	12.8%	15.3%	13.7%
7	Apple Inc.	United States	Technology hardware	6.0	8.1	10.0	11.6	3.3%	3.5%	4.7%	5.1%
8	Roche Holding AG	Switzerland	Pharmaceuticals, biotechnology	10.2	9.8	11.8	10.8	19.8%	19.0%	21.9%	18.9%
9	Johnson & Johnson	United States	Pharmaceuticals, biotechnology	8.5	9.0	9.1	10.6	11.4%	12.9%	12.7%	13.8%
10	Merck & Co., Inc.	United States	Pharmaceuticals, biotechnology	7.2	6.7	10.1	10.2	17.0%	17.0%	25.4%	25.4%

(continued on next page)

TABLE 9.1 Top 25 Companies with Highest R&D Expenditures (continued)

2018 RANK	COMPANY	COUNTRY	INDUSTRY	R&D EXPENDITURES (BILLIONS OF U.S. $)				R&D INTENSITY			
				2015	2016	2017	2018	2015	2016	2017	2018
11	Toyota Motor Corporation	Japan	Automobile and components	9.5	9.9	9.8	10.0	3.9%	3.9%	3.7%	3.9%
12	Novartis AG	Switzerland	Pharmaceuticals, biotechnology	9.7	9.5	9.6	8.5	18.0%	18.8%	19.4%	17.0%
13	Ford Motor Company	United States	Automobile and components	6.7	6.7	7.3	8.0	4.7%	4.5%	4.8%	5.1%
14	Facebook, Inc.	United States	Software and services	2.7	4.8	5.9	7.8	21.4%	26.9%	21.4%	19.1%
15	Pfizer Inc.	United States	Pharmaceuticals, biotechnology	4.0	7.7	7.9	7.7	16.9%	15.7%	14.9%	14.6%
16	General Motors Company	United States	Automobile and components	7.4	7.5	8.1	7.3	4.7%	5.5%	5.4%	5.0%
17	Daimler AG	Germany	Automobile and components	6.9	7.2	7.8	7.1	4.4%	4.0%	4.2%	3.7%
18	Honda Motor Co., Ltd.	Japan	Automobile and components	5.7	6.2	6.5	7.1	4.8%	4.9%	4.7%	5.3%
19	Sanofi	France	Pharmaceuticals, biotechnology	5.6	6.1	6.2	6.6	14.6%	14.6%	14.9%	15.1%
20	Siemens Aktiengesellschaft	Germany	Capital goods	4.8	5.3	5.8	6.1	5.6%	5.9%	6.2%	6.2%
21	Oracle Corporation	United States	Software and services	5.5	5.8	6.8	6.1	14.4%	15.1%	18.4%	16.1%

238

2018 RANK	COMPANY	COUNTRY	INDUSTRY	R&D EXPENDITURES (BILLIONS OF U.S. $)				R&D INTENSITY			
				2015	2016	2017	2018	2015	2016	2017	2018
22	Cisco Systems, Inc.	United States	Technology hardware	6.3	6.2	6.3	6.1	13.4%	12.6%	12.8%	12.6%
23	GlaxoSmithKline plc	United Kingdom	Pharmaceuticals, biotechnology	4.7	4.8	4.9	6.0	15.0%	14.9%	13.0%	14.8%
24	Celgene Corproatiom	United States	Pharmaceuticals, biotechnology	2.3	3.7	4.5	5.9	31.7%	39.9%	39.8%	45.5%
25	Bayerische Motoren Werke AG	Germany	Automobile and components	5.0	5.1	5.2	5.9	5.1%	4.5%	4.6%	5.0%

TABLE 9.2   Top 10 Most Innovative Companies According to Survey Among Executives

2018 RANK	COMPANY	R&D EXPENDITURES (BILLIONS OF U.S. $)
1	Apple Inc.	11.6
2	Amazon.com Inc.	22.6
3	Alphabet Inc.	16.2
4	Microsoft Corporation	12.3
5	Tesla Inc.	1.4
6	Samsung Electronics Co., Ltd.	15.3
7	Facebook Inc.	7.8
8	General Electric Company	4.8
9	Intel Corporation	13.1
10	Netflix Inc.	1.1

R&D departments have a tendency to work on problems from a certain size onward only. Polaroid was very proud that it was able to handle huge innovation projects involving half a billion dollars.[8] Such gigantic sums lead people to be so afraid to fail and make any changes that they continue the projects until they may cause a serious threat to the company. Project duration has no relation to the speed of innovation. Assumptions made at the beginning of a project may no longer be valid by the time it is concluded. And there is nothing worse than having to acknowledge that the final product has already been replaced by disruptive technology. One promising approach is to make innovation part of the requirements for every department and for every employee.

How Do We Recognize a Disruptive Idea?

Tesla has had an electric vehicle with sufficient range available since 2013. Uber has been up and running since 2009. Google has been working on autonomous driving for 10 years. This should have been enough time to catch up, you would think, and yet the traditional automobile manufacturers are lagging behind. Why is this?

The example of a delegation from the German furniture industry shows this most clearly. It was a group of 15 managers from furniture manufacturing companies and distributors, and they had begun their third pitch. As is usual procedure in Silicon Valley, the participants desperately wanted to introduce a potential startup concept in a brief presentation as proof that they could operate and think like a startup. A pitch is meant to convince investors to put money into a startup company or project. My role was to listen to the pitches from the point of view of an investor and provide feedback on their ideas.

The first two pitches presented ideas that sounded interesting but apparently would provide little gain because the market for them was too small and there was too little revenue. The third pitch was about an idea for a platform on which furniture dealers and manufacturers would be able to sell surplus or dead stock for a low price, "always with a focus on the customer who will be happy to get cheap furniture."

However, this was not really true, after all. The main beneficiaries were the dealers and manufacturers themselves, which would in this way make money from old stock and would not have to spend money on disposal. This is a quite legitimate desire, of course. Half the pitch was over when the speaker mentioned in passing what the real disruption was and the actual benefit for the end customer. This is what usually happens: the same furniture item is produced by the same manufacturer for different dealers and brands. The only difference is that such a chair—from the same manufacturer, the same specifications, the same design, and the same quality—is sold for about $200 a piece under a luxury label, whereas a furniture discounter offers it for $30. This fact is an open secret in the industry and is part of the usual procedure.

So what is the disruptive idea here? A platform that showed those differences and makes the prices transparent would be truly disruptive. This is what customers would most benefit from, saving a lot of money in the process. This, however, is precisely what was not mentioned in the pitch because this idea would have destroyed the business model and done lasting damage to the relationship between brands, dealers and manufacturers. This is something they did not want to risk or could not risk because those relationships had been built over many years.

And this is where the startup companies come in, outsiders to the industry, not involved in any relationship and not heeding industry secrets. They have nothing to lose and therefore step on everybody's toes. The more disruptive the idea, the more people will complain. This is what Uber, Google, and the others are doing, and this is what Daimler and BMW cannot simply copy. Uber works in parallel with the taxi services, but Mercedes is the main supplier of taxi vehicles in Germany and is not in a position to affront its customers. Traditional car manufacturers and suppliers are walking on thin ice.

Traditional car manufacturers have already recognized the sign of the times, but it is not altogether clear how quickly they can market their technology. Mercedes is known to have been working on self-driving vehicles since the 1990s, without having had any great impact so far. Audi announced that it intended to found a wholly owned subsidiary called SDS with technology partners that were to push the development of self-driving vehicles.[9] Porsche, in turn, builds on its subsidiary for the software area, called Digital GmbH.[10] Ford threw itself into the game of autonomous vehicles, but while it may talk the talk, you can feel it from the people responsible that their hearts are not in it. General Motors seems to be serious with its autonomous car initiative with GM Cruise. And now Volkswagen, led by the example of Porsche, seems to be very serious about EVs as well.

Why the Automobile Scandals Are a Chance for Germany and the United States

The diesel emissions scandal and the cheating software are a story that does not lack in drama. Millions of vehicles from Volkswagen, Audi, and Mercedes (and probably those of other manufacturers as well) are involved; recalls, dozens of possible solutions, penalties, legal action, falling sales, and criminal proceedings and investigations are happening on a global scale. Employees who were dismissed, quit, or even ended up remanded into custody are all over the news. Anyone who believed that it could not get worse just had to wait until news got out about a price-fixing cartel involving five German manufacturers that had existed for decades. Technical arrogance—the presumption of knowing better than anyone else what good technology is and what it should be able to do, even if one did move into a gray zone for that—may have been a reason for the diesel emissions affair, maximizing profit without caring for the loss of others. In the short term, VW's conduct certainly will have a negative impact on the entire German economy and its reputation, but at the same time, this may be the chance to finally enliven the desperately needed startup sector. Faced with the tendency in larger companies to sublimate innovation with the knowledge of one's own size and by processes that have been optimized for mass scaling and execution, the German automotive scandal may offer the unexpected opportunity to release one's own innovative potential or that of others.

This can be proved with an example from the past, namely, the Austrian wine scandal in 1985. That was the year when the authorities discovered that winemakers had illegally added a chemical called diethylene glycol to their wines to "improve" the taste and sell them at a higher price. This substance is usually used as an antifreeze component in car cooling systems. In addition, the practice of adding sugar to the wine was widespread. However, as soon as those practices were made public, sales just plummeted. The Austrian wine industry was finished, and the culprits ended up in jail.

Thirty years later, Austrian wines are at the top of the international wine charts—with no illegal tricks at all. This scandal forced the legislature to adopt more severe provisions and controls while the winemakers took some time to rethink their attitude and slowly but surely return to their basic business, namely, to produce high-quality wine, made from naturally ripened grapes and without additives.

In 2016, Volkswagen employed 610,000 people worldwide and had an annual turnover of $220 billion. In 2016, the company was the largest automobile manufacturer in the world, featuring 10.3 million vehicles produced, slightly more than Toyota with 10.2 million.[11] The company combines more than a dozen brands, such as Volkswagen, Bugatti, Audi, Porsche, MAN, and Bentley. With $13.2 billion, the company also has the highest budget for R&D of all publicly listed companies. Unfortunately, as we know, this did not boost its innovative power—unless you count variations on fraudulent practices as innovative.

In the wake of the diesel emissions scandal, VW was penalized in the United States, agreeing to spend $2 billion over the next 10 years to create a U.S.-wide charging station network for EVs. And $800 million alone is to be spent in California. What looked like a bummer for VW turns out to be an advantage for the company and American EV drivers. VW started advertising its coming EVs by pointing out the extensive charging station network it is building—or rather was forced to build. Electrify America, the name of the company VW created to build the network, suddenly turned into a competitive advantage. But while the penalty for cheating included having the charging station network open for any EV, the stations come predominantly with the plug standards that VW prefers. Well, once accustomed to bending rules, it's not easy to get out of that habit.

The automobile industry today is eerily similar to the petroleum industry, where we have learned about the "resource curse." The discovery of resources and raw materials is often both a blessing and a curse for a country. The raw materials suddenly bring a lot of money to the country, driving inflation up. There are exchange rate fluctuations. At the same time, the raw materials industry drains other industries because it lures qualified staff away by offering high salaries. It favors wastefulness and corruption, undermines democracy and legislation, provokes conflicts with landowners, and displaces some population groups. And all of this is put in motion for a rather "dumb" product that requires little intelligence and involves little additional value.[12] Admittedly, the automobile industry makes a much more intelligent product, but old technologies such as combustion engines bind qualified engineers who might be able to provide a much greater service to humanity if they worked on something else. It is a waste of human capital if someone wastes his or her entire career being responsible for a sealing ring or a piston rod.

Because accruals for damages from the diesel emissions affair amount to tens of billions of dollars, VW, Daimler, and the others now have to save money in other areas, making many employees redundant. Thousands of engineers and creative staff will (have to) leave sooner or later, not least due to the imminent fundamental changes in the propulsion system. Many of them will receive a settlement. And although the German labor market with its high number of employees is looking good compared with the European average, many former automobile engineers may jump at the chance of founding their own companies and realizing their own ideas. Just imagine the creative potential those people bring with their knowledge and their capacities and skills. Areas such as robotics, drones, electronics, portable technology, medicine, and digital transformation may create thousands of new workplaces and companies and turn it into Europe's most important economic engine.

The question is, is the German economy prepared for this? Is there enough venture capital available? How can the government and the institutions help those founders by providing favorable legislation? Many things will have to change in politics in order to bring about the next German economic miracle and remove bureaucratic obstacles as much as possible. The German automobile scandals are the best chance for Germany of taking its economy into a modern digital era. A combination of highly qualified former automotive workers, settlements, venture capital, and startup-friendly provisions can turn Germany into a European leader and generate jobs and industries.

To quote economist Paul Romer, "Don't waste a good crisis." Germany has the chance to grab it! The German automobile scandals paradoxically may be the best thing that has "happened" to Germany recently. Experts, however, as well as the public, are doubtful about whether the manufacturers are likely to grab the opportunity offered. So far, the reaction of management, workers' councils, and owners does not look good. Even though some recent steps taken are in the right direction, they may be too little and too late.

And let's consider just one more upsetting piece of thought. Volkswagen alone was struck with $29 billion of fines in the wake of the diesel emissions scandal. Compare that with the amounts that the two leaders in the new technologies have spent so far, Waymo in autonomous driving and Tesla in electric driving. If we are generous and assume that both companies spent maybe $12 billion or $14 billion to advance those technologies to today's state, that's just half the amount that Volkswagen had to spend on fines. Volkswagen could have built two Teslas and two Waymos just with the fines and advanced the industry (and the environment) in ways that would have been staggering. And now remember what the company chose to do instead? I would never have the idea to call that behavior "commercially responsible," as German auto managers called it while pointing to Tesla's losses. While it's not my money that German automakers had to spend in fines, it's still upsetting when you imagine the alternatives.

CHAPTER 10

Timescale

What Will Happen to Us and When?

The future is already here—it's just not evenly distributed.

—WILLIAM GIBSON

Once you have had the chance to digest this mountain of information, you will comprehend how backward the technology for vehicles really is, despite the fact that we are using it every day as a matter of course. Every year, millions of people die or suffer injuries in car accidents. Our mobility demands a high price, not least because we lavishly waste our resources. First, we have to build vehicles with high safety standards because we are lousy drivers. This additional weight of the safety features requires more fuel. We are building bigger cars for that rare exception in which we travel with more than one person on board. And the fuel chain is so inefficient that we eventually only use less than 1 percent of the energy required for the actual transport of a person. This is the breakdown for combustion engines:

- **Extraction:** 10 to 20 percent just for the extraction[1]
- **Transportation:** 5 to 10 percent for providing the fuel where needed
- **Refining:** 20 percent for refining[2]
- **Maximum efficiency in the engine:** 30 to 40 percent[3]
- **Percentage of human weight in a standard motor vehicle:** 5 percent

If you take a gallon of water for comparison, we are just using a sip of water for transporting people. Nobody can tell me or persuade me that we have a highly efficient individual transportation system in place.

Bearing this in mind, we should be impatient for new technologies to replace the old ones, rendering our mobility more sustainable, cleaner, safer, and cheaper. Legislators and corporations are in a tight spot here. All groups involved and all stakeholders have to understand that mobility will change in the future, how it will change, and which other everyday areas are affected by this. The question basically is not whether this happens, but when. The second question is whether we will be in the lead or continue to leave the position of leadership to others.

In earlier chapters we discussed the state of the art and the intensive efforts on all sides in some detail. The signals are there, and the options are available; now we have to combine the various technologies and make their potential available to us. In order to drive home the urgency of this step, I will illustrate the facts and developments on a timescale so that you can get an idea of the speed with which all this is happening.

The development of autonomous vehicles has already reached the point at which experts are imploring the manufacturers to share their findings so that objectives can be reached sooner and the technology can become safer.[4] Tesla, as mentioned earlier, has offered the Department of Transportation (DOT) access to all autopilot data to increase the safety of this technology by creating massive data sets.[5] This anticipates possible demands on the part of the regulating authorities that now have to obtain sufficient expertise together with legislators to improve their assessment of the risks as soon as possible and adopt appropriate provisions.

In the United States, official bodies are convinced that autonomous and electric vehicles will be available for the masses within the next 10 years. Infrastructure projects already have to include this into their planning. As mentioned earlier, Los Angeles has already more or less discarded any expansion of public transportation, whereas Florida is in the middle of planning road projects for self-driving vehicles. Two-thirds of all American cities that took part in the Smart City Challenge organized by the DOT in 2016 are definitely continuing their planning under consideration of self-driving cars.

How will we experience self-driving cars? Probably through a gradual introduction on certain tracks and various applications. First areas of application already exist in the world: the autonomous campus shuttle service at universities in the United States, the Netherlands, and Switzerland and autonomous trucks traveling in the Australian mining area of Rio Tinto.[6] Highway sections allowing autonomous trucks are probably next. Highways are actually comparatively easy for this technology to master, in contrast to city traffic: there is only one direction of travel, no traffic lights, relatively few road signs, and (normally, at least) no pedestrians. The next step would be to legally and technically prepare individual city districts that are undersupplied with public transportation for autonomous vehicles.

This slowly opens an increasing amount of public traffic areas for autonomous driving while excluding manual driving at the same time. Very likely there will be a transition period in which both types of vehicles coexist on the streets. It may be advisable, though, to keep this period as short as possible in order to be able to have the full benefit of the positive effects of autonomous EVs as soon as possible. The founder of the Uber rival Gett, Shahar Waiser, expects a ban on manual cars as soon as 10 years after the registration of the first autonomous car.[7]

The challenge is in how we deal with the manual cars and combustion engines that we do not need anymore. There may be a transition period, to be prolonged if the vehicles are refit as self-driving cars with a conversion kit so that cars that are already on the road and are manually controlled would be converted at a later point. We heard already about Kopernikus.auto and Comma.ai, which were both working on such a self-driving refit kit.

Because vehicles are already recycled to 100 percent, the cycle for this is already available. They will be taken out of service gradually. As an afterthought to the revelations we witnessed in 2017, it is much more important that millions of cars with diesel engines will have to be taken out of service after the Volkswagen emissions scandal, long before their natural end-of-service life. In this respect, the diesel emissions affair was a huge environmental pollution affair not just because of the higher emissions but also because of the sheer number of vehicles to be replaced. The European Union Commission is almost at the end of its patience because of slow progress on the part of the German government and the resistance the manufacturers involved are putting up. The Commission has even threatened a ban on those manipulated diesel engine vehicles in the entire European Union.

Generally, we can recognize two approaches to the situation with the traditional manufacturers, illustrating how they pursue the development of self-driving vehicles. On the one hand is the evolutionary approach, introducing new technologies and functions step by step. This includes driver assistance systems such as the Tesla autopilot, the Drive Pilot in Mercedes S class cars, and the central zFAS (term is a bit of a mouthful) in the Audi A8. On the other hand, we find the revolutionary approach that Google adopts. One can passionately discuss which approach is safer. However, both approaches drive progress and will probably converge in the end.

The timescale predicted (and partially already completed) by the experts for the introduction of autonomous vehicles is as follows:

2019

- Tesla and other manufacturers offer driver assistance systems that allow partially autonomous driving: changing lanes, parking, and automatic braking, if necessary.
- Manufacturers such as Google, Audi, and Uber conduct testing on vehicles in a controlled environment; this includes (partially closed) highways, test tracks, and inner-city zones with low speed limits. In these cases, there is still a driver on board who can take over control if necessary.
- The first regulating bodies adopt national provisions. In the United States, the authorities in California, Florida, Nevada, Michigan, Louisiana, North Dakota, Tennessee, Utah, and Washington, D.C., are the ones with the most far-reaching and generous interpretation.[8] In August 2016, there were already 16 states about to implement respective laws for autonomous vehicles.[9] France allowed tests with autonomous vehicles on public roads in August.[10]
- Showcase cities such as Taiyuan in northern China exchanged all their taxis for EVs in 2016, Beijing is planning to retire all its 70,000 taxis with combustion engines over the course of just a few years and replace them with EVs, and Shenzhen finished replacing 20,000 taxis with electric taxis in December 2018 and 16,000 buses with electric buses in December 2017.
- Driver assistance systems will undergo additional function extensions and automate an increasing number of tasks; that is, they will be able to autonomously handle driving on highways. *Corridors*—road sections on highways—are set aside for autonomous truck traffic.
- Traffic authorities will present regulations managing testing and operation of self-driving vehicles. California introduced regulations that allowed autonomous cars to travel without any human control on public roads in April 2018, with Waymo becoming the first company to be granted a license in October of the same year.

- Manufacturers are going to increase the number of test miles, get more information and experience in various scenarios under varying conditions, and improve the fundamental technology as well as its safety. Tesla has at least half a million vehicles with hardware presumably fit for self-driving on roads all over the world already in the hands of clients—all in one software update. Since the spring of 2017, Tesla has been collecting sensor data from tens of thousands of customer-owned cars in order to centrally accelerate machine learning for self-driving technology and return it to the customer vehicles via upload. This is the latest point in time for the big manufacturers to perhaps still close the gap.
- Prices for sensors and electronics are going to decrease even more and become affordable for the lower price segments as well.
- Tesla's Model 3 has been available for sale since the summer of 2017 and thus is the first vehicle that definitely marks the breakthrough for electric mobility.

2020–2023

- From now on, new productions from most manufacturers will have all the technology on board (admittedly in different stages of development) that allows for autonomous driving: sensors, cameras, processors, software, even LiDAR systems. These cars will recognize traffic lights and road signs and act accordingly.
- Ridesharing corporations are already planning on the introduction of commercial autonomous vehicle fleets that will have their place as a service in certain cities and districts.
- More and more autonomous vehicles will be found on the streets. This is simultaneously the turning point because legislators, regulating authorities, and city planners now really must react and introduce appropriate changes.
- Self-driving ridesharing cars cover hundreds of millions of miles in self-driving mode in a very short time. The conclusions drawn from these data will massively influence and possibly drastically change our understanding of how cities and traffic work.
- The first large manufacturer, Ford, has announced the market launch of autonomous vehicles without a steering wheel or pedals for 2021.[11] Volvo is expected to present a similar development of its own, and Fiat-Chrysler will offer similar vehicles together with Waymo. Other

manufacturers have also presented their concepts and published statements, but so far, all we have are vague promises.

- EVs will be cheaper than combustion engine vehicles both at purchase and during operation. It will therefore become uneconomical to buy and run a car with a combustion engine.

2025

- All taxis used in urban areas will be EVs run by fleet managers, and in many city districts, all of them will be autonomous cars. The vehicles are charged at their waiting stations if that should still be necessary during the day.

2030–2035

- This is the point where the last manually controlled cars for the mass market have already been produced, and there are now only special models with manual control.
- The last combustion engine vehicle is manufactured, and the last vehicle combustion engine plant closes for good.
- Transition regulations are introduced, banning manually driven vehicles from public streets and restricting them to closed tracks (e.g., mountain and coastal roads that are available for manually controlled cars on some weekends, much as we see today when there are classic car races). Manual control of vehicles will become a pure leisure time activity.
- The last person in the world obtains a driver's license.

2045

- The last manually controlled cars will have disappeared from public streets, with a huge impact on cities and traffic. Traffic areas that used to be reserved for cars in the past will be returned to humans.
- Public local transportation is deconstructed. Streetcars, bus lines, suburban train lines, and subways will disappear. Many more people will be able to participate in mobility and at a lower cost at that.

Are these predictions overly optimistic or even possible? Perhaps yes, perhaps not. There is a series of good arguments why those developments may not happen so quickly or happen at all. Human nature is prone to habits, behaviors, and irrationalities that have been known to throw a spanner in the works of many (allegedly) good ideas and (luckily) some bad ideas, too. Natural obstacles may be more difficult to overcome than we expect. *Black swan events*, those rare and unpredictable global events in human history that are unprecedented and unexpected at the point in time they occur, may occur and have an impact on many things. The terrorist attacks on September 11, 2001, qualify as such an event, as does the financial crisis in 2008. And let us not forget the election of Donald Trump as president of the United States. And then there are the German automobile scandals. Those events can push or drown individual technologies or shrink or expand the timeline for change.

What Manufacturers Have to Consider

> Any useful idea about the future should at first sound ridiculous.
>
> —JIM DATOR

Since the global financial crisis in 2008 and the ensuing insolvency procedures for General Motors and Chrysler (Ford barely escaped the same fate), the overall situation has improved again for automobile manufacturers. The last couple of years brought record sales results—love, peace, and happiness, you might say. As a matter of fact, however, the stock prices for traditional manufacturers are decreasing, and the entire industry has misgivings. Could it be that "peak car" is the swan song for the combustion engine, just as "peak horse" indicated the end of the carriage-drawing horse?[12]

Today, there are 43 million vehicles in Germany, and over 260 million in the United States. In total, cars in Germany cover about 380 billion miles every year, which breaks down to about 8,800 miles per vehicle annually.[13] The average American drives even more with 13,400 miles per year, which brings the total number of miles driven in the United States to 3.22 trillion. Let us assume for a moment that all those cars are self-driving electric Ubers with an average speed of 35 miles per hour; in this case, they would be used for much longer than the average 38 minutes per day in Germany and 54 minutes per day in the United States for combustion engine vehicles. They could be on the road for up to 20 hours per day, which would result in a performance of 700 miles per day. The remaining 4 hours would be used for charging and service. Such a car therefore would be able to reach an annual driving performance of about 270,000 miles. If we regard today's annual demand for driving, which is 380 billion miles in Germany and 3.22 trillion in the United

States, there would be just about 1.4 million vehicles needed to cover the current demand for driving services in Germany and about 12 million vehicles in the United States.

Even if we calculate generous reserves and assume that people drive more with those vehicles than they used to and that public transportation is replaced by them, and even if each vehicle may only be on the road, say, 10 hours instead of 20 hours per day, and even if we add some private cars to that electric robotaxi fleet, we will never reach the vehicle volume we have today. Between 10 and 25 percent of the current total vehicle stock would be sufficient. In the United States, we would have to add about 30 percent to this amount because of the calculated average driving time of 54 minutes per day per car.

Obviously, not everyone will be ready to do without a private car, even if it is autonomous and would be cheaper in a sharing model. There are too many other factors that also play a role. A craftsperson will need a car to transport his or her tools. Jamie Carlson, for example, senior director of autonomous driving for Chinese startup NIO, does not believe in the sharing model for personal reasons because he always keeps child seats and toys in his car and cannot imagine having to rearrange everything every time he starts out from home. He clearly states that NIO is making cars for private owners, not as shared vehicles.

So is this basically bad news for car manufacturers? Perhaps yes, perhaps not. A car that is driven more will also have to be replaced more often, just as taxis today are replaced every year or two. This further increases the quantity, and quality becomes even more important in order to prevent faster wear and tear. The sales for premium vehicles might increase if such vehicles are also used in the sharing model and new customer groups can be acquired. The life cycle of each model would no longer be calculated to be four to seven years, but a lot less, perhaps like that of a smartphone. If vehicles in the fleet are exchanged every year or two, there is also a chance of introducing new models and completing updates much faster. Technological innovations are put out on the streets more quickly. Tesla shows how this can work. Software updates will soon be made every month, and moreover, models will be adapted every few months, with hardware parts that are included but that will only be activated at a later time, as described earlier. The authorities, in turn, can adapt the legal framework to changed conditions more quickly and implement them practically in the national vehicle fleets.

Increased use of cars that are robotaxis also implies that vehicles can be serviced quickly and easily and that a component exchange can be completed quickly. This last aspect is most likely to become crucially important to fleet operators. A car that is in the garage earns no money.

Good news for the users? With the sharing model, car travel would definitely become cheaper. The cost could be reduced by up to 90 percent. In the United States, owners have to plan on spending almost $12,000 per year for loss in value, gasoline, insurance, maintenance, and other expenses. In Germany, the same expenditure for a midrange car is about $6,300 (€5,600) annually.[14] Even hard-core car owners who could not imagine ever being without their own car may start to rethink that attitude if they discover that they would only have to pay about $1,200 in the United States or $630 (€560) in Europe per year and would have more flexibility in terms of available vehicle types as well. If we translate this to the situation in the United States in which the average annual travel performance of each of the 260 million vehicles is 13,400 miles, the savings will be even more dramatic.[15] In the United States, the number of miles traveled was 3,220 billion in 2016 according to NHTSA.[16]

Remember, this is only a reference to the cost; we have not even started on the convenience factor yet—no more searching for parking, no maintenance, no cleaning to organize or do yourself, no charging and/or refueling, and last but not least, no more stress while driving. The car just runs, and the passengers can deal with other things.

We can safely say that the traditional automobile industry will have to change its business model. In the future, the aim cannot be an increase in sales to individuals, but it must be to ensure provision of mobility services in fleets. Other industries are good examples of such a change. Energy producers no longer try to sell as much power as possible but instead offer a lighting service, a heating service, a cooling service, an entertainment service, or a mobility service. As a consumer, I want my house to be warm, my fridge to be cool, my TV and computer to be running, and my EV to be driving. Once the calculation of cost is no longer based on the amount of power used but rather on the individual service, the power plants will be motivated to make sure that the same service can be provided using less energy.

Car brands will become less important. Admittedly, they are a synonym for quality, but as a matter of fact, in the time of robotaxis it does not matter at all even today whether I am traveling in a Mercedes taxi or a Toyota Prius taxi. What is important, however, is that I reach my destination quickly, safely, for a reasonable price, and conveniently and that I can recharge my phone even if the trip is just a short one. I am not really concerned about whether I have an "Audi experience" at the same time—regardless of which city I am in.[17] I will be riding in a Waymo or Uber, in much the same way as I fly United or American Airlines without knowing whether it is an Airbus or Boeing plane that I am flying in.

The Automotive Ecosystem: Welcome to the Valley

In the past, those who wanted to see the latest automotive trends had to go to one of the big trade fairs. The North American International Auto Show in Detroit and the Frankfurt International Motor Show IAA are just two of them. Meanwhile, however, they have been upstaged by the Consumer Electronics Show in Las Vegas. Chrome and the sound of motors are no longer trendy; silent EVs are. And those EVs are simply better suited to presentation at an electronic trade exhibition. Any top-level car manufacturer has to show something there nowadays. Tesla, for example, does not participate in the Detroit show at all, not least because the company is not allowed to do any sales in Michigan because of the dealer legislation there.

The shift in automobile power in the western states is mirrored in the automotive ecosystem that has developed in the Silicon Valley and in California as a whole. Car manufacturing plants are growing like weeds in one of the most expensive locations in the world. Tesla produces in Fremont; Gigafactory 1 in Reno is just three hours north of Fremont. Lucid Motors is based close to Fremont in Newark, the electric bus manufacturer Proterra is in Belmont, and Karma Automotive—formerly known as Fisker—has its home in Irvine near Los Angeles.

Detroit, Stuttgart, Wolfsburg, and Munich no longer set the agenda or the speed: all the manufacturers had to be present in Silicon Valley in recent years. Those who want to play in the "World Series" cannot wait until their own country finally grants permission to test autonomous vehicles and until test tracks can be built. In Silicon Valley, all of that is already available—the ecosystem of technology companies, research institutions, legal environment, experts, and qualified staff with the right mindset are there.

Anybody who ever visited Silicon Valley may have attended some meet-ups, informal events with presentations and discussions that usually take place in the evening after work and are organized by private people and supported by companies. Startup company founders, investors, and interested parties meet with managers from the big corporations: Daimler and Bosch, Tesla and Peloton, NVIDIA and Kleiner Perkins, BorgWarner and Stanford, Udacity and Nissan. Prominent representatives from the industry all joined together. Sebastian Thrun, George Hotz, and others regularly are found among the speakers. In a full display of the Silicon Valley mindset, everyone opens up to public discussion, a marked contrast to the reserved communication culture favored by traditional manufacturers, which prefer not to talk about their activities (except when it concerns illegal price fixings or technological trickery).

Alison Chaiken, development manager with Peloton Technology, is one of the most active organizers of automotive issue meet-ups.[18] This is one more factor that contributes to Silicon Valley being so much ahead of the traditional manufacturers in many areas of the modern automobile industry. Something that keeps surprising me are the fascinating and well-founded contributions to the discussion following the presentations and panel discussions. It is noticeable that the people asking the questions are involved in pushing the limits further and driving the development forward.

The impact of Tesla on the growing automobile manufacturing know-how in Silicon Valley cannot be overestimated. There are 40 suppliers for Tesla parts that have their production facilities in this region, and dozens of others of the total of 300 suppliers worldwide are planning to open offices there. They want to, and need to, stay close to the "parent ship."[19]

Classic car manufacturers used to move an increasing number of their parts production to tier 1 suppliers. Corporations such as Bosch, Continental, and Magna provide entire component systems to the manufacturers, which then proceed to assemble everything into the finished automobile. Car doors and the trunk units are delivered to the manufacturer as a finished part. Up to 70 percent of a vehicle is produced by the suppliers.[20] In simple terms, the competence of a manufacturer is limited to the design, the chassis, the engine, brand building, and branding. These are areas in which Volkswagen, Mercedes Benz, BMW, and Opel have excellent skills.

The same principle applies in other industries as well. Nike, for example, does not produce any sneakers or sports pants. Nike is responsible for the design, communicates the exact specifications for production to its suppliers, and takes care of logistics, branding, and sales in the stores. Apple develops and designs the iPhone, hands over production to the contract manufacturers with strict requirements to be met, is responsible for distribution and branding, creates the software, and sells the final product.

The example of Apple clearly shows changes in electronic goods production—they allow a glimpse of what lies ahead for the traditional car manufacturers. Until the late 1990s, Apple did build its own computers. As a matter of fact, Apple owned production plants as well. When Tim Cook became the man responsible for logistics and production, he radically changed all that. The production requires expensive machinery, which has a negative effect on a company's cash position. Just as with perishable goods (e.g., milk), electronic goods must have a fast turnover. Machinery ties up capital. The objective was, however, to have an inventory turnover within just a few days, not over the course of months. This is the reason why Apple sold all its factories and outsourced its production and storage facilities.

Producers such as Foxconn, in turn, produce for other clients, not just Apple. The company is able to offer lower prices for its components because of its higher volume than any single client corporation could hope to bargain for. Even if you include additional cost, the total expense is still lower than if you tried to produce the product yourself. This model turned out to be perfectly scalable when Apple started to develop iPods in addition to computers and eventually iPhones as well. Car manufacturers are moving in the same direction, although their pace is slower. However, those of the traditional car manufacturers that have the intention of surviving the next automobile revolution will have to "digitize" their expertise. Engine production will disappear, and software becomes a central component, the driving force and central sales pitch topic.

Of "Scouts" and City Mobiles: Even Buses Can Drive Independently Now

Matters are changing for self-driving buses as well. Daimler is working on something, as are new startup companies such as EasyMile and Navya from France and SB Drive from Japan. Actually, field trials are happening everywhere already:

- Milton Keynes (Great Britain) is testing 40 autonomous small vehicles for two passengers with a maximum speed of 12 kilometers per hour—the so-called Lutz Pathfinders.
- CityMobil2 is an international research project with autonomous shuttle bus systems in La Rochelle (France).
- Meridian Shuttle is active in Greenwich (Great Britain) and Singapore.

- EasyMile EZ-10 operates in Wageningen (Netherlands) and Bishop Ranch (United States).
- Daimler is doing tests in Amsterdam (Netherlands).
- Navya runs trials in Las Vegas.
- The Swiss postal service is testing Navya buses for passenger transportation in Sion (Switzerland).

And the list could go on. Self-driving buses are running in Germany, Austria, and Switzerland in cities such as Lausanne, Salzburg, Berlin, Dusseldorf, and Hamburg, although those are currently test operations and have yet to take up regular operation.

Heavy Trucks on Tour All by Themselves

The recipe for success was relatively straightforward for the top dogs among the truck manufacturers, such as MAN, Mercedes, Scania, and DAF. Trucks were supposed to use as little fuel as possible and work reliably. Comfort was an extra to keep the drivers happy. In the United States, there was also the chrome-blinking almost Baroque touch to be considered. The manufacturers played the same game, just exchanging a couple of percentage points of market share every now and then.

But suddenly, there was turmoil. Newcomers such as the startup company Ot.to, acquired by Uber, are attacking traditional truck builders and sending drivers into unemployment with its self-driving trucks. Peleton Technology from Mountain View, Embark, Kodiak Robotics, Plus.ai, and Starsky Robotics in the Silicon Valley, TuSimple in San Diego (and China), and the Chinese company Baidu are some other newcomers in this field. Volvo, Peleton Technologies, and Embark are making electronically controlled coupling as a convoy happening, and Tesla wants to launch an electric tractor unit, a project with which Mercedes has also run trials. Waymo has been testing its self-driving technology on trucks. FR8, Trucknet, and Uber Freight, by contrast, are creating new competition for forwarders because they are combining freight space and freight tests. This reduces the risk of having half-empty trucks on the road—which, in turn, leads to the demand for trucks increasing to a lesser degree than the growing share of freight transportation would lead you to think. This might spoil the fun for the manufacturers.

A survey among 2,000 European forwarders conducted by the business consultants at Bain & Company shows that brand loyalty is decreasing. The manufacturer's brand is no longer the decisive factor; it has been replaced by cost and reliability. Additional services are also important, ranging from digital services to vehicle maintenance.[21] Once again, digital transformation strikes. Trucks differ very little from one another, but digital services might carry the day. And this is where the newbies with their digital expertise have an obvious edge.

In the United States, 22.8 percent of all emissions are caused by trucks and 42.7 percent by passenger cars. Because there are fewer than 3 million trucks compared with 260 million passenger vehicles, this share is substantial.[22] Although only 1 percent of all the vehicles in the United States are trucks, they are responsible for a quarter of the pollutant emissions, cover 5.6 percent of the total miles driven, and are the reason for 9 percent of all traffic fatalities.[23] Overtired drivers, lack of attention, and stress or road rage are good arguments to replace humans behind the steering wheel—at least to some degree. During the first phase, trucks are likely to be used as self-driving vehicles only on straight stretches of highway. The driver would take over at the exit, on smaller roads, in town, and to approach the loading bay.

Looking at the numbers, it does not appear quite as impossible to believe that big corporations and startup companies are working on autonomous electric trucks, although some of us may feel a shiver down our spines thinking of a 40-ton robot moving on our roads. Truck drivers in the United States are allowed to drive a maximum time of only 11 hours per day and 60 hours per week and need breaks. During those breaks, the truck is just standing around and does not make any money for the forwarder. The driver also accounts for a third of the total cost of a truck. Another third of the cost is for fuel, which would be approximately $90,000 (€80,000) per truck in Europe, because it seems impossible for most drivers to go easier on the accelerator and save up to 30 percent of fuel. Autonomous trucks, by contrast, can simply continue driving, and they can do that so efficiently that they save quite a bit of money.[24] PricewaterhouseCoopers estimates that the annual operating cost for trucks decreases by 28 percent when those trucks are traveling autonomously. The cost would therefore be reduced from $130,000 (€115,600) to $93,250 (€82,800) per year.[25]

This new development also has an impact on job profiles as well as business models. Instead of employing 3.5 million truck drivers in the United States and 540,000 in Germany, there is going to be increased demand for computer-savvy logistics experts in the head offices. Forwarders may miss out if truck manufacturers and customers start collaborating with each other directly.

The now defunct startup Ot.to had already passed its first commercial test. Together with the large U.S. brewery Anheuser-Busch (founded by German immigrants more than 150 years ago), they delivered 50,000 cans of beer in October 2016 with an autonomous Ot.to truck that covered a distance of 120 miles without any help from the driver. I showed the video Ot.to published to an audience more than once, and the spectators' reaction when the driver unbuckles his seat belt and leaves his seat to go to the rear of the vehicle is telling.[26] Suppressed outcries and widened eyes betray their surprise. This is going to be quite the opposite in the future. We will have gotten used to driverless cars so much so that we are more likely then to be worried when we see a driver behind the wheel, not when there is nobody, a little like seeing someone brandishing a gun in public.

The startup company Embark did the first commercial freight hauling with autonomous trucks from Long Beach, California, to El Paso, Texas.[27] Peleton Technology, from Mountain View, by contrast, concentrates its work on *platooning* trucks, backed by $18 million of venture capital. Its first products have already been delivered. They permit, among other features, V2V communication and maintain the distance to the preceding vehicles by means of radar.[28] And then there is Starsky Robotics from San Francisco, another company aiming to launch autonomous trucks in the market.[29] The company did its first real driverless delivery in Florida in 2019, with the help of remote controlling technology.

Meanwhile, the Chinese internet giant Baidu presented the first self-driving truck in collaboration with Foton Motor Group.[30] Baidu was interested because of its telematics and entertainment applications, CarLife and CoDriver, which are already being used by 60 automobile companies, mainly in China. Baidu presented self-driving vehicles together with three other companies: Chery, BYD, and the Shou Qi Group. Startup companies such as TuSimple from Beijing, for example, store the driving style of truck drivers to be able to offer autonomous trucks.[31]

Traditional manufacturers, however, are not going to give up this business without a fight. Volvo, MAN, and Daimler are also busy developing autonomous trucks. Volvo already tested "road trains," and Daimler has tested driverless trucks on the German Autobahn.

Small and Manageable: Pretty Pods

Some time ago, my Facebook feed was full of comparisons of popular car models and their previous versions. Two Porsche 911s, one from 1963 and the other built in 2013, were posted next to each other. The one from 1963 looked almost tiny against the massive rear of the younger model that appeared to have almost doubled the size of the older one. I observed the same for the Mini Cooper. The model built in 1959 fits almost entirely onto the rear seat of the current Mini model. The size of the current VW Polo is more like that of a Golf/Rabbit 30 years ago, whereas today's Golf model is as big as the old Passat model series. This means that our cars have become bigger, although usually there is only one person inside.

Autonomous electric taxibots would offer the opportunity of designing vehicles in a more compact way with fewer resources used. Several manufacturers have adopted this concept. In 2010, General Motors and Segway presented a two-seater pod called EN-V in Beijing. In Milton Keynes, England, plans are under way to use the LUTZ Pathfinder in urban traffic.[32] And Adaptive City Mobility presented the CITY eTAXI at the CeBIT, a means of transportation that weighs all of 1,200 pounds (550 kilograms) and has an exchangeable battery, offering room for three passengers or—as was successfully demonstrated—for one passenger with a Euro palette. This is useful if you think about how often we have had problems in the past finding a taxi that would fit our palette as well.[33]

Self-Driving Motorcycles: Is That Even Allowed?

There are some strong arguments for the autonomous movement of vehicles with four and more wheels. But what about those with two wheels? Should those also be computer operated? Do we want that? Can we do that?

The answer to the last question is definitely yes. There was a motorbike at the DARPA Grand Challenge as well. Anthony Levandowski, who later worked for Google and then went on to found Ot.to, was the mastermind behind the first autonomous motorcycle. He was also very unlucky because his motorbike tipped over at the startline: he had forgotten to activate the equilibration stabilizer.

Lit Motors, a startup company from San Francisco, works on the first steps of a self-stabilizing motorcycle that should then also drive autonomously.[34] Such a vehicle could be particularly efficient as a taxibot offering individual transportation services for short distances in town and a fuel-efficient one at that. Lingyun Intelligent Technology, a company from Beijing, has a similar objective, as does BMW too.[35]

The Bavarian car manufacturer presented a future design model of an autonomous electric motorcycle that is supposed to reflect the zeitgeist: the company's Concept Link. The modern zeitgeist is exactly what the iconic U.S. brand Harley-Davidson is struggling with. Harley drivers are getting on in years, and there are fewer customers every year. Today, the average owner of the American dream of freedom is 50 years old. Not long ago, the average age was 35. Some older drivers have to give up their ride for health reasons. Self-driving technology is intended as a means of preventing errors and accidents, not as a means to spoil the fun of riding a motorbike. Of the percentage of bikers who have traffic accidents, almost 30 percent of all fatal motorcycle accidents happen without any external interference. Self-driving technology that recognizes errors or other vehicles early may save those drivers and give them a real chance to survive and tell their story.

Things Manufacturers Should Expect

Automatization went after blue collar jobs, Artificial Intelligence goes after white collar jobs.

—BEN LEVY, BOOTSTRAPLABS

What is the background image on your computer or iPhone? A photo of your family? Your last holiday at the beach? Your dog? Whatever it is, it probably is not a picture of your workplace. That is what really struck me many years ago during a visit to the BMW plant in Dingolfing, Bavaria. Whirring and humming surrounded us as we made our way through the assembly hall. The apparently chaotic but perfectly choreographed dance of people and machines around sculptures of glass, metal, and plastic showed at close range how one of the most desired vehicles is produced. We woke from our dream only when we had eventually passed the shelves with engines, three stories high, and closed the door of the IT manager's office behind us. He had a very sparsely furnished office where he sat "enthroned" above the shop floor to ensure that all the computer systems on the assembly line worked smoothly. When we turned out attention to him, we noticed something straightaway.

"Is that the design for the new BMW 5 series?" asked one of us. The IT manager looked at the background image on his computer. "Yes! The marketing department released the photos to the public today," he replied. And after a long pause, he sighed romantically without looking away from the photo: "It is such a beautiful car."

And that is what is so special about the automotive industry: the passion the staff feel for their work. Cars induce emotion. We link them to feelings and memories in a way that is almost irrational. For many people, their first car was the fulfillment of an old dream. It stood for independence and freedom. A car was also for many the first place where they kissed somebody else other than their parents or relatives, a place where almost all of us had sex and where many children were conceived. Entire movie categories exist around cars. TV shows and movies have cars as their central characters. Who is not familiar with KITT, Herbie, Chitty Chitty Bang Bang, the Batmobile, or Bumblebee and the Transformers?

We don't say, "My car is parked over there," but rather, "I am parked over there." The car is part of our person, a physical extension, perhaps a transcendental, extrasensory experience. Many people give names to their vehicles. My VW Polo many years ago was called *Flitzi*, German for "speedy." We anthropomorphize our cars and recognize their special quirks. My old Renault started to "sing" when it reached a certain speed, just like the Tesla one of our friends had. The loss of their car or of their driver's license has hit many people hard, so hard that they lost their will to live and saw no sense in going on—some committed suicide.

And yet this emotional connection is about to come to an end. For an increasing number of young urban people, a car is just a burden, a monster that needs space and costs money, something that keeps their attention away from their smartphone.

Those who are directly affected more or less know the threatening and exciting events that are approaching, even if not everything has been tried and tested yet. The change will come, and it will be massive. A head-in-the-sand policy of the "It can't be that bad" or "It won't happen quickly" is not helpful. Even people who are not directly affected by the automobile industry will have to rethink their lives—city planners, hospital administrators, and road sign advocates, to name but a few.

A study prepared by the McKinsey Global Institute estimates the economic effect of the new autonomous vehicle technology at up to $1,900 billion by 2025 (see Table 10.1).

TABLE 10.1 Estimated Economic Impact of New Technologies Until 2025

TECHNOLOGY	LOW ESTIMATE ($U.S. BILLIONS)	HIGH ESTIMATE ($U.S. BILLIONS)
Mobile internet	3.7	10.8
Artificial intelligence	5.2	6.7
Internet of things	2.7	6.2
Cloud	1.7	6.2
Robot technology	1.7	4.5
Autonomous vehicles	0.2	1.9
Genetics	0.7	1.6
Energy storage	0.1	0.6
3D printers	0.2	0.6
Modern materials	0.2	0.5
Oil and gas exploration	0.1	0.5
Renewable energy sources	0.2	0.3

Source: McKinsey Global Institute.

Not a Single Profession Should Feel Immune

You, too, may have to face the fact that your job will be subject to changes. It is more likely that future generations will change their jobs and even their general professional areas more than once in their lifetime. The education system already faces the challenge of having to train people who need to adapt to new job situations or can create their own jobs. And unions have to stop reacting to the demands of the twenty-first century by offering models from the nineteenth century.

While social upheaval is not a new phenomenon in the history of humanity, the breathtaking speed of change we are witnessing today is unprecedented. Let me say it once again very clearly: yes, we are right in the middle of the second automotive revolution; this is happening *now*!

Every successful revolution drastically changes societies and hierarchies. The beginning of the end for nobility and the rise of industrialists and workers began with the industrial revolution. Innovation has revolutionary impact. Mechanical looms made weavers obsolete, containers reduced the number of workers needed at a port, agricultural machines replaced farmhands and dairymaids. At the same time, the demand for good training for skilled workers in an increasingly mechanized world grew. Our current education system basically was established at the same time as the industrial revolution. Factories need workers who can read instructions and operate machines.

In comparison, mechanization in the agricultural sector happened quite slowly. About 200 years ago, the majority of people were still working in the agricultural sector compared with less than 2 percent in our so-called industrialized nations.[36] In the United States, for example, the number decreased from 80 percent around the year 1800 to 1.5 percent today. The average job loss ratio at the time was 0.5 percent annually. Farmers, farmhands, and dairymaids did not suddenly face job loss; the events unfolded slowly, and everybody had time to prepare. Their children were able to go to school and became mechanics, electricians, and industrial workers.

Nowadays, even the number of people employed in the industrial sector is 30 percent at most. The service sector in countries such as England and France, by contrast, employs up to 80 percent. The traditional employment model is on the retreat as well, however. In the United States, 34 percent of employees are already considered self-employed, and the percentage is expected to rise to 40 percent by 2020.[37] Although the share is comparatively low in Germany with just 3 percent, it has doubled since 2000.[38] The number rose from 705,000 to 1.344 million in 2016.[39] In Austria, the share of self-employed people is 7 percent.[40]

With its almost 800,000 workers, the automobile industry is the most important industry in Germany.[41] The German Association of the Automotive Industry (VDA) estimates that 5.4 million people depend directly on the automobile industry, which generates about $456 billion (€405 billion) of economic output per year. The biggest impact will be on the petroleum industry once the majority of cars on the streets are EVs. This is likely to trigger a political change in the oil-producing countries that could dwarf all other changes so far. Those numbers show clearly how massively the automobile revolution can impact job markets if traditional manufacturers let the versatile newcomers from Silicon Valley and China take precedence and do not stand up to them. This, however, is not all. As you will find out over the next couple of pages, many more professions are also influenced by the way we travel.

The younger truck and taxi drivers, garage operators, and car mechanics are probably going to find out about the impact during their lifetime. The changes in the automobile industry and the related sectors of the economy that we already discussed can be expected to take place in less than one generation. Some of the jobs we are going to lose require high qualifications and were considered to be safe jobs. We are not going to experience a slow slide into a new technological era—we are going to fall head first into it.

One hundred years ago, it was horses that lost to technological change. You could say that they could no longer be employed. If horses had the right to vote, they would certainly have led very forceful political campaigns. Given today's technological progress and the rise of artificial intelligence, humans today are in a similar situation to the horses 100 years ago. We are no longer needed. A large part of the population suddenly will not be able to find gainful employment anymore, and it is not their fault at all. Even highly specialized and renowned professions are not exempt. A medical diagnosis of cancer can be made more accurately by IBM's Watson than by a team of doctors nowadays. Fighter pilots lose in combat against AI-controlled fighter jets, and computers play Go better than humans do.

The University of Oxford conducted a data analysis that predicts the probability ratio of automation for a total of 702 professional groups.[42] Forty-seven percent of employees today would be in some small way to very massively affected across almost every industry and sector. This calculation does not even include the progress in AI to the necessary extent. Tesla, Uber, Google, and Apple are not just "happening" to those working in these professional groups or the automotive industry; they are the future that cannot be stopped. Tesla and the others are just their tools. Publisher Tim O'Reilly says:[43]

> The most exciting companies think of technology as a tool to create more opportunities for people, not reduce them. About the least interesting way to think about self-driving cars is merely as a means to cut payroll expenses. Instead we should be thinking up all the ways they will empower new economic activity: cheaper, smarter public transportation networks; better access to medical care. We should think the same way about all new technologies. Take Zipline, one of Silicon Valley's hottest startups, which is running a pilot program that uses drones to deliver medicines and blood for transfusions on demand. It's starting in Rwanda, a country with sometimes impassable roads and poor health infrastructure.

At least five dozen professional groups and sectors are most likely to be affected by changes in the automobile industry either directly or indirectly. They are going to either disappear completely or survive in a "lean" form. Let us take a look at the effects on different industries and professional groups in detail.

Drivers of Any Kind

Jobs of people driving automobiles are most affected by all this, including drivers of ambulances and vehicles with hazardous goods. In Germany, around 540,000 professional drivers are working in freight traffic.[44] In the United States, there are 1.7 million truck drivers, and the U.S. Bureau of Labor Statistics expects an increase to 1.89 million drivers by 2022.[45] If you add drivers of vans, the number increases to 3.5 million. In 2014, trucker was the most frequent job in 28 of the 50 U.S. states.[46]

This extrapolation is logical if you consider the increase in freight transportation resulting from the internet and the increasingly close connections between production processes. The problem is the omission of changes in the transportation industry. Autonomous trucks will slowly make professional drivers redundant. The number is more likely to tend toward zero than toward 1.89 million. And because truck drivers also have to take breaks, sleep, eat, and be entertained, there are hundreds of hotels, service stations, and small towns that depend on them. Those places are going to lose their loyal customers. The business consultants at McKinsey think that every third truck will be semiautonomous by 2025.[47]

In Germany, there are 36,000 taxi companies, and we have a total of 250,000 licenses for passenger transportation.[48] The number of business licenses for Austria was 16,447 in 2014, with 7,469 being taxi licenses and 174 being licenses for horse-drawn carriages, the famous Fiakers loved by tourists. This latter group—I think I can say this much—is not going to be endangered by autonomous taxibots.[49]

Forwarders are generally threatened if truck manufacturers and clients such as supermarkets or electronic goods chains decide to handle transportation themselves. This concerns the following occupational groups:

- Taxi drivers
- Uber/Lyft drivers
- Chauffeurs
- Bus drivers
- Truck drivers
- Courier drivers
- Valet parking services
- Emergency vehicle drivers
- Forwarders

Jobs in Automobile Production

One-third of all the employees in the traditional automobile industry are in some way connected with the engine and the surrounding ecosystem. Thousands of employees (the estimates range from 120,000 to 180,000 in Germany) will have to make room for battery engineers and chemists, but especially for computer scientists, experts in AI, robot engineers, and computer vision and electronics experts—just as horse breeders, saddlers, bridle makers, veterinarians, and stable owners had before. Once the transition to EVs is complete, the number of staff needed by the suppliers will decrease from 310,000 to about 220,000.[50] Of the 810,000 workers in German automobile production, 210,000 will lose their jobs in a best-case scenario, up to 270,000 in a worst-case scenario. The industry will not be able to provide sensible work for about a third of its employees. Let me take two sites as examples representative of many plants in Germany and Austria: the VW engine works in Salzgitter has 7,500 workers, and the BMW engine plant in Steyr employs 4,100 workers. These and other sites will see a massive reduction of jobs.

However, the comprehensive (manufacturing) expertise for motor production will not become completely superfluous. Autonomous electric cars require an infrastructure that will benefit from this manufacturing expertise: cells, battery packs, charging infrastructure, power supply, maintenance, vehicle interior design, and special vehicles are all part of this. What is, however, especially required is a different mindset and the necessary readiness to develop new fields of knowledge recognizing the respective opportunities and chances and not just focusing on the risks.

The workers' councils in German corporations and union representatives are slowly becoming aware of the consequences. Michael Brecht, chairman of the workers' council for the Daimler Corporation therefore warns against ceding competencies in electric mobility to suppliers.[51] Representatives from the labor union IG Metall, usually known as the hardliners in the Volkswagen corporation, regard China's aspirations of building its own electric mobility industry as a direct danger for German manufacturers and therefore demand a quick change away from today's combustion vehicles.[52] This concerns the following occupational groups:

- Engine builders
- Exhaust gas experts

- Fuel experts
- Transmission builders

Whereas there are jobs at risk here, new jobs will develop in other areas. The manufacturers of sensor technology and software are expecting an additional $20 billion to $25 billion annual revenue by 2020.[53] This breaks down to an expected $10 billion to $15 billion per year for road maps for navigation and collision prevention systems and $9.9 billion for cameras, radar systems, ultrasound sensors, and LiDAR systems.

Jobs in Traffic Control

Every car in Germany generates about $56 (€50) to $68 (€60) in fines for districts and cities. Considering the 43 million combustion engine vehicles operating today, the total sum is substantial: more than $2.25 billion. Every day in the United States about 125,000 traffic citations are issued at an average of $150, leading to an annual total amount of traffic fines of almost $7 billion. What happens once this is gone? Currently, American highway patrol officers are busy securing accident sites in 80 percent of the incidents to which they are called, but those will become less with self-driving vehicles. Those cars also keep to traffic rules and the speed limit, and they are on the move for most of the day, thus not parking (illegally), so there is nothing to punish. Traffic police and parking enforcement officers can be assigned new tasks. This concerns the following occupational groups:

- Traffic police
- Parking guards
- Towing services
- Traffic reporters
- Traffic court judges
- Lawyers specializing in traffic offenses
- Producers of breathalyzers

Jobs in Traffic Area Infrastructure

Who are the *users* of traffic lights and road signs today? Humans and human drivers. Not self-driving vehicles, because they can obtain the required information from suitable expandable and adjustable road maps.

In Germany, there are 20 million road signs and 4 million signposts. Each one costs between $90 and $225 plus installation cost. A traffic light—there are 1.5 million of those in Germany—costs $40,000 to $280,000, and the operating cost amounts to $5,600 every year excluding power. This concerns the following occupational groups:

- Traffic light producers
- Producers of road signs

Jobs in Driving Instruction

If there are no drivers anymore, who needs driving schools and tests? The driver's license itself will be obsolete, as will be all the additional certificates drivers of special goods or special vehicles have to obtain nowadays. In Germany, there are currently 21,485 driving instructors employed in just over 11,000 driving schools, although numbers are already decreasing and the remaining instructors get older and older.[54]

Perhaps this is already an indication of the slow decline of the profession of driving instructor. After all, EVs have no transmission, and autonomous vehicles do not need a driver. This concerns the following occupational groups:

- Driving instructors
- Driving examiners
- Driver's license administrators

Jobs in Research Facilities and Universities

When a scientist says something is possible, they're probably underestimating how long it will take. But if they say it's impossible, they're probably wrong.

—RICHARD SMALLEY

Many research institutions and universities today are concerned with optimizing fuels and engines. Forty percent of the entire R&D effort in Germany is funded by the automobile industry.[55] These institutions will have to redirect their efforts to remain competitive. Whereas traditional institutions for vehicles will lose their importance and may be closed, others will experience a boom. Institutes working on sensor technology, electronics, data processing, and battery chemistry are going to see more money coming in. This concerns the following occupational groups:

- University professors
- Researchers

Jobs in Vehicle Maintenance

EVs do not need an engine, a transmission, a radiator, an exhaust gas system, or a fuel system. Consequently, they do not need an oil change, do not need any replacement of spark plugs and sealing rings, and, at least in part, do not need brake replacements because such cars mainly brake using their motor. Autonomous automobiles cause fewer accidents and hence need less repair work. Up to 70 percent of the work completed by car mechanics will be eliminated.[56] In the United States, 739,000 people worked in garages and body shops in 2014. In Germany, there were 390,000 employees.[57] And if private car ownership decreases and is replaced by fleets, automobile clubs will have to morph from member clubs to business-to-business (B2B) service providers. This concerns the following occupational groups:

- Car mechanics
- Vehicle emissions testers
- Roadside mechanics
- Operators of car wash facilities

Jobs in Vehicle Sales

If an increasing number of people decide to forgo their own cars and instead rely on transportation service providers, fewer cars will be sold to consumers and more to fleet operators. This business with end customers, or business-to-customer (B2C) businesses, changes and largely turns into B2B transactions. This means that cars are no longer sold to many individual customers but to some few large corporate customers demanding corresponding discounts. The buyers have more leeway in price negotiations.

Nevertheless, current sales figures do not yet reflect this change. German car dealers employ about 70,000 people. But we will need fewer sales representatives in the future. In the United States, about 2 million people are employed in the motor vehicle and parts dealer industry.[58] This concerns the following occupational groups:

- Car loan providers
- Used car dealers
- Custom car shops
- Car sales staff

Jobs in Construction and Traffic Area Planning

The good news first: apartments and houses are going to be more affordable. The fact that we can use the space now used for parking for other purposes offers a range of new options. And let us not forget the space required for access ramps to garages. Municipal parking garages increase the distance between individual buildings and make it more difficult for pedestrians to reach their destinations easily and comfortably. The solution today? More cars, with the consequent need for even more parking space, and larger cars requiring more space, too.

Fewer vehicles will require less space on the roads and for parking. Self-driving vehicles furthermore are able to use the space more efficiently than human drivers. This concerns the following occupational groups:

- Construction workers for parking lots and garages and parking lot attendants
- People working in road construction
- Parking lot security guards

Jobs Connected with the Provision of Fossil Fuels

Everyone knows that hundreds of billions of dollars are spent on petroleum, mainly filling the wallets of regimes not always known as staunch defenders of democracy and liberty. Every minute, the United States transfers $612,500 to such countries, just by paying for gasoline. Per year, the total sum is $300 billion.[59]

The United States is spending more than $600 billion annually for its armed forces. It is estimated that 10 to 25 percent of the military budget is required to secure the gasoline supply.[60] This is a hidden tax America and the rest of the world are willing to pay for the prevailing dependence on petroleum—and one that could become largely void. Oil has to be extracted but also transported and refined before it can be burned in car engines. More EVs on the roads make for an alternative and more environment-friendly power mix.

In the United States, the coal industry employs 174,000 people, compared with fewer than 10,000 in Germany.[61] These numbers are clearly below the record numbers seen during the 1950s, when more than 300,000 workers were employed in German coal mining companies. Now the oil industry will follow suit.

In 2016, there were 14,500 gas stations in Germany, employing a total of 100,000 people.[62] If we do not need gas stations anymore, the little supermarkets attached to them will also disappear. Mind you, this is not necessarily a bad thing. In the United States, for example, about half of all cigarettes sold are sold in gas stations.[63] In 2015, cigarettes amounted to 35.9 percent of the sales in gas station stores.[64]

Anyway, just think of how anachronistic it actually is to ship millions of tons of liquids around the globe so that we can just burn the majority of it. This concerns the following occupational groups:

- Employees in oil production
- Employees in refineries
- Drivers and operators of tank trucks
- Employees working on oil tankers and platforms
- Employees on oil pipelines and storage tanks
- Gas station owners and staff
- Staff in environmental impact assessment
- Soldiers

Jobs in the Insurance Sector

So far, insurance companies have not made any policies for commercial self-driving vehicles, but if the predictions hold true and 90 percent of all accidents can be avoided with self-driving vehicles, the insurance premiums also have to go down. Fewer claims for damages mean less revenue, and that does not look good in the annual statements. Experts are expecting a drop in vehicle premiums of at least 40 percent.[65]

The revenue for German insurance companies includes more than $26.6 billion from car insurance; in the United States, the respective amount is

$200 billion. In all of Europe, the amount is $135 billion, and globally, car insurance premiums amount to $700 billion.[66] Almost half of all insurance premium income is from car insurance. The dramatic decrease in this item on the balance sheets will certainly stun shareholders. Today, car insurance is also ideal for the industry because it serves as a gateway to customers for other products to be sold. This access would also be largely lost and therefore acquiring new customers would become very expensive.

It is a small wonder, then, that manufacturers and fleet operators such as Tesla and Uber applied for insurance licenses themselves. Tesla, for example, has introduced its own insurance plan under the name of InsureMyTesla in its service portfolio in Australia and Hong Kong. This insures the vehicles as well as the charging stations.[67] At the same time, the company plans to directly include the cost for the insurance and maintenance plan in the sales price. Once this is accomplished, buyers of Tesla's cars would then not even have to start looking for an insurance company.[68]

Changes in premiums are quicker to occur in the wake of safety-relevant equipment than any vehicle fleet can actually be updated, so it can take up to 30 years until all the vehicles are brought up to the latest standard and the oldest models have been decommissioned. Airbags, for example, were introduced in 1984, but it took until 2016 for 95 percent of all vehicles on American roads to actually have one. However, the insurance premiums had been reduced by 25 to 40 percent long before then and had been adjusted to the improved safety standard. This concerns the following occupational groups:

- Vehicle insurance representatives
- Claims auditors and settlement clerks
- Insurance call center agents

Jobs in the Healthcare Industry

In the United States, 1.2 million people are injured in car accidents every year, and almost 40.000 were killed on the roads in 2016. In Germany, traffic accidents are fatal for more than 3,200 people every year, with 394,000 people injured in 2018. In addition, approximately 400 people die in Austria every year. The healthcare system is designed and dimensioned to accept and treat traffic accident victims. As the number of accidents decreases, the capacities in this sector will decrease as well.

Would you have imagined that sectors such as organ donation also would be affected by a decrease in traffic accidents? There are many donors among traffic fatalities. In the United States, 12.3 percent of all organ donations come from traffic fatalities.[69] More than 123,000 people in America are waiting for a heart or kidney, and every day, 18 of them die because they could not have that lifesaving surgery in time despite the 28,000 transplants every year.[70]

It would definitely be cynical to argue that we should keep accident numbers at their current level (or even increase them) in order to maintain our supply of organs for donation from accident victims and maintain jobs in the medical sector. However, other technologies may fill the gap and provide solutions. Perhaps this situation will increase our readiness to develop alternative methods and create organs through bioengineering and 3D printing. This concerns the following occupational groups:

- Emergency services
- Surgeons
- Nurses
- Therapists
- Support and care personnel for the disabled

Jobs in Local and Long-Distance Transportation

If self-driving vehicles are easily available at an affordable price, and if they are also comfortable, nobody will go to the trouble of using a bus or a train. Why should people carry their luggage to and through the station to find their train car and seat or walk down those dark flights of stairs in subway stations if they are already seated in a self-driving electric Uber? Women especially would welcome the increased feeling of safety.

Actually, the railways and public transportation are under pressure from various competing means of transport: autonomous trucks do away with the need to reload onto railway cars, and long-distance buses are cheaper. Subway trains can already travel without a driver. It makes economic sense to abolish public transportation, too. It is very expensive and time-consuming to build a subway. This concerns the following occupational groups:

- Train drivers
- Conductors and inspectors
- Railway staff

- Counter clerks
- Streetcar drivers
- Subway and bus drivers
- Security employees

Jobs in Overnight Accommodation

An overnight trip in a car from Munich to Hamburg or San Francisco to Los Angeles that I spend sleeping saves me the cost of a hotel. Perhaps hotels then extend their service line and offer a shower and breakfast only, instead of a room. And as we saw earlier, we will not need any service stations and restaurants for truck drivers anymore either. This concerns the following occupational groups:

- Receptionists
- Cleaning staff
- Service station workers

Jobs in the Entertainment Business

If we can watch the latest movies during our trip, as during a flight now, why, then, should we still go to the movies? This concerns the following occupational groups:

- Movie theater operators
- Ticket attendants

What's My Line? Guessing Professions in the Future

New technologies create new industries and new professions. This also holds true in the case of the second automobile revolution. Discussions about job loss and future operating fields flare up periodically. Other industries show us how this works: Airbnb, the accommodations platform, is competition for hotels and guesthouses but also generates income for people who previously would never have thought about offering rooms, apartments, or houses. Landlords earned $3.2 billion with Airbnb in the United States in 2015; landlords in Europe earned $3 billion.[71] Service industries developed around the platform, dealing with cleaning and management on site. Smartphone producers pushed aside producers of simple cell phones but also made room for app developers. Companies such as Uber and Airbnb would be unthinkable without smartphones.

Just look around you: the fact that there are service stations on highways and hotels everywhere and that there are holiday magazines and travel guides about sights all over the world shows clearly how much the automobile has changed our way of life. Some elements could have been anticipated: infrastructure planners, roads and road construction workers, parking lots, mechanics and gas stations, as well as rest stops. Other things were less predictable: shopping malls on the outskirts of towns or goods offered at a low price. Marc Andreessen, founder of Netscape and venture capitalist, sums it up beautifully:[72]

> The biggest economic influence of the automobile was not the automotive industry itself—the biggest economic impact was outskirts of towns, and commerce, and parcel delivery and movie theaters, and hotel chains, and adventure parks, and highways, and service stations and all the rest of it. To put it simply, the way we live today is a consequence of the invention of the car. Before, people did not actually travel anywhere, so every destination you go to is a consequence of the automobile.

Perhaps this short list of potential future jobs inspires you to think about this a little, perhaps even doing a "jolly" society guessing game: what other jobs can you think of? Let your imagination roam freely.

- Battery engineers
- Experts in AI
- Electronics specialists for road construction and connected cars
- Producers of in-car entertainment systems
- Charging station managers
- Power suppliers
- In-car app developers
- In-car advertisers
- Advertising route optimizers
- Traffic planners
- Data service providers

CHAPTER 11

Wave Effects and Leaps of Faith

Forward, March!

Report: 98 Percent of U.S. Commuters Favor Public Transportation for Others

—THE ONION

Millions of employed people travel to their place of work every day. Every commuter spends about 50 minutes of his or her life in traffic every day. Taking the number of working people in the United States, this means a total of 6 billion minutes traveling every day. If you assume an average lifetime of 70 years, then basically 162 lives are wasted in the United States every day just by commuting.[1]

Everybody who has completed military service knows this phenomenon: many soldiers march in step in long columns, one after the other. At the command of "Forward, march!" the entire squad starts moving together. Every soldier can rest assured that the person ahead of him or her makes the first step at the same time. In this way, the group moves like a single body. If, however, only the first row starts marching, the second row then starts moving a little later, then the third row, and so on, the entire squadron would be in a wave-like motion. It would be almost impossible to fall into step again.

Honestly, did you never wish for such a command at traffic lights? When the lights switch to green and the first cars start moving, all the others behind are still standing still. The technical term for this phenomenon is *startup lost time*.[2] The first vehicle must ensure that nobody else drives into the intersection when the light is red. This waiting time is the biggest time killer, but the waiting time is less for each following vehicle. Still, you have to plan in about 2 seconds. SUVs take an additional 20 percent of time because of their size, weight, and lower acceleration. The startup time in the end should not be neglected.

Self-driving cars and connected cars take us quite a bit closer to being in step; all the cars together would only require a single (common) startup period. Or they do not even have to stop because they synchronize their speed with the traffic light settings. At intersections without traffic lights, the vehicles would also adjust their speed and coordinate their maneuvers so that none of the road users would have to stop. Researchers created a concept video to show how nicely traffic could flow compared with intersections controlled by traffic lights[3] (see Figure 11.1).

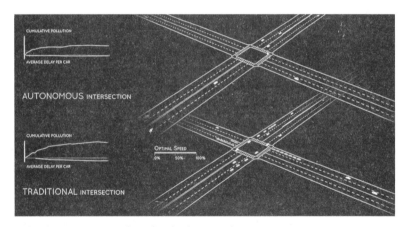

FIGURE 11.1   Simulated vehicle coordination at intersections

The difference is amazing. Instead of wave effects at crossroads where vehicles slow down, stop, and are arranged in rows one after the other, the distance between the vehicles remains almost unvaried. Braking and accelerating, along with the respective energy consumption and wear and tear on the brake pads, are almost completely eliminated. Together with the increase in traffic flow, the drive experience increases as well.

The simulation in Figure 11.1, however, does not consider pedestrians and cyclists, in contrast to a video that spectacularly combines all road users at an intersection.[4] This popular short movie uses film editing tricks to give the impression that the road users just barely avoid collisions and demonstrates traffic flow without any significant deceleration (see Figure 11.2).

FIGURE 11.2   Traffic participant coordination at an intersection

The real scenario always considers self-driving vehicles, pedestrians, and cyclists and adapts the respective speeds in such a way that everyone can travel safely, if necessary, including coming to a complete stop. Once pedestrians trust the reactions displayed by a self-driving vehicle, we are going to experience similar events as those often demonstrated by the late Dutch traffic planner Hans Monderman. During a conversation or while giving an interview at an intersection known to be "shared space," where all traffic signs and signals had been removed, he stepped into the traffic walking backwards. In this way, he demonstrated how the road users showed consideration for one another. Our future will be somewhere between the vision in the video and Hans Monderman's work.

Wave effects can also be observed on highways, not just at intersections. A car that is slowing down forces all the other cars behind it to do the same. Once everybody accelerates again, it frequently happens that people have to stop again because there is congestion (and they ask themselves what on earth could have caused it).[5] The first car that, for whatever reason, had to or felt the need to brake passed that spot perhaps about 30 minutes ago, and still the wave effect persists. The resulting stop-and-go movement is often very frustrating for everyone on the road (see Figure 11.3).

FIGURE 11.3   Traffic wave effect (© The Mathematical Society of Traffic Flow)

Whereas road operators currently are trying to soften wave effects with a dynamic adjustment of speed limits, such an effect is going to disappear in the future because of the communication between individual autonomously driving traffic participants and their faster way of capturing the traffic situation and its implications. The vehicles move like an intelligent swarm. Another behavior that is going to disappear is braking because of curiosity, when drivers pass a crash scene in the opposite lane. As a matter of fact, researchers show that even a few autonomous vehicles among those controlled by human beings may lead to a drastic reduction in such wave movements.[6]

Deliveries Order Ubers, Too

The entire delivery and transportation industry is still in an early technological stage. Many startup companies focus on it in order to cause disruption there as well. Communication is still very much telephone based, bills of lading are still actually printed out on paper, and only occasionally is (outdated) software used, but it is never fully integrated. Many standards make communication between the software systems difficult.

In the United States, 532,000 forwarders handle a total of 400 million complete loads (i.e., full truck loads) commissioned by 1.7 million customers. The total volume forwarders manage is valued at $700 billion, $600 billion of which involves complete loads, and the rest consists of parcels.[7] Ninety percent of the forwarders have fleets consisting of fewer than six trucks, so the market is definitely very fragmented. Also, 13,000 freight managers assist clients in finding the right forwarders for their delivery. Their mediation fee is 15 to 20 percent of the freight value, and their services are very much in demand because of the many small companies in this sector.

The interest on the part of Chinese manufacturers and startup companies is no surprise, either, with 7.2 million trucks and 16 million drivers traveling on Chinese roads.[8] There are 130 Chinese metropolises with over 1 million inhabitants, which creates such a huge demand for transportation services that we find it difficult to get our heads around it. The transportation industry in China is estimated to be valued at as high as $300 billion.

A negative aspect is the industry's inefficiency. First of all, surely you are aware of those long truck lines at border stations where drivers are forced to wait until their freight is processed. Then they often have to sit around for hours doing nothing because they have to wait until the next truck load has been negotiated. The waiting time alone is estimated to cost $26 billion of wasted economic output in the United States.

It is a small wonder that a couple of startup companies smell a good business coming up. Uber appears to be toying with the idea of a freight mediation service with its division Uber Freight. The company has already gained some experience with this with its mediation platform for taxi rides. The freight loads themselves now look for and order an Uber instead of individual passengers.[9] Perhaps the concept of surge pricing can also be applied to goods transportation so that prices rise in times of greater demand. In any case, the retail giant Amazon has no intention of leaving this space to the competition. Amazon announced the launch of its own freight mediation service in 2017.[10]

Big "Apple" and the Battle for Big Data

> Without data, you're just another person with an opinion.
>
> —W. Edwards Deming

What do BMW, Mercedes, and Ford have in common? The fact that all of them make cars? No, that would be too easy. They all refused to collaborate with Apple on a car, the mystical iCar. Actually, the negotiations failed not so much because of physical details but because of the question of who would own the data generated by the car—the all-important data, the real "golden treasure" of the second automobile revolution. Cars already generate huge amounts of data. Hundreds of sensors create thousands of measurements every hour—per sensor. But what do they really measure, and for whom?

First of all, operational data are collected by cameras and LiDAR and radar systems, as well as ultrasound sensors. Then there is the kind of data generated by every other sensor in a car, for example, for tire pressure, coolant temperature, and remaining battery capacity. The drive data allow improvements in the algorithms and decision making that permit a better and safer driving performance for self-driving cars. A connected car collects data because of the communication with other vehicles and objects. The drive scenarios that every car experiences can be analyzed and routed back for the entire car pool to improve fleet performance. Personal data provide information about who is in the car, the selected route, and the activities the passengers are involved in during the trip. In addition, data about whether the car is in an unsafe neighborhood and whether it is driving conservatively or more sportively could be of interest to insurance companies. And finally, a huge amount of data is received by providers such as Uber and Lyft via transportation networks.

Former Intel CEO Brian Krzanich estimated that every self-driving vehicle will generate up to 4 terabytes of data every day if the current mileage also applies in the future.[11] As a layperson, you might ask yourself where all these data can be stored and how they can be saved to an external data storage device.[12] Well, not all the data are stored; some are deleted once they are no longer required. Then there is information you probably want to keep for longer, or perhaps you have to store it for a certain period of time for various legal, maintenance, or incident-related or administrative reasons.

In the end, any single vehicle probably does not need to save more than 1 terabyte of data locally; the rest of the information is uploaded to the cloud. Again, this may just be a fraction of the data volume stored in the manufacturer's cloud. The transmission bandwidth may quickly prove to be a bottleneck. How fast are the networks that allow me to upload data and communicate with other objects.[13] After all, we want to quickly be able to extrapolate sensible information from the immense data volume,[14] and this requires appropriate computer capacity.

So what do the data contain, and what makes them so valuable? The commercially most lucrative data describe user behavior. How is the car used? Those who know a lot about this and other aspects can offer additional services or make improvements.

In Germany, this kind of information is mainly regarded with distrust: it is a risk factor. Data protection in Germany has a very high priority and is a particularly sensitive topic. In light of the two totalitarian regimes that violated citizens' private data space regularly and sometimes with fatal consequences, this attitude is understandable: basically, everything related to passing on data is immediately analyzed critically. Consequently, people even exhibit a kind of self-censorship, in which not only are data to be protected, but their collection or creation should be avoided in the first place. Engineers aim to avoid generating any kind of data that might get them in trouble.

However, I feel that this is a misconception of data protection. One creates the respective problem even when it is not really necessary. After all, this should mainly be about personal data, not machine data, but everybody is trying to avoid any risk at all, the engineers as well as the corporate legal departments: the issue simply is too hot.

German car manufacturers have defined the data generated in a digitized car as personal or individual-related data as follows:[15]

Personal nature of data: Any use of a modern motor vehicle involves the continuous generation and processing of a large range of data. Data can especially be traced back to the owner or the driver of the car or the passengers, especially if other information is taken into account as well, containing information about personal or factual relations of an identifiable person. The data generated by and during the use of the vehicle are definitely to be considered as personal data as per the definition used in the German Federal Data Protection Act (BDSG) if a link to the vehicle identification number or the license plate is included.

Even before the first autonomous vehicles are on the road, Germany is already concerned about drafting regulations without any understanding of the kinds of data measured and how they might be used. Nevertheless, what this country so urgently desires and calls *digitalization* or *Industry 4.0* is nothing less than the exploitation of that potential. Because the people in charge are not focusing on chances and opportunities but only on dangers and potential abuse, Germany and the EU in general may fall behind many other countries.

It is a fact that we are in a rapidly changing legal gray zone. As with the intolerable European cookies' regulations, legislation is so slow and technically behind the times that regulations are adopted that neither are suitable to address the original intention nor take the actual importance into consideration, not to mention the fact that they are technically obsolete by the time they come into force. The United States is no exception. The hearing of Facebook CEO Mark Zuckerberg, appearing before the House and Senate committees in April 2018, showed how little the members of Congress understand crucial internet technologies and the business models behind them. Without understanding them, it becomes tough to properly regulate them. However, there is no doubt whatsoever that action is required and that the right to information self-determination must finally be given an appropriate legal framework. As a user, I want to be able to determine what happens with my personal data. Facebook and Google have to provide tools that allow European users to view their collected personal data and, if required, delete such information.

Newcomers such as Apple and Google have their core competencies precisely in this sector: recognizing the users' behavioral patterns and catering to them. For example, Apple's ecological system shows a comprehensive view of its users, including their Apple devices, and media preferences around iTunes, apps and their uses, preferred music and films, and the users' shopping habits are saved. Similarly, Google's comprehensive free offers are all about transparent customers. Which customers were looking for what, and where were they at the time? Who are their contacts, and what are the users chatting to others about? These kinds of data make Google and Apple the most valuable corporations in the world.

It is small wonder that the battle for vehicle data is so fierce. Self-driving vehicles allow more media consumption; the entire "infotainment" sector alone makes $31 billion every year.[16] And Apple, Google, and others are perfectly positioned in this field. Today, users are already using mobile systems featuring Android and iOS operating systems. So why should I use the vehicle manufacturer's operating system instead of my smartphone, especially when the former is far inferior in terms of familiarity and user friendliness? Accordingly, the car is the next territory to be conquered by Silicon Valley. The traditional manufacturers may already have lost the battle, even if they are in the process of developing their own systems, as Ford has with AppLink or GM with MyLink. SmartDeviceLink was developed as a kind of industry standard by Ford. It is intended as an interface for the link from smartphone to car.[17] However, all of this starts and ends with the number of external developers that are occupied with those things.

Their many years of experience in designing operating systems and the existence of an entire ecosystem of developers and tools grant an almost unbeatable competitive advantage to Google, Apple, and Microsoft. Drivers already tend to rely on Google Maps rather than on navigation systems built into their vehicles or stream their music directly from their smartphones instead of using the installed hard drive.

The control of passengers' motion behavior allows business opportunities of unimagined scope. The consulting firm Accenture estimates the additional revenue from data processing obtained during the life cycle of a single vehicle at $6,265 per car.[18] Chris Carson, former CEO of the now defunct driver assistance technology startup Caruma, thinks that for every year on the road a car generates a data value of $1,400.[19] According to Accenture, almost two-thirds of all future sources of income will come from the offer and sale of additional products. Estimates from the consulting company McKinsey say that the entire data service sector in the automotive industry may grow to up to $750 billion by 2030.[20] The market for smartphone apps looks almost tiny compared with that: about $100 billion.

Regulatory authorities can already prepare for a lot of headaches but also chances—in any case, a lot of work for them. First drafts from the U.S. NTSB also include the issue of access to the generated data.[21] The authorities demand access in case of accidents to help determine the cause and the question of fault and adopt measures to make self-driving cars safer. For this purpose, everybody involved has to agree to a standard so that data can be exchanged between different manufacturers as well. What is not yet clear, however, is how this information can be shared without anyone having to divulge trade secrets.[22]

On the other side of the coin, this means that (investigating) authorities and other official entities such as the police or technical offices such as TÜV Rheinland have to acquire the know-how to execute digital forensics on the cars using such data and be able to determine the causes of accidents from the available data. They have to understand how to measure and assess the safety of such a mobile data generator. Automobile clubs and repair shops must be able to read and interpret the respective data to recognize and repair defects. This work could be contested by the manufacturers themselves, which, in turn, means that the authorities just mentioned should already offer such jobs and positions now to obtain the knowledge for their own organizations. Data analysts, database specialists, the internet of things and AI experts, and software programmers would be a good start.

A look at Fiat shows that this company seems to have fewer qualms about handing itself over to the digital industry depending on your view of things. The Italian company closed a cooperation agreement with Google regarding the construction of 100 self-driving Waymo minivans in a joint pilot project in May 2016.[23] It was the first time ever that Google shared details of its previously top-secret technology for self-driving vehicles with an automobile manufacturer.

The first vehicles were delivered as early as October 2016. Since then, Waymo has received several hundred vehicles and has the option to order 62,000 more.

You Are Right Here!

Australia had a bit of an unusual problem in 2016. It had to shift its lines of longitude and latitude and adapt its data to GPS satellites. The reason for this measure is the continent's tectonic activity. Every year, Australia moves 3 inches (~7 cm) to the north. Because local measurements are based on fixed points on the continent, they obviously move together with the country, but the longitude and latitude measures also obviously do not, so eventually there is a discrepancy. The deviation amounted to 5 feet (1.5 m) since the previous data were obtained in 1994. A difference like this is simply too much for navigation systems and in particular for self-driving vehicles. As Chris Urmson pointed out, "5 feet of deviation on a road would be an unhappy day indeed." Australia therefore had to adjust the local coordination system until the deviations that continue to occur will be automatically considered in a new system from 2020 onward.[24]

GPS signals are good enough for us humans, but not for cars. "Exact to a couple of feet" for a car means that it may already be heading into oncoming traffic in the wrong lane. This is the reason why high-precision maps that are exact down to an inch, such as the ones provided by TomTom, HERE, Google, and Apple, are not just important for navigation in general but are an essential requirement for self-driving cars. This is why all the manufacturers are busy working on their own road maps, because they have to measure and digitally prepare more than 4 million miles of road in the United States alone.

With its maps application launched on February 8, 2005, Google transferred the virtual world to the real world—before that, there was simply a list of information from the internet. Today we think that this is perfectly normal, but at the time, it was a revolutionary step. Over time, Google managed to pack its maps full of information and the Street View function so that we today can virtually explore locations and streets all over the world. We look at the district and the street of our holiday home before leaving our own. Disabled people can check from their home whether a store or restaurant is accessible to them. We find out about opening hours before starting out. And the use cases just keep coming. Researchers analyzed changes in Canada's capital Ottawa by training machine learning systems with Google Street View data from past years to pinpoint the areas that had undergone gentrification.

In 2013, Google bought the Israeli startup Waze, allegedly for $966 million. This company started out as a crowdsourced street map project for Israel but quickly evolved into a free navigation app that simply communicated the user's speed of motion anonymously. Waze used this information to determine whether there was traffic congestion ahead on the route. This allowed the company to suggest alternative routes to its users. Today, some elements of Waze are an integral part of Google Maps, for example, notification of traffic obstructions. There are other apps as well that integrate Waze, for example, the taxi apps Lyft, Cabify, and 99Taxis.[25]

Apple had to discover the hard way just how difficult it is to create high-quality road maps for navigation. When the iPhone manufacturer replaced Google Maps with Apple Maps on its smartphones in 2012, the ever-successful technology giant was subject to an equally giant disaster. Its road map solution was full of errors. Within a few weeks, Google Maps was back on the iPhone. Since that fiasco, Apple invested a lot of money in Apple Maps and has dozens of mapping cars driving around the United States and other countries to obtain precise and up-to-date information.[26] It was just this intensified activity that fueled the rumors about a potential proprietary Apple vehicle project (code name Titan).

A consortium consisting of the German manufacturers Audi, Mercedes, and BMW purchased the road map service HERE from Nokia for $2.7 billion in 2016. HERE also started incorporating user data into its service. Vehicles with advanced assistance systems will be able to communicate road condition information to provide a current overview of the traffic and road situation to other drivers.[27] In this way, the information collected by 200 HERE mapping cars in more than 1,000 cities and 100 countries is supplemented.[28]

Toyota, in turn, holds shares in the Japanese map provider Zenrin, a company generating data for Google, among other activities.[29] Uber wants to break free from its current dependence on Google Maps, not least because of its efforts to enter the market of autonomous vehicles. Uber, for instance, bought the Bing road map solution from Microsoft, also taking over the 100 people employed there, and has now invested another $500 million into road map surveying.[30]

Mobileye, a manufacturer from Israel, employs 800 people to annotate images that support the interpretation of road markings for the system and has recently announced its collaboration with HERE.[31] The company is using a reduced 3D model employing orientation points. Apple has 4,500 employees all around the world working on similar tasks. Ford invested $6.8 million in the 3D map startup company Civil Maps.[32]

There are a good many startups in the wake of the large manufacturers. The German company TechniSat based in Daun is working on navigation systems, and Mapbox offers a software kit that allows developers to access road map data and integrate the information into their applications.[33]

The easily readable road maps offered by Google, Apple, and TomTom today are easy to read for us humans only; self-driving cars cannot use them: they are too imprecise and full of gaps. The data that can be used by the cars have to include much more detailed information than any human would ever need. This includes the height of the curb, the width of sidewalks, bike lanes and their exact position, dividing lines, and special road markings. Because road conditions actually change with surprising frequency, not just because of the weather or the season but also because of construction sites or incident light depending on the position of the sun or the time of day, the information must be updated continuously. This is where traffic participants can join in crowdsourcing projects to assist the special mapping cars used by Apple and Google to capture the entire San Francisco Bay Area.[34] Autonomous vehicles compare the digital with the actual conditions on their route and communicate deviations. For this to work, they have to be able to transfer a sufficient amount of data without using up their entire capacity. Consequently, the data have to be compressed and limited to the absolute minimum. As an additional requirement, the aim is to do without any additional sensors and only use the ones already installed.

Tesla is benefiting from this already. Its system can, for example, remember a speed bump and signal it back to headquarters. In the next step, all the other Tesla cars receive a current map so that the vehicles can automatically adjust their floor height as they approach the speed bump and instruct the driver to reduce speed. There was a winter video recorded by a Tesla driver that shows just how well this works. The road was snowy, and no road markings were visible anymore. The Tesla autopilot could still keep to the lane perfectly because the vehicle was able to rely on the treasure trove of previous trip data and knew where it was on the road without any visible lines.[35]

Digital Experience

The fastest way to my heart is the answer to my question: What's your WiFi password?

—MODERN DATING

In Europe, everyone is talking about *digital transformation*. This topic has been hotly debated for a number of years. Many people are trying to understand what exactly that means and how Silicon Valley proceeds. They are desperately looking for possible applications. The result usually is an app simply added on top of one's own product. However, this is a small aspect of digital transformation at most. It is nothing more than, as Americans like to say, "putting lipstick on a pig." It may look nicer now, but it is still a pig. The term *transformation* also would indicate that one state of being transitions to another, leaving us with another final state. Changes are, however, continuous and never stop. They change from shape to shape to shape. Once a change is completed, it is time to tackle the next one. Digital transformation therefore is not a project; it is a process, and an endless one at that.

Let us not forget that the term itself—like the German catchword *Industrie 4.0*—is quite unknown in Silicon Valley. The use of digital tools is not regarded as a separate field to be added to an existing one. It is often the central element, in fact. Let us take a look at the car-sharing sector. Uber's initial question was not how it could challenge existing taxi service providers. The company was trying to find out what an app to provide improved transportation services would have to look like. The company first created the app. Then it convinced car owners to offer their services and unused resources. The company welcomed the taxi drivers who signed up in addition to the drivers who before had had nothing to do with transportation services, but the company's main concerns were the individual drivers and limousine services. The company also never tried to acquire cars itself or hire drivers. The core competence was the app, the database behind it, and the algorithms offering efficient service.

Whereas companies dealing with digital transformation mainly look at how they can use their own existing resources more effectively with digital tools, companies from the sharing economy mainly focus on how they can offer the best service for customers. This is a massive difference in the economic approach and may develop its own dynamics. This approach allows one to question everything without regard to any social sensitivities. Established corporations are so closely linked to other industries and interested parties because of their history that a kind of internal censorship applies. Certain views are not permitted out of fear that well-established relationships may be endangered and damage one's own company.

There is one detail I want to provide at this point, one that is not unimportant and may help to comprehend Germany's problems with digital transformation. One reason may be the use or rather nonuse of digital tools by company CEOs. Elon Musk is well known for his tweets in which he directly interacts with customers and the press and publishes statements, sometimes to the chagrin of shareholders or the Securities and Exchange Commission. Steve Jobs also continued to pursue customer contact by showing up in Apple stores unannounced or working on customer messages in the system himself. Not a single one of the top three executives in the German automobile industry has an active Twitter account or is otherwise known for his—and yes, it's a male-dominated industry—use of other digital tools. This just confirms the overall impression: digital transformation is not set as an example at the top.

Let us take a look now at some areas of the second automobile revolution that will make the digital experience an important factor.

One-Click Updates: Over the Air Instead of Back to the Workshop

The announcement from the U.S. traffic agency NHTSA regarding a software error in Volkswagen's e-Golf was technical and sober: "Due to an over-sensitive diagnosis in the battery management system, an overvoltage may erroneously be indicated, and the electric motor may unexpectedly shut down."[36] In the United States, more than 5,000 vehicles were recalled back to the dealer so that mechanics could install a software update. Apart from the cost incurred by VW, the recall was inconvenient for owners.

When events such as this occur, you start asking yourself why this sort of thing still has to be such a hassle. Why is it not possible to do it via the internet? Just imagine that you had to go to a service point with your smartphone for every software update or error, plan your time, and potentially wait in line until a technician could install and test the appropriate software patch. This is a procedure that no manufacturer of smartphones and computers could hope to enforce on its customers, but that is still standard in the automobile industry.

This still occurs despite the fact that Tesla with its over-the-air updates demonstrated how to do it without problems. The new software is installed overnight, and errors are resolved or new functions become available. Tesla owners were ecstatic when they suddenly had a "summon" function available (independent parking and maneuvering of the vehicle without the driver), an autopilot mode, or the "dog mode" that allows you to leave your dog in the car with the air conditioning on, keeping the car cabin at a comfortable temperature. The internet was flooded with videos of how a car comes rolling out of the garage and picks up the driver waiting at the roadside. Also, after a critical mention by *Consumer Reports* magazine of the effectiveness of the Model 3 brakes, Tesla just made a software update available, and a few days after, the critical issue was fixed. Just like that.

While this kind of updating would not be worth turning your head for in the digital industry, quite the contrary, it is revolutionary in the automobile sector. Opportunities leading to the faster development of new functions thus are missed. The autopilot update that worked for the latest versions of the Tesla Model S gained 10,000 test vehicles for Tesla in one go. The data collected improved the company's database, and all Tesla owners benefited.

However, those software updates must be analyzed critically. Some videos showed drivers putting too much trust in the autopilot, although Tesla had pointed out the limitations. In reaction to their overenthusiasm, some other updates followed that once more limited certain functions until additional data were available.

The chairman of TÜV Rheinland in Germany believes that a Tesla vehicle after such a software update would differ too much from the original vehicle for which the circulation permit was issued and that, consequently, the permit would have to be revoked temporarily, as it would be for a tuned vehicle.[37]

Two worlds collide: the digital industry, on the one hand, where quick and regular updates are standard, and the automotive industry, which is focused on risk avoidance. And both of them are right. Nobody would want to risk his or her life in a car that could cause an accident because of a software error in an update or a program that is still a beta or test version. And yet we would like the convenient solution that adjustments can be made without a visit to the garage and have a safe vehicle. Because digital functions increasingly gain importance for vehicles and are the driving force behind the upheavals in the industry, we must find measures that ensure safety and convenience. Car manufacturers have to learn how to program and manage highly failsafe software.

They might find inspiration in the procedures adopted in the nuclear industry or in space travel applications. Methods are available, although they still entail enormous cost.

Software probably will be installed not just by the carmakers but also by third parties as well. Platforms similar to Apple's app store or Google Play open up new dimensions for independent software providers. Automobile manufacturers will have to obtain the necessary competencies if they want to have the full benefit of this, including, for example, building a community of developers using the programming tools and interfaces offered by the car-makers. And once again, Google, Apple, and others are ahead on this path. They have already gained a lot of experience with operating systems for mobile devices, computers, and, most recently, the internet of things.

Winning the "Battle" with Dashboard and Entertainment

The fight for dominance in the race toward the car of the future is apparent on many obvious and some less obvious fronts. The dashboard is one of the bat-tlefields. My former Volvo S60, built in 2014, already had a digital display. My new Tesla Model 3 comes with no dashboard, but a touch screen in the middle of the front with all data displayed there. Meanwhile, each vehicle pre-sented or announced—regardless of the manufacturer—has digital indicators. The question that is completely unanswered, however, is who creates the oper-ating system for them and eventually controls the ecosystem.

Google is at the forefront with Android Auto.[38] The company already offers thousands of applications such as Google Maps and the music service Spotify, so the driver will no longer expect or need such apps from the auto-mobile manufacturer. Even vehicle-specific data are transferred to smartphone apps via dongles nowadays.

Google, Apple, and Microsoft have a pool of developers available for their millions of apps. Even a car manufacturer with an excellent financial background would not be able to match that, apart from the fact that car manufacturers have their core competence in building cars, not in software programming, the creation of operating systems, or the setup of developer ecosystems. The software companies themselves have a number of advantages. First, they gain access to the data generated during a trip. At the same time, they increase customer loyalty to their smartphones. It is not so attractive to take the leap and use a competing product such as the iPhone every two years when upgrading to a new smartphone if your own car is exchanged for a new one every five and a half years on average. You have grown comfortable with the setting, and exchanging your smartphone means loss of the familiar car dashboard.

This is another reason why Apple is wasting no time. Apple Carplay is Google's counterpart. Car manufacturers such as Ford nowadays try to play it safe by integrating both.[39] But what will the long-term solution be eventually? If the value of a car is in the data, do we really want to leave the field to the technology giants without a fight? Is there any chance whatsoever to keep up with them?

The Taxi Lobby Complains: Sharing Is Against the Rules

As soon as automobile manufacturers enter the car-sharing business together with disruptors such as Uber and Lyft, conflicts with taxi companies are inevitable. Mercedes, traditionally providing more than half of all the approximately 90,000 taxis on German streets, is subject to the effects of this fact now. Taxi companies are starting to wonder whether it would not be a good idea to switch to a different brand if their so-far preferred manufacturer turns into a competitor. This is an important signal for Mercedes to reconsider priorities: it may be more profitable in the long run, not to mention an investment to the company's future welfare, to set up a proprietary taxi service instead of just delivering vehicles to taxi companies.[40]

Wherever Uber arrives, the company meets with resistance, a not wholly unwelcome phenomenon for venture capitalists. The number of open court cases is huge. Germany, the country that likes to parade around carrying the insignia of digital industry and Industrie 4.0 and even put the topic on its political agenda, is the place where the discrepancy is most visible. There is this innovative and revolutionary company that offers a much better service to customers, and government agencies such as courts have nothing better to do than to protect the profit interests of the traditional taxi service providers. In this way, the taxi lobby got rid of an annoying competitor, thanks to its good relationship with regulators and legislators.

A particularly schizophrenic statement is this: "Uber's aggressive tactics even put off potential customers," said Andreas Müller, a financial analyst who tried the company's Frankfurt service after first using Uber on a business trip to Chicago.[41] Mr. Müller said that he liked the convenience of paying through his smartphone but soon turned against the company after reading that it had continued operating in violation of court orders and did not directly employ its drivers, who are independent contractors. "That might work in the U.S., but that's not how things are done here in Germany," said Mr. Müller. "Everyone must respect the rules." Here we have someone who is happy with the improved service but never asks himself whether the laws that the court decisions are based on are actually still valid for Uber or whether they should be modified, and he conveniently also forgets that taxi companies also have freelancers working for them.

People like to talk about the digital revolution, but woe when it really comes: it is beaten down instead of serving as an incentive to offer better service, to the disadvantage of the customers and damaging to the digital efforts as a whole.

Business Models

Self-driving electric Ubers call many different business models into question. Technological innovation can cause disruption, but so can the full force of various innovations in different fields: how and through which distribution channels is said technology sold, how is it presented as a brand, what changes does it bring to the services previously provided, what new services and models does it allow, and what types of clients has it not yet reached and is it addressed to? I could continue this list endlessly.

One model gives an impressive example of how much this might change the industry. If we assume that an average car drives 200,000 miles during its life cycle, and that car companies earn $2,000 on every vehicle they build and sell, then the profit for each mile is 1 cent. That's what car companies basically have been counting over the past 100 years. But if you can earn 20 to 30 cents per mile driven with operating robotaxi fleets, then this changes the profitability equation quite a bit. This explains why a small and young company such as Tesla has a higher market capitalization than General Motors, Ford, BMW, and even Daimler, companies that build ten to hundred times more cars.

"I'd Like to Buy an Electric Car": Dealers and Their Clients

If you buy a car in Germany, any visit to the dealer usually just consists of a test drive and configuration of the desired vehicle. You wait two or three months, and then either the vehicle of your choice is delivered to your door or you pick it up at the dealer's. In the United States, by contrast, you usually drive home in the car on the same day. Dealers usually have a large range of vehicles with different features ready to be sold. Any cars that are standing around for too long are often sold at a huge discount. In comparison with German car sellers, who would be more aptly called *vehicle configuration consultants*, Americans appear to be *aggressive sales strategists*.

Nevertheless, both sales models are under pressure. Who needs a German dealer if you can configure your car on the internet and order directly from the manufacturer? In fact, that is what I did with my Model 3. I was on a business trip in Germany, when configuring and ordering my car. I made a down payment online, uploaded the required documents, and two weeks later the car was delivered to my home address and my trade-in was driven away. The aggressive sales techniques of American dealers, by contrast, are the reason why their customers' trust in them plummeted to the second to last of 22 positions in a Gallup survey.[42] The technological changes pressure the dealers because the newcomers no longer want to or cannot rely on traditional sales models in the automobile sector.

Tesla has its own sales centers. The car dealers are all Tesla employees and do not get the usual commissions, as is the practice in automobile sales. The result is a more relaxed sales talk for customers. Anyway, Tesla prefers direct sales because traditional dealers are not keen to sell electric cars. First of all, it is easier to sell what you know. Second, the lucrative business segments are financing and service. According to the National Automobile Dealers Association (NADA), American dealers earn 2 percent on each sale but more than 10 percent of the sales price through car services over the next few years.[43] But as we all know, EVs need less maintenance, and this threatens this calculation. In fact, Tesla doesn't even have a service plan. Nothing that says to come in every 10,000 miles or once a year. There is not much to service, with the exception of refilling the windshield wiper fluid and checking the tires. There is no oil or oil filter to change or spark plugs to clean.

Many car dealers still do not know how an EV works and continue to prefer their combustion engine cars. There are a number of stories on internet forums published by people interested in purchasing EVs. There are tales of missing test vehicles and nonexistent charging stations. A survey in the United States based on visits to 308 dealerships shows a devastating picture.[44] The following quotes from clients will give you a somewhat bizarre insight into everyday (non)sales practices:

I couldn't do a test drive because the key was lost. I was encouraged to purchase a non-electric vehicle instead. [Nissan dealership, Connecticut]

I called the dealership and was told that they weren't certified to sell EVs and that their sales department wasn't equipped to handle them. [Ford dealership, Maine]

There were only two EVs on the lot, and neither were sufficiently charged for a test drive. [Mercedes dealership, California]

They didn't have any electric vehicles on offer, and he said he wasn't interested in selling them. . . . The only way to get him to sell electric vehicles would be if Volkswagen forced him to. [Volkswagen dealership, Maine]

The respective internet forums of German groups of EV fans are similar. This is a post dated December 30, 2016, from the Facebook group Elektroauto:[45]

Germany, no-service country: Today we wanted to find out more about the i3 by BMW and potentially buy one. . . . We waited for 20 minutes; then the receptionist asked us to wait until the afternoon as the only contact person was out to lunch. We heard his voice from the background talking to another colleague: "I sure won't do any overtime for an i3."

So we went to Tesla, bought a Model 3, and made a down payment. We immediately felt good and welcome there and received competent advice.

This snapshot shows that it will be difficult for the competition to win over customers, let alone keep them.

Here is a similar report from another user in the same Facebook group:

I once went to see the BMW dealership to look at the i3. Ten minutes into the conversation, the sales rep says: "Do you know that I could also sell you a real car for that kind of money?" They're not in the least interested in selling an electric vehicle. The story is the same with VW.

The accounts are all similar:

Had exactly the same experience with the Audi A3 Hybrid. . . . My e-mail enquiry was answered six weeks later, and they wrote, "If you are still interested, we could try to arrange a test drive next year." And they added as an aside something to the effect of "but we'll only do a test drive if you are really interested.". . . So I went to Tesla, got three one-hour test drives and ordered a Model S. Have never been more satisfied. Pity for those German premium brands.

Interestingly enough, several U.S. states have banned the direct sales of vehicles by their manufacturers. There are historical reasons for this. During the economic crisis of the 1930s, Ford, for example, kept its production at pre-crisis levels and forced the dealers to buy the cars, knowing full well that they would not be able to sell them. The dealers, in turn, were worried about being canceled by the dealer list and then not getting any deliveries once the times were better. The dealers were also completely dependent on the manufacturers in terms of product features and market positioning.

Consequently, the dealers created an association and successfully lobbied in their respective states, persuading legislators to adopt regulations that granted dealers territorial protection against the car manufacturers on a local level, including a ban on manufacturer sales points in order to exclude any direct competition. You need a license in all the states to sell cars. This is the reason why Tesla is in dispute with many local governments with its direct sales model. In some states, such laws have already been abolished, in Massachusetts and Maryland, for example.[46] The U.S. Department of Justice estimates that the sales figures for dealerships might fall by more than 8 percent as soon as manufacturers are granted permission to sell directly.[47]

The Nürtinger Institut für Automobilwirtschaft (IFA) notes that the dealer network in Germany is also subject to marked changes. In the year 2000, there were 18,000 dealer groups, but their number shrank to 7,400 in 2016. The IFA expects the number to decrease further to 4,500 in 2020. Cars are increasingly bought online today, and smaller dealers have disappeared. A quarter of the dealerships post losses, and half of them work for less than 1 percent of return. Although the revenue from repairs is increasing, the profit is immediately swallowed by more technical complexity and the accompanying investment.[48] In addition, many cars today seem to be computers on wheels, and the software service is provided directly by the manufacturer (see the earlier discussion of over-the-air updates).

The dealers are also under pressure from other directions. Business models are changing. Tesla, for example, has its business premises in top locations in city centers. If the scenario comes true that people will have fewer cars and switch over to sharing models, the dealers' business will consist less of sales to end customers and more of sales to fleet operators, provided that dealers will actually still be part of the sales chain and fleet operators do not buy directly from the manufacturers (or the manufacturers become fleet operators themselves). BMW's DriveNow and Mercedes' Car2Go are good examples of this. Who will then still need a dealer, especially if—as in the Apple stores—you could buy products for several thousand dollars in under five minutes without any salesperson involved at all (you just open the Apple store app on your iPhone, scan the barcode on the box, confirm payment, and leave the store with your product).

With this possibility, it becomes completely absurd to spend three to four hours at the dealer's office to buy a car. Naturally, the amount of money involved is higher, but if much of my information has to be checked repeatedly and input into the system manually, and the salesperson has to go into the next room to obtain dozens of pages with antiquated matrix print copies, then you understand why buying a car is not at all a positive experience.

Lower Taxes Without a Steering Wheel: How on Earth Do We Accomplish This?

Vehicles are a source of constant income for towns, municipalities, states, and the federal government, for example, motor vehicle taxes, diesel oil taxes, road tolls, fines for traffic violations, parking fees, etc. It seems that there is no limit to the authorities' inventiveness. We should avoid looking at drivers as the nation's pitiful cash cows, however. Parking fees in cities are lower than the market price level would be, whereas apartments are 20 to 30 percent more expensive because of building regulations that still demand the inclusion of parking spaces. Those prices are cross-subsidized by all of us, including people with no cars. The economic damage to the country because of exhaust pollution, accidents, noise, congestion, and general wasting of resources is higher than the income generated from duties, taxes, and fees.

The annual tax revenue in the United States from the production and use of automobiles amounts to $206 billion.[49] More than 1.5 million people are employed in the automobile sector, 322,000 of them directly with the large automobile manufacturers, more than 500,000 with the supplier companies, and more than 700,000 with the dealers and workshops.[50]

Every German car generates $56 to $68 for the municipal coffers every year from fines for illegal parking, speeding, and other traffic offenses. Electric robotaxis stick to the rules and do not park. And voilà, we have a revenue shortfall of $2.26 billion.

One consequence of more EVs on the roads is a decrease of revenue from fuel taxes and diesel oil taxes. As mentioned earlier, more than $44 billion of revenue was generated for the German state from energy taxes.[51] In Austria, diesel oil taxes amounted to slightly over $4.5 billion for the state in 2014.[52] And in Switzerland, the amount was $4.79 billion altogether in 2015.[53] A lot of tax income is earmarked to maintain the traffic infrastructure. As a replacement for the loss of income resulting from the decrease in the number of combustion engine vehicles, the EU Commission is contemplating a tax on the energy content instead of the volume.

However, electric robotaxis also generally reduce the national economic damage that our current habits of private transportation have caused and that the cities, municipalities, and federal and state governments have to consider in their budgets. Fewer accidents, improved ecological footprints, noise reduction, and a decrease in infrastructure costs and public local transportation costs are the main possible sources for savings.

Not So Sure: Who Pays for Accidents?

If a robotaxi causes an accident, the passengers are insured, as they would be if they traveled by taxi. The insurance must be paid by the owner of the vehicle—a company in many cases. The question of who has to pay in such a case, the vehicle manufacturer or the other party involved in the accident, depends on who caused the accident. The situation is slightly different when semiautonomous vehicles are involved, that is, those below level 3 that cede control to the driver under certain conditions. Did the driver cause the accident because he or she was at the wheel and had a delayed reaction? Or was it the car that did not react appropriately or alert the driver in a timely enough manner? Or was the accident caused by something else altogether? With such cases coming up, some insurance situations will become more complex, and others will slowly disappear.

In the United States, NHTSA believes in a conservative estimate of traffic fatalities at half of today's total, thanks to the new technology.[54] If self-driving vehicles can manage this because of their safe style of driving and quicker reactions, the income in the U.S. car insurance market would drop by as much as 90 percent from a ridiculous $200 billion to $20 billion; the German equivalent would be a drop from today's $22 billion to $2.2 billion. Now try to explain this dramatic reduction in the balance sheet to shareholders. The decision of who has to pay also depends on the manufacturers.

Volvo has already declared that it would assume liability for accidents caused in the future by self-driving Volvo cars.[55] At the Tesla Autonomy Days in 2019, Elon Musk was also asked about liability, when laying out the road map for a taxi service with Teslas. Not only would the fleet be supplied by Tesla-owned vehicles, but Tesla owners could also add their cars to the fleet. Musk stated that the liability in case of an incident with a car joining temporarily the robotaxi fleet may be with Tesla, not the individual car owner.

At the beginning, we are likely to see the following effect: with the first self-driving vehicles, the number of crashes and claims for damages will decrease and ensure higher profits for the insurance companies. They might try to include the high-priced technology in their policies and premiums while sensors and computer units in those cars are still comparatively expensive. Once we have the first numbers and calculations for risk assessment of autonomous vehicles, we may expect the insurance premiums to become less expensive, with the respective impacts on revenue and profit margins.[56]

As a countermeasure, the insurance companies might reassess the risk of human drivers compared with computers and make insurance premiums for human drivers so high that the costs are basically prohibitive. Just imagine if the risk for a human being causing an accident were 10 times higher than for a self-driving car by 2025. In that case, there would be an immediate effect. Instead of paying $50, you would have to pay $500 for insurance. The startup company Root was the first insurance company to offer a reduced car insurance premium for Tesla owners. If the car is used in autopilot mode, Root charges only 50 percent of the premium.[57]

If fleet operators such as Uber or Lyft dominate the market in the future, insurance companies will mainly have to deal with companies instead of individual clients. Currently, private clients account for a third of the revenue in the insurance market in the United States.[58] This must necessarily change negotiating positions. Larger insurance companies can offer higher discounts and more services than smaller ones. We can therefore expect smaller insurance companies to be pushed out of this market. And there is another thing: fleet operators may generally want to forgo any individual insurance for their vehicles because of the low number of accidents and opt to cover damages directly from the revenue from the operating business.

Some insurance companies are already experimenting with similar new business models. The first approach is the result of the fact that cars today are basically parked for 23 out of 24 hours. Providers such as the startup company Metromile, which has $200 million of venture capital available, started out with recording the time the vehicle is in motion and limiting the insurance premium to that period plus the driver.[59]

Autonomous linked vehicles equipped with many sensors may manage insured events quickly and adjust the premiums. As soon as the vehicle moves into a district with a higher crime rate, the premium goes up. In the case of an accident, the damage can be settled on the basis of the sensor data while everyone is still on site.

But there is a chance that insurance companies will have a place in the age of the autonomous car, too. Because of the necessary cross-linking, there may be cases of mega-accidents with thousands of vehicles involved if there was a defect. In cases such as this, the question of who or what caused the accident becomes almost secondary. Hackers might seize control and make all cars under their control turn right simultaneously, or perhaps a faulty algorithm suddenly wreaks havoc. The latter scenario almost came true in another sector already: in 2010, there was what is called a *flash crash* at the stock markets, and the Dow Jones Index plummeted by 10 percent in just a few seconds. As a matter of fact, there have been more than 18,000 such minicrashes at the various stock exchanges since 2006. Today, most of the "traders" are actually algorithms that humans do not even notice.[60]

The British government was the first to work on regulations for insurance policies for autonomous vehicles.[61] The laws not only were extended to include the vehicles but were rewritten to the effect that the insurance policies apply to both a human driver and a computer. Insurance companies are obliged to pay damages to innocent accident victims but can be held harmless through the manufacturer of the autonomous vehicle in the case of technical defect.

Cars as Bank Customers

U.S. banks loaned the incredibly high sum of $1,027 billion for car purchases in 2015.[62] This amount does not even include the huge sums for car leases. The automobile market comes in third in terms of loans granted, preceded only by real estate and student loans. Of all the car sales in the United States, 86 percent are financed by loans compared with only 26 percent in China. In Germany, 50 percent of all new cars are financed with a loan, as are 28 percent of all used car purchases.[63]

In the sharing services, Uber offers loan financing to its drivers via Xchange Leasing so that they can buy new cars and maintain the quality of their vehicles as well as the experience of passengers. One gladly accepted side effect is a greater (product) loyalty of drivers to the company. And Uber came up with another idea, too: 30 percent of its drivers do not have a bank account. Taxi services in the United States are strongly based on cash payments, so many taxi drivers did not find it necessary to have a bank account.[64] This makes it more difficult to pay the money earned in a sharing industry that is dominated by digital payments. Uber therefore offers an option to open a business bank account when it recruits drivers to make Uber more attractive for the drivers. This brought Uber 300,000 small customers overnight, more than all the large banks would bring in together.[65]

Once autonomous cars start earning money with extra trips after they have dropped their owners off at work, they might need a bank account, too, into which payments for fares can be deposited and from which the costs of fuel and maintenance can be paid. Such a bank account would not be linked to a person, but to the vehicle. At the Consumer Electronics Show, the German gears manufacturer ZF together with UBS and Innogy presented eWallet, which is exactly this.[66] This step makes for other interesting questions. How can a vehicle identify itself and prove its identity with a bank? Will it have to pay taxes on its earnings?[67]

Smart City Challenge: Onward to the City Without Cars and Parking Lots

We already discussed traffic in the cities in detail and the question of how traffic planners aim at keeping traffic flowing if at all possible (except traffic planners in Switzerland, who are trying to do the opposite and are proud of it). There is indeed a *general theory of walkability* for a city as defined by Jeff Speck, an American city planner.[68] He differentiates among four main characteristics a city can be measured by from a pedestrian point of view: a city must be useful, safe, convenient, and interesting. A useful city will allow inhabitants to have all the aspects of everyday life within reasonable walking distance. Streets must be planned to be sufficiently safe so that pedestrians are not endangered by vehicles. The surroundings and all the buildings should give pedestrians a feeling of convenience and comfort as if moving through an outdoor living room. Jeff Speck lists large plazas as contrasting examples: those squares usually are not able to attract pedestrians. And eventually, public spaces must be interesting and offer some variation along the sidewalks: popular shopping streets, varied architecture, and small parks scattered across the city that are within easy reach of pedestrians in their respective quarters.

What would our cities look like if the space that is nowadays reserved for parking lots and streets would be available for other purposes? Historical photographs showing street scenes before the rise of the automobile are striking in their depiction of how humans use the entire road space quite naturally: children play, dogs nose around, and adults stop for a chat in the middle of the road. For us today, such things are unthinkable. Conceptions of the potential impact of autonomous vehicles on our city life almost seem like science fiction, but they are actually just a return to a previous way of life. Humans take public spaces back again from the automobiles.[69] Autonomous vehicles in sharing models offer a chance to utterly reshape our cities. They need fewer parking spaces because they are mainly moving around. Consequently, we need fewer individual vehicle units that take up less space. This newly freed space can now be used for other purposes, and some pioneers among the cities are planning for this kind of future now.

Whereas cities simply sort of "happened" in the past, they are usually planned in detail today. Roman border towns, Alexandria, Haussmann's Paris, and Barcelona's Eixample housing blocks are exceptions from the unplanned chaos many cities thrived on for thousands of years. At a river; near a bridge, a mill, or a mine; places to sleep, cook, and wash; and eventually housing accumulated. Today we have urban development planning and make sure that everything is "in order"—which is not always the same as obtaining a high quality of life.

Urban planning is more than just thinking about which street should lead where, determining places for apartment blocks and shopping centers. There is public local transportation to consider, the sewer system and electricity grids, nursery and grade schools, and street objects such as street lights and benches, too. And we have new technology here as well. Los Angeles, for example, started trials with smart street lights that light the streets but also send feedback to the head office about whether they are switched on or off and street lights that communicate with road users and warn about emergency vehicles or dangerous situations with light signals.[70] They can also potentially serve as charging stations for electric cars. One such provider is the British startup company Telensa. The company already completed installation of such solutions in Moscow and Shenzhen thanks to a venture financing input of $18 million that it received.[71]

And don't underestimate the effects of a good location on the success of the enterprises active in the area. It may be crucial to whether a company is able to have offices in the city itself or only in the surrounding urban area.

A central location means a larger pool of talent and customers, whereas a location outside the city offers more affordable rents. A survey compared the market prices of 38 comparable companies in New York City that left town and moved to the suburbs with those of 35 companies that decided to stay within the New York City limits. The market price of the companies that left was only half as good in the years after the move.[72]

The distance that commuters have to travel from their homes to their workplaces also has an impact on economic output. The U.S. Environmental Protection Agency determined that the farther commuters had to travel to reach their workplaces, the lower was productivity in that state.[73]

Cesare Marchetti, a physicist from Venice, analyzed the average time people calculate for their travel time to and from work. He noted that the time seemed to regularly "level off" at about 1 hour, regardless of whether the worker traveled on foot, on horseback, on a bike, in a car, or on a train. On average, workers accept half an hour each way for work. In the relevant literature, this finding is known as *Marchetti's constant* and has remained the same over several centuries and independent of the means of transportation used.

Although forms of urban planning and transport may change, and although some live in villages and others in cities, people gradually adjust their lives to their conditions (including location of their homes relative to their workplace) such that the average travel time stays approximately constant at one hour per day.[74]

It is interesting to note that humans also try to increase their commute time on purpose. People living close to their place of work often stop at a café to prolong a minicommute.[75] I do this as well. As long as I am not traveling for my consulting activities, I stay at home and write books. My way to work is just a few feet: from my bed to my desk. So boring! I therefore have the habit of dropping into a nearby café and working there for 3 or 4 hours. In this way, I create a little commute to work, 20 minutes in all, but time for me to think through the day's work and ideas so that I then can do the work or write things down.

It is impossible to predict all the developments in the wake of the introduction and adjustment of transportation. It was quite possible to forecast an increase in the number of streets and parking lots 100 years ago, but it was surely a surprising effect of the new technology to find that once there were cars, there were also huge malls, shopping centers, and big box retailers such as Walmart, Real, and Ikea on the periphery. We should therefore be aware that the actual changes shared cars will bring us cannot all be predicted yet. The only thing we can predict is building regulations!

You have to take a walk through a city and look closely to understand how much it is dominated by cars. In some cases, parking garages, parking lots, and private garages use up to a third of the available space. The reason for this is the local building regulations that mandate the number of parking spaces to provide in relation to an apartment, a store, or offices. Researchers analyzed four small American cities to see how multistory car parks and lots impact urban planning and marked the space on aerial photographs.[76] The results speak for themselves.

In Cambridge, Massachusetts, home of Harvard University, the regulations mandate 0.09 parking space per 1,000 square feet of built-up area, so basically a little under one parking space per 1,000 square feet. Those spaces marked in red are hardly visible on the city map. In the California university town of Berkeley, offering three times as many parking spaces as Cambridge (factor of 0.25), the red spaces are already clearly visible but do not dominate the city. In New Haven and Hartford, Connecticut, the situation is drastically different. They have 0.6 and 0.86 parking space, respectively, per 1,000 square feet. The cities are riddled with multistory car parks—half the map is red. Paradoxically, this leads to urban sprawl. Because there are structures intended for parking between the apartment blocks and business premises, the distances to be covered on foot are longer, so the demand for parking rises. If you happen to know those four small cities, you will immediately be able to decide which one offers the highest quality of life. It is definitely not the one with the most parking spaces.

The Chinese city of Wuhu near Shanghai with almost 4 million inhabitants has announced that it will make a complete transition to self-driving cars in the next five years.[77] In collaboration with Baidu, the city plans to introduce autonomous buses, trucks, and passenger cars throughout the city.

In the United States, only 6 percent of the cities had considered autonomous vehicles in their transportation planning until recently.[78] The U.S. Department of Transportation (DOT) therefore organized the Smart City Challenge, an ideas competition asking for submission of new concepts for urban traffic and the linked use of autonomous vehicles, sensors, and data.[79] Seventy-eight American cities submitted their ideas and concepts. Mark Dowd, former deputy assistant secretary of research and technology with the DOT and now executive director of the Smart Cities Lab, noted in particular that two-thirds of those cities considered autonomous vehicles as an integral component of future urban traffic solutions, including the winner of the competition (with a prize of $40 million), Columbus, Ohio. This city plans to use autonomous city buses to provide improved traffic connections for inhabitants in the poorer districts so that they have better access to healthcare centers in other parts of the city.[80]

In other words, not only do most urban administrations regard autonomous vehicles as inevitable, but they are also impatient for their arrival to solve urgent traffic issues. It does not matter where, in which town or municipality, autonomous vehicles complete their test drives; representatives of local governments are already competing for their presence and are ready to push legal obstacles out of the way.

By now there are smaller communities as well that are planning with autonomous vehicles in mind. Babcock Ranch, a town in Florida, not only built a new suburb with environmental-friendly energy supplies, but also projected the use of autonomous minibuses and vehicles as a local means of transportation and especially took care to organize the necessary infrastructure within walkable distance to the respective accommodations and housing.[81] This, in turn, changes the construction of houses: no garage and no obligatory space for parking on the street.

The traffic planners in Los Angeles, a city that particularly suffers from automobile traffic, have practically given up on planning new local transportation. Although the city still had a functioning tram network in the 1920s, it was bought and closed down by dummy companies financed by the automotive industry—you may remember the movie *Who Framed Roger Rabbit*, in which this is mentioned in passing. Anybody who ever had to move at a crawl on one of the eight-lane highways in LA knows that the city has a huge traffic problem. The city therefore decided to simply skip public transportation after years of neglecting it and move forward to autonomous vehicles immediately.[82] By 2035, Los Angeles plans to have zero traffic fatalities as a result of its Vision Zero initiative. Yes, zero, that's the number. Self-driving cars are the most crucial component for achieving this ambitious goal. The state of Wisconsin is also thinking about reducing its investment in expensive road infrastructure in favor of autonomous vehicles.[83] The savings alone might be incentive enough for administrations to hurry along permission for self-driving cars and a ban on manual ones. We are facing the unique situation that governments will adapt to new technologies faster than most citizens want them.

Autonomous cars also play a huge role in the NHTSA initiative Road to Zero, shared by several road safety organizations.[84] Within just 30 years, the number of traffic fatalities in the United States is supposed to drop to zero: people are counting on the safety offered by driverless cars. Once again, this is an example of how a vision or the related emotional trigger might move authorities and companies to push disruptive innovations and market them. The same is going to happen in Germany, at least with regard to organizations such as the Deutscher Verkehrssicherheitsrat (DVR) (German Advisory Board for Traffic Safety) or Verkehrsclub Deutschland (VCD) (German Traffic Association).[85]

While the use of personal narratives or emotional stories is perfectly normal in the United States, Germans find it more difficult to accept narratives in that kind of framework. A delegation from Lower Saxony, led by Olaf Lies, former Lower Saxony's Minister of Economy, Labor and Traffic, returned home from a visit to Google/Waymo full of excitement. The element that had moved the participants more than anything else was a video about Steve Mahan, former CEO of the Santa Clara Valley Center for the Blind and blind himself. He traveled to his appointments by himself in a Google car.[86]

And what is the answer from the German manufacturers? Slogans! "Das Auto" ("the car"), "Vorsprung durch Technik" ("progress through advancement"), "Freude am Fahren" ("joy of driving"). Disgraceful. Stories and narratives are what inspire us. Gene Roddenberry's *Star Trek* series inspired the flip phone with its communicator. It also inspired a whole series of startups working on the tricorder, a medical device that measures several vital parameters such as body temperature and pulse. The TV series *Knight Rider* is behind many a dream of self-driving, communicating cars like K.I.T.T., and the wristwatch you can talk into evolved into programs like Siri. Unfortunately, neither the stories nor the respective technology is from Germany.

Marc Andreesen, former founder of Netscape and now head of the Andreessen-Horowitz investment fund, has discovered an even more radical approach to developments:[87]

> There are mayors who would, for example, like to just declare their city core to [ban] human-driven cars. They want a grid of autonomous cars, golf carts, buses, trams, whatever, and it's just a service, all electric, all autonomous.
>
> Think about what they could do if they had that. They could take out all of the street parking. They could take out all of the parking lots. They could turn the entire downtown area into a park with these very lightweight electric vehicles. No pollution, no noise, no nothing.

It would be almost like going to an airport, where you could drive and then you drop your car off and then a self-driving golf cart would take you into town. There are cities that want to do that, and not just in the US—there are cities internationally that want to do that, including in some countries where the government can order that to happen. College campuses, retirement communities, amusement parks, industrial campuses, and large office complexes in some cases are places where this stuff can get rolled out in a top-down way. I think you'll see a hopscotching kind of thing, as opposed to sudden mass adoption.

Some cities have already adopted certain drastic measures as a result of the diesel emissions scandal and pollution caused by fine dust. Anne Hidalgo, mayor of Paris, for instance, announced a ban on particularly "dirty" vehicles in the entire city after a first test involving a complete blocking of the magnificent Champs Élysées and another section of road along the Seine. The number of cars is to be cut in half, and public transportation and bicycles have priority.[88] Diesel vehicles are to be prohibited in all of Paris by 2020.

Other cities suffer even more. When I was in Bangalore in 2014 and, ingeniously, wanted to explore my surroundings and the hotel complex while jogging, I was soon exhausted and had problems breathing. The extremely dirty air took its toll. Whereas more than 50 percent of all vehicles in India were equipped with diesel engines in 2012, the government meanwhile has lifted the tax breaks that served as incentives for buying those vehicles. Since then, the share of diesel engine vehicles has been cut in half.[89]

Oslo, capital of Norway, prohibited entry into the city for private diesel vehicles in January 2017 for the first time in its history—with a share of diesel engines that is at 45 percent. The reason given: danger of smog.[90] In a similar vein, Athens, Madrid, and Mexico City also want to ban diesel engine vehicles from their streets by 2025. With their latest findings regarding exhaust and fine dust pollution of gasoline-powered engines—their actual values are no better than those of diesel engines—a general ban of combustion engines seems to be even more likely.[91] Perhaps it is also time to think about even more drastic (although very costly) measures and general removal of manually controlled combustion engine vehicles (i.e., buying them back from their owners) in order to accelerate the general introduction of autonomous EVs and keep the transition period to self-driving electric and manually controlled automobiles as short as possible.[92]

Some experts warn against losing sight of other vehicles because of the concentration on driverless passenger cars. We also need to find solutions for garbage trucks and delivery vans that also require rather a lot of (parking) space in a city and today frequently double-park and block all traffic.[93]

Autonomous Trips to the Countryside

We often mention cities as the first field of application of autonomous cars, but driverless cars actually offer a lot for people living in rural areas as well. Between a quarter and a third of the European population live more or less remotely in the countryside. Upkeep of public local transportation in this case is costly and work intensive. Often a bus line services a stop only every couple of hours. At the same time, the issue with the last mile remains unsolved with public transportation, too. In some cases, passengers have to face a walk of a few miles after getting off the bus to reach their final destination.

This situation could improve with autonomous vehicles. The advantages speak for themselves. First, autonomous vehicles are smaller and therefore require fewer resources than the usually half-empty or empty intercity buses. Second, one can "save" on the driver. Third, the cars do not have to keep to a fixed timetable but instead come when ordered. And they stop right in front of one's door. This grants mobility to everyone, even those who do not own a car.

Even agricultural machines such as tractors and harvesting machines could be running day and night without any human control, and the farmers would become managers and remote controllers of their machines. There is a video of a farmer in Canada who uses an autonomous tractor with a trailer that is filled by a harvester that drives alongside it.[94] It is only a matter of time before the harvester drives autonomously as well.

The Braess Paradox and Other Tales of the Road

Intersections are magnets for accidents. In the United States, 50 percent of all accidents occur at intersections, compared with just 16 percent at traffic circles. An interesting fact in this context is that there is a total of 56 points of conflict at intersections—variations conducive to crashes. Thirty-two of these conflict points are of a vehicle-vehicle type; twenty-four are vehicle-pedestrian conflict points.[95] Traffic circles reduce accident numbers by 37 percent according to the Washington State Department of Transportation; injuries can be reduced by 75 percent and fatal accidents by as much as

90 percent.[96] Traffic circles force vehicles to slow down, there is only one direction, and there are no red traffic lights someone might try to cross at the last minute.

Contrary to general expectations, traffic situations that feel dangerous are safer than those that appear harmless. We are more careful approaching the former, whereas we just stumble into the latter. Accordingly, there is a rule of thumb for road planners: build highways in such a way that there is a little bend to negotiate every minute or so—this will reduce driver carelessness. In the United States, rumble strips are a standard installation, both in the middle of the road and alongside it. They help to avoid 70 percent of the accidents caused by fatigue and inattention,[97] and either they are milled into the tarmac or the asphalt there is somewhat elevated. If the tires come into contact with these strips, small but very loud vibrations penetrate into the interior of the car. They are literally meant to shake the drivers awake and remind them that it is time to take a break.

One argument presented in favor of road construction is that it ensures jobs. However, we should take good care to differentiate what exactly is meant in each case. Highways and cross-country roads connecting urban conglomerations and factories create jobs indirectly because they allow an exchange of goods and services. Road construction inside cities makes the city centers less pedestrian friendly and reduces the quality of life for the people living there—in this case, the negative aspects prevail. The building of highways with large machines and small construction teams is no match for work on thoroughfares, sidewalks, or bike lanes, where about 60 to 100 percent more workers are required.[98]

Despite all their statements in favor of public transportation, politicians still allocate four times as much money to road construction than to any other means of transportation. In 2011, the figure reached $40 billion, which was supplemented by an additional $65 billion to $113 billion in open and hidden subsidies, according to the California Environmental Protection Agency.[99]

Martin Wachs, retired professor of infrastructure at the University of California, dryly noted, "A good 90 percent of our roads are not congested 90 percent of the time."[100] Paradoxically, more roads do not reduce traffic—they create it or else generate even more traffic jams and thus increase travel times. Dietrich Braess, a German mathematician, proved this phenomenon as early as 1968. It is now known as the *Braess paradox* and shows that an additional road increases the travel time for all traffic participants if the traffic as such remains the same.[101] The underlying reason is partly that we—if we have a choice—put our own well-being above the general good. If we optimize the first, the situation for everyone deteriorates. In scientific terms, we often speak of *latent demand*. If you build more streets, more people will use them.[102] You are more likely to visit people, more parcels are delivered, and more babies are put to sleep by tucking them into the car and driving them around until they fall asleep.[103]

You can see the Braess paradox at work in several examples. Interstate 405 in Los Angeles once had to be fully blocked for several days because of urgent repair work, and beforehand, there was talk of "carmaggedon." Yet quite the opposite happened; there were fewer traffic jams and a drastic fall in environmental pollution in the surrounding communities.[104] In the same way, traffic jams in neighboring streets were reduced when 42nd Street in New York City was blocked. New roads make traffic conditions worse, as one instance showed in Stuttgart in 1969.[105]

If you would like to test these facts for yourself but have no car, there is an alternative method: use an escalator. A London-based consulting firm analyzed how standing or moving on an escalator impacted throughput of persons.[106] If 40 percent of the people were moving on the escalator, the transport times for those who were standing was 138 seconds on average, compared with 46 seconds for those who were walking. When everyone was just standing, the average time for everyone was 59 seconds, equating to an improvement of 79 seconds for those who were standing and a tiny disadvantage of 13 seconds for those who were originally walking. The group as a whole would benefit if everyone just stood on the escalator, but that is just what people will not do.

The number of traffic options, and especially how we are using them, is a question of cost. The value of municipal roads in Germany is estimated at $227 billion.[107] In total, the network of roads in Germany is about 400,000 miles long, with 8,000 miles making up the Autobahn and 143,000 miles of interregional roads.[108] In 2013, the traffic areas in Germany amounted to 5,600 square miles.[109] The annual expense per square mile would be $1.46 to maintain the surface, but the actual expense is just 84 cents.[110] This means that road maintenance should be about $21.4 billion annually, but only $12.4 billion is actually spent. However that may be, it is definitely a huge amount that could be better used elsewhere, especially if you also consider that 90 percent of our roads are not being used 90 percent of the time.

Roads are also sources of income, actually generating three times as much money as they cost. In 2010, revenue in Germany was slightly over $56 billion compared with expenses of just under $19 billion.[111]

Looking for the Lost Parking Time Zone

Every year, the 43 million vehicles on German roads cover 380 billion miles. This can be broken down to 8,700 miles for every single vehicle every year. Assuming an average speed of 37 miles per hour, the conclusion is that every vehicle is used for a total of 38 minutes every day, which, in turn, means that for more than 23 hours per day, it is just standing around. Automobile? Perhaps "autoparker" would be more suitable. The situation in the United States is not much better. The 260 million registered vehicles are driven a total of 3.185 trillion miles per year, corresponding to an average mileage of 12,240 miles per year per vehicle, or 33.5 miles per day.[112] The average speed is about the same as in Germany, and every car spends 54 minutes per day in motion. American cars are just standing around uselessly for 23 hours and 6 minutes every day.

And parking is not cheap either. The cheapest type of asphalt-covered municipal parking lot in the United States costs about $4,000, and the most expensive one costs $60,000. This last one is in a shopping center in Seattle. The usual cost of a parking space in a surface lot is about $20,000 to $30,000; underground parking is at about $40,000. In Germany, real estate managers also estimate the construction cost of one underground garage parking space at between $34,000 and $45,000.[113] If you then consider that there are at least four parking spaces for every vehicle (some sources even counted up to eight parking spaces for every vehicle), the value of the parking space exceeds that of all the cars in the country.[114]

Because every parking lot additionally requires access roads and possibly ramps, the space used is more than just the space for one car. Parking space may increase the cost of an apartment by up to one-fifth. In Seattle, a survey showed that the cost for living space is at least 15 percent higher if parking is obligatory, despite the fact that during the night, that is, the time period where most parking is required, about 37 percent of that space remains empty.[115] What is more, we cannot do without garages and parking lots. Today, building requirements include detailed information on the number of parking spaces to be provided with new apartments or commercial premises. This can become very expensive and make houses or stores unaffordable for many people. In some cities, up to 15 percent of the built-up area is used for parking. The irony of it all is that the authorities are stricter in punishing violations of the parking regulations than in punishing violations of regulations for subsidized living for low-wage earners.

However, the truly perverse fact is yet to come. First of all, every tax-payer has to pay for parking, regardless of whether he or she has a vehicle or not. You may be using public transportation only, walk, or take your bicycle, but you will still have to pay for car-centered infrastructure. In short, all of us are supporting those who drive cars. Studies suggest that goods cost about 1 percent more because of the parking regulations. From low-income workers, money is taken and passed on to groups of the population who are better off. In addition, those parking spaces are "rented" (through parking fees) at prices that are far below their actual value. This holds true for both garage space and roadside parking.[116] This makes driving cheaper for drivers, which makes them drive more, which makes cities less pleasurable to walk or ride a bike in. American "parking guru" Donald Shoup made the following comparison:

> If cities required restaurants to offer a free dessert with each dinner, the price of every dinner would soon increase to include the cost of a dessert. To ensure that restaurants didn't skimp on the size of the required desserts, cities would have to set precise "minimum calorie requirements." Some diners would pay for desserts they didn't eat, and others would eat sugary desserts they wouldn't have ordered had they paid for them separately. The consequences would undoubtedly include an epidemic of obesity, diabetes, and heart disease. A few food-conscious cities like New York and San Francisco might prohibit free desserts, but most cities would continue to require them. Many people would get angry at even the thought of paying for the desserts they had eaten free for so long.

Researchers believe that 8 to 74 percent of the vehicles in moving traffic are searching for a place to park.[117] The average time lost for this is between 3 and 13 minutes; in the city center of Los Angeles, an incredible 96 percent of all moving cars between 1 and 2 p.m. are just looking for a place to park. If you consider that each parking space is used by an average of 10 vehicles every day, the time for searching in a best-case scenario is 30 minutes every day per parking space. At the average speed of 9 miles per hour, driving around easily adds up to 4.5 miles per parking space, naturally with all the respective side effects, such as gasoline consumption and exhaust emissions.[118] Per year and per parking space, this easily results in a distance that is double the distance between San Francisco and Los Angeles.

It is small wonder that startup companies such as ParkWhiz are trying to solve this problem. This app, provided by ParkWhiz, shows the user where there are free spaces in public parking facilities and guides the driver to those spaces.[119] In San Francisco, there is an initiative attempting to control the supply of parking spaces on the street with flexible price lists. The SFPark initiative uses so-called smart parking meters that adjust the price on the basis of the day, time, and place. The idea is to keep about 15 percent of the available parking spaces free in order to reduce the time for searching and thus to simultaneously reduce gasoline consumption.[120]

In addition to the search itself, drivers looking for parking behave differently from others. They drive more slowly and reduce the speed for all the other road users as well. One estimate indicates that almost one-fifth of all collisions are caused by people performing parking maneuvers.[121] On top of that, there is also a gender-specific difference. Women are willing to search parking near their destinations for a longer time, whereas men are ready to park some distance away. They both are bad judges of distance: women overestimate the walk from the parking lot to their target destination, and men underestimate it.[122]

One survey analyzing the commuting data in Lisbon, Portugal, found that the same degree of mobility could be reached with only one-tenth the number of vehicles actually used. A fleet of 26,000 taxibots could cover the driving mileage of 203,000 vehicles currently moving through Lisbon. This would leave an area equal to 210 soccer fields free to be used for other purposes.[123] A similar study conducted for Singapore found that one could handle the entire traffic in the city-state with one-third of the number of vehicles.[124] And for New York City, the MIT Institute for Artificial Intelligence used a simulation to calculate that 3,000 self-driving taxis could replace the current taxi fleet of 13,000 cars in 98 percent of all trips.[125] Ann Arbor, Michigan, by contrast, where 120,000 cars are moving, could do it with 18,000 shared autonomous vehicles.[126] And a study in Munich, where currently 700,000 cars are registered, concluded that 200,000 private vehicles could be replaced by 18,000 robotaxis without any loss of mobility.[127]

For the United States, where an average household has 2.1 cars, a decrease by 43 percent to 1.2 cars per household has been forecast, with the number of miles per car increasing from today's 11,200 to almost 20,500 miles per year.[128] The related cost for the average American company would amount to about $18,000 per year according to the American Automobile Association. This means that 1 mile traveled costs about 60 cents, four times the amount of a mile traveled in an autonomous vehicle.[129] Overall savings in mobility could be more than $3 trillion for American households, 19 percent of the American gross national product.[130]

Worldwide, the parking industry today generates about $100 billion every year, with the revenue from parking garages amounting to two-thirds of the revenue in the United States. The rest comes from parking fees on city streets.[131]

The question is what we will do with all those unused parking lots and facilities? According to one survey, this free space could amount to as much as 25 percent in San Francisco.[132] Demolishing them all is a costly solution. Perhaps living in a parking garage will become the next hot thing? You know, like those cool lofts today, apartments created in old warehouses and factory halls, empty parking garages could become an "in" place to live.

The future is promising, at least in terms of parking spaces. It tells us that cars will occupy much less space than they do today and that cities and their inhabitants will finally get some air to breathe again.

Lost in the Jungle of Signposts, or Why Traffic Lights Are Stealing Our Life Away

The place: a suburb in Vienna, mid-1970s. We are leaving the school building. After a stressful day at the elementary school, my friends and I are on our way home. My school friend Gerald and I are clowning around as we all walk to the tram station, trying to outdo the others by walking faster. We are watching each other closely while making wide movements with our arms. Suddenly, there is a loud clanging noise. Gerald had whacked fully into a signpost on the sidewalk. Fortunately, a small dent in the signpost was the only damage, and Gerald was fine.

Something like that was inevitable. On Germany's streets, we find a maze of 20 million traffic signs and 4 million other roadside signs. In Austria and Switzerland, the total is 2 million each. The law distinguishes between 500 different traffic signs.[133] The results of a study in Maryland show that on an average stretch of road, a driver has to take in a new bit of information every 2 feet. This makes 1,320 messages absorbed at a speed of 30 miles per hour, corresponding to 440 words per minute, or about three paragraphs of text. Consequently, in addition to the other tasks drivers have to juggle, there is also the rapid processing of said information.[134]

Despite courageous attempts to diminish this bureaucratic nightmare of signs, it keeps growing. And it comes at a cost! Each sign costs about $100 to

$200, plus transportation, installation, operation, and possibly lighting.[135] In the Swiss canton of Aargau, one in eight signs was removed as an economy measure a few years ago. The amount saved was 1,000 Swiss francs per sign (~$1,000). There were pilot projects in other cities, too, in order to get rid of the forest of signs altogether, for example, in the Dutch town of Drachten, where late traffic researcher Hans Monderman had all street signs and traffic lights removed and defined the traffic areas as *shared spaces*.[136] The result was a dramatic decrease in the number of accidents. The concept of shared spaces was so encouraging that it has been adopted by a number of other towns in various countries.[137] We may expect to see more shared spaces with the increase in self-driving vehicles in our towns.[138]

As a matter of fact, the idea behind the installation of the first signs was actually a good one. Signs were meant to facilitate orientation and to increase traffic safety. The opposite effect occurs today owing to the sheer number of signs. Many of them are confusing or contradictory, which interferes with the drivers' concentration on the actual traffic.

In addition to the signs, we have 1.5 million traffic lights in Germany.[139] The cheapest ones cost $40,000; more complex ones, such as traffic lights for bicycles or a streetcar, may cost up to $300,000.[140] Operating cost per year is up to $5,700, plus the cost for electricity of $900. These high sums are one reason why a growing number of towns and municipalities opt for traffic circles. And there are more expenses for road users. In England, drivers spend up to one-fifth of their total driving time waiting at red traffic lights.[141] This corresponds nicely with the period of time researchers calculated: assuming a daily travel time of 38 minutes over 50 years, each of us spends almost 2 weeks of our lives waiting at traffic lights.[142]

Because self-driving vehicles and connected cars won't need those pretty expensive street installations anymore (because they receive their information electronically or digitally), traffic signs and traffic lights will disappear from the physical world and reappear in the digital/virtual world in a more modern and less expensive shape. So let's look forward to a clear view!

From the Assembly Belt to Vertical Integration to Artificial Intelligence Design: Production Through the Ages

Order or disorder depends on organization.

—SUN TZU

Automobile production has been optimized to the extreme since the introduction of the assembly line by Henry Ford in 1908. *Vertical integration, just-in-time,* and *just-in-sequence* are just some of the buzz words. Each bolt, each lamp, each door is delivered by the suppliers at the exact time and in the exact order required. We do not see 100 black BMW 3 Series in a long row on the assembly line being put together but instead the most diverse types such as station wagons and convertibles in sequence, in any color, and with any number of equipment variations. Work no longer follows a strictly linear, unchanging line but actually appears chaotic at first glance. Calling this an assembly line would do no justice at all to this highly complex machinery. In more than 100 years, the automobile manufacturers have managed to build infinitely precise machines that, in turn, build other machines. There are robots at almost every point on the assembly line, welding, inserting windows, installing seat assemblies, and transporting chassis and parts from station to station without human intervention.

Audi toys with the idea of abandoning the assembly line altogether and using only robots to obtain even greater flexibility and save money. In this way, one could easily continue to work on cars with specialized equipment requiring longer process steps while vehicles requiring less production time could "overtake" them.[143]

A machine to build machines—this is Tesla boss Elon Musk's dream. The complexity of an automaton building products is much higher than that of the product itself, even in the case of such a versatile product as an automobile. Musk suggests that there is much more potential in the optimization of a factory than in the product design itself. He estimates this potential to be in the ballpark of a factor 10.[144]

Artificial intelligence systems already write some code themselves.[145] On the whole, this process develops certain similarities to a process we would call *reproduction* in biological terminology. Genetic code is sequenced, and a new life begins.

AI in the production process does not start with production: it starts with the automotive design itself. In the Autodesk Gallery in Market Street, San Francisco, you can find an airplane design that was commissioned by Airbus and resembles a biological organism—a body crisscrossed with tendons, bone structure under a microscope, and leaf veins—rather than a regular sequence of structural elements we expect to see in human-made objects. Designs that are calculated and produced by an AI system, for example, components of hook elements, bicycle parts, and car chassis you may have seen before, appear as if created accidentally. They are both surprisingly attractive and "alive" but can also be produced more easily and at lower cost. Achieving the same degree of rigidity, the designs require only a quarter of the material used for previous designs, allowing a weight reduction of up to 75 percent.[146]

High-Performance Battery Cells to the (Electric) Forefront

Whenever people talk about battery storage and alternative power generation, we do not often hear the names of the energy providers that dominated the market in the past. We do not speak about E.ON, Vattenfall, RWE, or EnBW or Pacific Gas and Electric or Florida Power and Light Company in the context of "green" electricity. There are new companies in the energy sector, too. Automobile manufacturers such as Tesla with its Gigafactory 1 and internet companies such as Entelios drive the development. In the beginning, people were happy to leave this area to the newcomers, even though their visions appeared as too unrealistic and expensive. Now that a change seems imminent, the energy giants are almost too late for the party. Traditional utilities offer just about 12 percent of the green electricity produced in Germany.[147]

Automobile manufacturers and internet companies present new approaches and models and have an edge in knowledge. They are able to make better use of the enormous amount of data generated by the production and consumption of electricity than the builders and operators of large power plants. Young enterprises such as Opower and Gridcure insert themselves between power companies and customers, offering services based on the data produced—from better utilization and planning of the grid, to proposals of where to install new charging stations, to the use of the batteries from EVs as interim storage for production peaks.

Ford is another company besides Tesla that believes in the necessity of keeping battery cell production in the company. As a matter of fact, Ford regards the development of batteries and the understanding of the chemical processes as core competencies that should not be ceded to other companies.[148] This is not an easy task at all. Nissan experimented with its own batteries as early as 1992 but then opted for a joint venture with an external manufacturer because its cells were cheaper. Unfortunately, cheap cells are not necessarily good cells. Consequently, Nissan has selected a different producer of battery cells for its latest Leaf model. Toyota was not successful with the production of its own lithium-ion accumulators either. Even Tesla was only able to get battery development to a level unparalleled by any other manufacturer because of its cooperation with Panasonic.

Smart Traffic Management for "Smart" Cities

Traffic forecasts are a little like the story of the chicken and the egg. If the forecast says that some highways will be completely free tomorrow, this might tempt more people to travel than originally planned. After all, they said the roads would be free, right? And then it's all totally different. A traffic forecast that is worth listening to therefore has to consider the human reaction as well.

Could it be that we will have to pay more in the future to be allowed to use faster road segments? Will we get a multiclass society with autonomous vehicles? There were attempts in Europe and the United States to make users pay extra for the preferred (i.e., faster) processing of their own internet traffic. The plans were shelved for the time being because such preferences would limit the freedom of opinion and information. In Moscow, for example, we already have special lanes for "officials" on the wide boulevards. This is often a middle lane that is reserved for government officials, emergency vehicles, and cars with flashing blue lights attached. Naturally, a shadow economy quickly emerged, driven by wealthy payers who wanted to outsmart the notoriously bad traffic situation in the Russian metropolis.

Smart traffic management not only is intended to keep the traffic flowing but also is helpful in avoiding accidents. The number of traffic fatalities in urban areas grew steadily in the United States from 1977 to 2015.[149] Los Angeles and other cities therefore announced their Vision Zero goal, an objective to be reached with autonomous vehicles and data from thousands of sensors. A car that almost runs over a pedestrian would inform a central database of this incident to help mitigate critical areas.[150] The idea of such a "smart" city is not completely farfetched or unreasonable. Rio de Janeiro, Santander, Singapore, and other cities already use a multitude of sensors.

Google Maps and Waze also generate data to keep traffic flowing, to calculate travel times, to notify of congestions, and to suggest alternative routes.[151]

Goodbye, Automobile Club?
From Member Club to Fleet Club

A delegation from a European automobile club arrived for a one-week visit in Silicon Valley. Once the participants had met representatives of two dozen automobile companies, suppliers, venture capitalists, and experts, one member of the managing board admitted that they had known about a lot of those "puzzle pieces" before but had been unable to put them together as a complete picture: self-driving cars, EVs, the internet of things, and over-the-air software updates. During the tour, the delegation had become aware of how the individual parts come together—in that moment they really saw what it all meant. And they realized that everything their club stood for was suddenly open to question.

Your typical automobile club is an organization offering services to its members such as roadside assistance and support after an accident, group insurance, return transport for injured members from abroad, and legal, contractual, and technical consultation. Like everyone else, it has to keep up with the times—something is added, something else is abolished. The National German Automobile Club (ADAC) discovered the camping program as a surprisingly important growth market. The camping guide and accompanying service offers sell like hotcakes. Having said that, the free regional road maps provided by the American Automobile Association used to be very popular, too. They became obsolete with the rise of digital maps from Google, Apple, and others and the introduction of navigation devices.

But basically this is not the decisive factor if you remember that fewer cars and more self-driving EVs in a sharing model will automatically result in fewer holders of driver's licenses and owners of cars. Automobile clubs will likely experience a massive exodus of private members, whereas companies and fleet providers are going to increase their activities. This has an impact on the price models and the scope and types of services offered because a larger volume of vehicles provides a better basis for negotiation. The self-image of automobile clubs as member clubs has to change completely when corporate clients replace individuals.

FashionTech: Cars Make the Clothes

Antenna-like spider's legs spike out from the shoulders of a model wearing a corset full of electronics. The antennas are not intended as a fashion accessory; they move. As soon as someone comes close, they rise in the air. The faster and more aggressively someone approaches the model, the more threatening the antennas become. This, in short, is called *FashionTech*. Dutch fashion designer Anouk Wipprecht experiments with electronic components in her designs. The corset just described allows outsiders to see what the threatening approach of the stranger means to the woman.[152]

Audi asked Wipprecht to create four individually designed dresses that were similar to an Audi A4. The young designer was inspired by the shape of the headlights and the sensors included. Ultrasound sensors in a dress determined approaching objects and reacted with a sound that turned shrill if the object moved faster and closer to the wearer of the dress. Light clothes, by contrast, reacted to surrounding colors.

Today we carry a key or a smartphone with us to connect with our cars. Our great-grandfathers wore a leather hood and an eye mask to protect themselves against the weather. In the future, we might come to use biometric sensors or clothes or even rely on bionic humans with electronic implants in their bodies. The Knight Rider watch on David Hasselhoff's wrist has long since disappeared into history. A car that recognizes your face and opens the door will more likely be all that you will need.

Before the Law All (Cars) Are Equal?

The legal framework must be prepared before we can use autonomous vehicles on a large scale. More than 70 countries ratified the Vienna Road Traffic Convention in 1968. It standardized traffic regulations and created a general acceptance of driver's licenses issued in other countries. Most traffic signs we see in European countries today still follow that standard and therefore mean the same everywhere. It is this convention that we should be grateful for when there are no difficulties for us when we have to present our German license to traffic police officers in Italy, Vietnam, or Saudi Arabia.

In 2016, this agreement was extended to include autonomous vehicles.[153] The use of driver assistance systems was deemed acceptable as early as 2014, as long as the driver was able to switch off or override the systems at any time.

In addition to the Vienna Road Traffic Convention, there are other regulations regarding the equipment of vehicles, issued by the United Nations Economic Commission for Europe (UNECE). Some of these regulations need to be adapted for autonomous vehicles to be acceptable in regular operating mode (not only in test mode)—but this has not happened yet. On a practical level, ECE R 13 concerns braking systems, ECE R 79 concerns steering systems, and ECE R 48 deals with lighting and light signal equipment.[154] ECE R 79, for example, states that the driver must be able to override the steering system at any time by making a steering motion (i.e., the driver must be able to maintain control of the vehicle). Automatic steering so far has only been accepted up to a speed of 6 miles (10 km) per hour.

The United States is a member of this convention, but the regulations are not automatically valid in the United States. In Europe, by contrast, the need for action is urgent. Regulations and legislation must be adjusted to include autonomous vehicles unless Europeans want to give up that market altogether and rely on America and China. I am talking about now while this technology is still being developed. Without goodwill from the authorities who give permission to manufacturers to test such vehicles in public spaces, Europe's domestic industry might lag even more behind the leading countries in the future.

The U.S. DOT presented a rework of the current guidelines in late summer 2016. Until that time, the regulations demanded that a driver be present to control a vehicle. Officials realized that the technologies are still being developed and cannot be conclusively assessed and evaluated yet. The 15 points listed in the guidelines refer to safety and validation measures, data protection, cybersecurity, ethical questions, and the protection of passengers in case of an accident.[155] And as mentioned earlier, starting in April 2018, autonomous vehicles may be on the roads of California without any driver on board. Three U.S. senators therefore drafted six principles for proposed legislation to govern autonomous vehicles, trying to strike a balance between safety and the quick introduction of this new technology:

- Safety first!
- Progressive innovation is to be supported, and existing obstacles must be removed.
- Remain technically neutral.
- Reinforce the different roles of the federal government and states.
- Strengthen cybersecurity.

- Inform the public in order to support responsible acceptance of self-driving vehicles.

In addition to a description of how autonomous systems may be used, a lot of careful thought will have to be put into the question of guilt and liability in the case of accidents. It is possible that autonomous vehicles will be assigned separate lanes and road segments, like the carpool lanes in the United States, where only vehicles with more than one person inside or having a special sticker may travel. And what about the sensors? Is the use of cameras and LiDAR systems on parked autonomous vehicles for surveillance or passing on information about a pothole compliant with data-protection legislation?

Why the Rail Company No Longer Keeps Passengers Mobile

The Deutsche Bahn (DB), the German national railroad corporation, also has to address some questions regarding its inherent objectives and the epic importance of the company. The company says that it is offering a solution for transportation or mobility. But the company name itself may be a block to grasping what the real issue is.

Jonathan Ive, British-born chief design officer with Apple, was invited by the British children's television show *Blue Peter* to assess design proposals that schoolchildren had made for their lunchboxes. His first comment was that one had to be very careful when wording this task. The simple use of the word *box* might already affect and limit the flow of ideas for solving the task. When you hear the word *box*, you usually think of something rectangular, something that must have a certain shape. Luckily, the children had not been bothered by that and let their imaginations run free.[156]

The word *railway* may tempt us to stay in the mental box of that word. We see images of railway tracks, locomotives, and stations before our inner eye. Someone coming from a railroad worker family like me may also vividly remember the overhead lines and that metal-tar railroad smell I was used to as a child. The object itself—the train—is always at the center, and the needs and goals of the customers come second. Despite the fact that European railroad companies are very much concerned with providing increasingly sophisticated services to their customers, the entire view of the matter is too mechanistic.

People choose means of transport to connect with other people, goods, or places. Several options are available. On the way from Munich to Stuttgart or

San Francisco to Los Angeles, I could take my car, a bus, a train, or a plane, and if I had enough time and was fit enough, I might decide to walk, ride my bike, or saddle a horse. In addition to that, an (economically) sensible combination may be an option. A typical journey would comprise, for instance, a taxi ride to the station, a train connection to the destination city, and a short walk to my final destination.

If I happen to be a top manager with a railroad company, long-distance trains will always be a central part of mobility for me. Even if buses and collaboration with taxi service providers are part of the standard service, they are less important and mainly intended to solve the problem of the so-called last mile. But what happens if I, the customer, do not really want to ride on a train? Changing trains when you have luggage is difficult and exhausting, there are often delays, my connecting trains will be gone, and the seating position indicators are broken. The railroad (as well as airlines and taxis) allows for too many fracture points in the transportation experience. An autonomous vehicle I order via an app is incomparably more comfortable, and it can take me all the way from Munich to Stuttgart or San Francisco to Los Angeles instead of just to the train station. And that would then be the real disruption for the railroad, the real competition.

This is connected with the very valid question of whether in the future it will still make sense to divide transportation companies according to their means of transport. Why should we separate airlines according to planes, railroad corporations according to trains, and taxis according to vehicles when all the customer wants is a comprehensive transportation solution with seamless service? Apple provides an example of this in its own field. Hardware, software, and content such as music and movies are all offered by the same provider. This is why a couple of years ago the company changed its name from Apple Computers to Apple in order to represent the entire product line. Tesla builds cars, sells them via its own dealers, and offers an extensive network of charging stations and mechanics, power storage, and solar roofs. Naturally, this is offered at a higher price, but you do have the benefit of a seamless service. As a consequence, other industries find themselves under pressure, too. The level of integration of the services Apple and Tesla offer has come to be expected in other industries as well.

As a first step, DB, SBB, and ÖBB should delete the "B" for *Bahn* (German for "railroad") from their names and actually remove it completely from their mental box, as should the National Railroad Passenger Corporation known as Amtrak delete the word *railroad*. Why should I have to suffer and drag my belongings through the crowds and the bustle at a station, upstairs, downstairs, on escalators, to finally get to the taxis or public transportation? Why isn't my Uber or self-driving car directly where I get off the train?

The situation is even more alarming for the railroads if we take a look at self-driving trucks at present. The railroads do have the advantage of being able to transport heavy goods at a relatively low price across large distances. Again, however, there is the problem of that last mile when you have to reload from rail to truck. Self-driving trucks will not employ drivers, will not have to take breaks, will be able travel energy efficiently in slipstreams, and will not have to reload the goods from one means of transport to another. The economic advantage of the railways would be lost. Volvo, Ot.to, Scania, and Peloton have already put road trains on the streets.[157]

The issue was quite obvious to the former chairman of the board of DB, Rüdiger Grube. He was wondering why there was no autonomous train yet. If the train loses its advantages—speed and comfort—then who would come to the stations anymore? Who would buy in the stores there? So how can the railroads retain their importance?[158]

A Network Without End: Public Transportation Under Scrutiny

In Europe, public transportation in cities usually has a good network, although people there have a different impression and there is always a reason to complain. The Swiss especially are a model of punctuality and predictability. People will probably always find a way to disagree about the interval between the connections and the acceptable distance to walk to the next means of transportation. But anyone who has ever used a service provider such as Uber won't ever want to go back. In San Francisco, for example, where public transportation is old and unreliable, Uber promises waiting times of just one to three minutes in most quarters of the city—and you spend them right where you were when you ordered your Uber! Uber is also a hallmark of safety. Its digital data trail is recognized as protection, especially by women.

Electric robotaxis force us to reevaluate the purpose and performance of public transportation. We will also have to reconsider the lanes and tracks reserved for public transportation as well as maintenance and development of the infrastructure provided. In contrast to buses and rail-bound transportation, autonomous travel service providers do not have to keep to set routes, do not have to follow any schedule, and optimize waiting times. Members of the city government of Charlotte, North Carolina, a city with 800,000 inhabitants, are currently debating whether they should really spend $6 billion on the planned extension of light rail and high-speed rail tracks. Both would not be ready before 2025, and by then, they might be outdated and need to be replaced by self-driving vehicles.[159]

Despite the well-developed public transportation systems in Europe and Asia, the number of cars in the cities is huge. Vienna struggles with the impact of 700,000 vehicles that cover an area of about 2,200 acres (900 hectares)—which is half the city's leisure park island in the Danube.[160] New York City, by contrast, provides 102,000 public parking spaces south of 60th Street. The area covered corresponds to half of Central Park.[161]

Just in Time for the Meeting: The Autonomous Sleeping Car Makes It Real

The hotel industry today survives because (business) travelers have to attend meetings, which makes it necessary to stay in a hotel for the night. However, instead of arriving the evening before to make a visit or attend a meeting in the morning, one could sleep in an autonomous vehicle without stress while traveling to one's destination and then just take a shower at the hotel and have breakfast. The distances covered could be greater, which, in turn, would create more competition for the railways and the airlines.

Much Ado About Oil, but Electricity Changes the World

The most obvious loser in the transition to electric robotaxis and robotrucks will be the petroleum industry. After more than 100 years, the mass acceptance of EVs results in a decreasing demand for diesel fuel and gasoline. Oil and gas will still be used in power plants for a while, to be transformed into electricity, but Denmark, Norway, and Germany have already demonstrated that on some days they can meet their countries' entire energy demand with alternative energies. China, with its nationwide efforts to replace diesel buses with electric buses for public transportation, is expected to save 270,000 barrels of diesel fuel a day by the end of 2019. For every 1,000 electric buses on the road, 500 barrels of diesel fuel are displaced each day, as Bloomberg estimates.[162]

Platinum resources also would be saved. Almost half of all platinum deposits today are used for automobile catalytic converters because this precious metal reduces poisonous pollutants.[163] However, at the same time, the demand for rare earth elements for the production of batteries increases. And now it gets political very quickly. The largest deposits of such resources are found today in China. Because of its huge domestic demand, the country reserves the bigger parts of its resources for itself, which makes them more expensive for the rest of the world. However, China's attempt to reduce the export of rare earth elements to cover its own demand has led to a contrary effect. First, Chinese companies circumvented the export ban, and second, the artificial limitation of the offer led to such incredible price increases that the exploitation of their own deposits became an economic option for other countries.[164] Chinese companies, however, are also very active in other countries. China has initiated great enterprises in Africa especially, thus securing China's political influence.

How much cost savings electric cars can bring was demonstrated by New York City. The city fleet of EVs has been shown to have a lower total cost of ownership over a period of 9 years by 21 percent. And this includes the higher purchasing price of EVs compared with hybrids or combustion engine cars and with the requirement for installing charging stations. But the overall maintenance costs and fuel costs are significantly lower.[165]

We already took a look at the cost development for electricity earlier in this book, and we found that things are not looking good for power producers. Between 2020 and 2025, solar power will become so cheap that transport from the power plant to the end consumer will be more expensive than the power produced on site with a solar plant. Tesla and other manufacturers are already producing battery packs for households, too, not just for cars. Besides, Tesla's acquisition of SolarCity creates the ultimate horror scenario for any utility. Households receive everything from one provider only and become independent in terms of energy by disconnecting from the public power grid.

With the growing distribution of EVs, there are new challenges waiting for power producers. First, there is the grid load resulting from a large number of vehicles being charged, and second, the opportunity of using the accumulators as a grid reserve in case of power shortages is reduced. So what are the new technologies and energy management systems required for this? Even established energy providers are increasingly "eaten" by software, in the same way digital components create the actual added value in other industries as well.

Consequently, power providers are more than ever forced to question their own companies' purpose (some of them started out as manufacturers of household and electric appliances). Their purpose certainly is not only to supply energy. When some providers changed their strategy and preferred regional wind-, solar-, and water-powered generating plants instead of building and operating large thermal and nuclear power plants, the entire industry was turned upside down. The massive losses German utilities such as E.on and Rheinisch-Westfälisches Elektrizitätswerk (RWE) suffered in recent years are proof of the fact that many found out too late which way the wind was blowing. In the future, the value of an energy provider will be less determined by the power plant performance than by the digital customer services with regard to energy management, which unfortunately is precisely where know-how and qualified staff are not available. The field of gathering and processing data is left to others barging their way in, such as Opower, which was recently acquired by Oracle.

In Europe, wind power already comes second after natural gas on the list of types of energy generation. In 2016, the total capacity was 153.7 gigawatts. Coal and hydropower came in third and fourth. Even solar energy is far ahead of oil, at sixth position. And nuclear energy is still at position five.[166]

We Can Do Without: Effect and Impact in the Wake of Petroleum

We should definitely not underestimate the political effects and changes. Most of the oil-exporting countries were able to finance their various regimes thanks to the very generous revenue from oil production. Price fluctuations time and again demonstrate just how unstable those countries are. One of the biggest oil producers worldwide, Venezuela, is suffering from a famine the country brought on itself. While its oil price was high, Venezuela was able to make up for the weaknesses in its general economy. When prices dropped, this house of cards simply caved in. Russia, whose national budget is also very dependent on the revenue derived from natural resources, tries time and again to cover political inadequacies with foreign policy coups and popular measures.

We can predict with some certainty that we will see even more instability in the oil-producing countries as the transition to electric drive systems proceeds. Even stable oil-producing countries such as Norway are worried about their country's future options. On the one hand, Norway supports the use of alternative energies and the purchase of EVs with generous state subsidies, and on the other hand, many Norwegians work in oil production, oil transportation, and the processing of the "black gold." Setting up alternative industries needs time and requires different qualifications.

Decreasing Energy Prices: The Other Side of the Coin

We Drive More Than Ever Before

As the price of power drops, each mile driven in an EV becomes cheaper. And as soon as self-driving vehicles have become the standard, it will be easy to spend your time productively doing something else. A Swedish study showed that we could then cover more miles. The Swedish government supported "clean technologies" aggressively. Swedes therefore bought more fuel-saving vehicles but drove more miles with them than ever before, which practically ate up any fuel-saving effect.[167]

We Expand Further

Self-driving vehicles and cheaper energy may lead to greater urban sprawl. It does not matter at all to me whether I am on the road for one or two hours because I can use the time differently. Therefore, I may want to build a house further out in the suburbs or beyond, and with this comes all the environmental impacts of such a move.[168] Self-driving electric Ubers allow a lifestyle that may have an adverse effect on the environment.

We Increase Emissions

We looked earlier at the energy expense required for the production of an EV and the batteries used. In order to draw a more balanced picture, we also have to include a lot more in our considerations, independent of the kind of vehicle, namely, the emissions created during the construction and provision of the energy carrier and traffic infrastructure, as well as during the operation and maintenance of the EV. All of this may increase emissions by 50 percent, experts say.[169]

Remote Controlled and Mugged: Cybercrime Meets Cybersecurity

The long-legged blond in her black power suit taps on her tablet, and chaos ensues. She just took control of all autonomous cars in the vicinity and has instructed them to hurl themselves from their multistory car parks onto our heroes. Like a pack of hungry wolves, hundreds of cars pursue members of the "family," overturning, crashing into houses, ruthlessly running down anybody and anything. Charlize Theron's character in the movie *Fast & Furious 8* unleashes a nightmare scenario. Really well imagined, but still, it is just a movie. And yet cars phoning home, cars connected always and any time—such cars will always provide some point of attack for ill-intentioned hackers.

Car manufacturers have to anticipate cases in which other participants in traffic and external people might try to impair or abuse self-driving vehicles. From cheap ultrasound sensors to expensive radar jammers or laser pointers and light-emitting diode (LED) flashlights with fast lenses to blind cameras, virtually everything has already been tried. What if hackers try to confuse the vehicle's sensors or gain access to them? During a hacker conference, researchers presented ways in which sensors could be blinded and incapacitated.[170] Strong lasers make them blind for a few seconds, and a delayed laser signal lets objects appear farther away than they actually are.[171] More elaborate attacks can fool the sensors in that they alter the perception of the environment and, for example, indicate to the car that a track is passable even though it is not. Such attacks on our safety must be punishable.

Safety measures range from access to the car to protection against unauthorized control. The latter can happen from outside, but also from within, when a passenger places a dongle in the OBD2 port normally used by mechanics reading vehicle data for maintenance purposes.[172] *Cryptography* is the key word in this context, but at the same time, the users want everything to work quickly and be convenient. Financial service providers were involved in more than 50 percent of all cases of identity theft reported in 2008.[173] Thanks to enhanced measures and legal regulations, their share decreased to 5.5 percent in 2014. At the same time, other industries became more interesting for hackers. In 2014, the medical and health sector was at the top of the list of identity theft cases, at 42 percent. Attacks on self-driving vehicles may be the next extremely lucrative activity. For this reason, Tesla CEO Elon Musk regards cybersecurity as a top-priority topic for all Tesla vehicles, and even hands out monetary rewards to hackers for identifying security loopholes.

However, first of all, we should be somewhat lenient with manufacturers of any kind. They all focus on their field of expertise, and for automobile manufacturers, that expertise regards innovations in the mobility sector. Identity theft initially is a secondary issue for financial service providers, medical organizations, and car builders. Still, while a hack in the field of financial services simply makes money disappear—which is bad enough—lives are at stake in other industries. None of us would want to be stuck in a hacked car or encounter such a vehicle in traffic and be at its mercy. For criminals, attacking a car may be attractive for several reasons. First, the car itself constitutes a value, plus the data generated with and by the car. Second, if each car should have a separate bank account, as we have seen in the eWallet concept, then every car becomes a gold mine on wheels.

As a matter of fact, the security agencies are critical of such scenarios. I mentioned this issue in a presentation I made at the Center for Homeland Defense and Security in West Virginia, speaking to sheriffs, firefighting commanders, highway patrol officers, antiterror experts, and other people from similar jobs. Here, as elsewhere, we will require specialists who know about the digital forensics of vehicles and defense against hijacked cars used to go on a killing spree or packed with explosives. Extreme scenarios should not, however, deter us from taking advantage of the possibilities the new technology offers—but we do need to make the cars digitally safe. Just as the Technischer Überwachungsverein (TÜV), Germany's technical inspection service, and car magazines today do crash tests with the latest models, we will have to do digital crash tests as well.

The U.S. Department of Justice has organized work groups for all kinds of threatening scenarios in all sectors, even exploring potential threats via medical devices such as a hacked pacemaker or other objects we subsume under the category of the internet of things.[174] Most of these devices were not made with the goal of protecting them against cyberattacks, just as many websites first of all looked to demonstrate their core competencies and were not specialized in defending themselves against digital attacks. Work groups for automobile cybersecurity, regardless of whether they are set up by state agencies or companies, look at the following items in particular:

- How can the car be protected against unauthorized access? People don't want their car to be stolen or for someone to enter their car or control it from outside while it is in use.
- How can we prevent a self-driving car from becoming a bomb on wheels, set to explode at a specific target (similar to the terrorist attacks in Nice or the Berlin Christmas market, where assassins drove stolen trucks randomly into the crowds of people)?
- How can we protect the generated data against abuse? First approaches consider anonymization, encryption, and transparency.

Legal regulations, such as the General Data Protection Regulation (GDPR) in the European Union, are just as important as the technical solutions themselves. It is hardly surprising that the first ideas for solutions came from startup companies such as Karamba Security from Israel, a country that is famous for its competence in cybersecurity on the part of its legendary secret service and military.[175] The company installs its own technology directly into the vehicle and then continuously monitors the vehicle's internal network with all its electronic control units (ECUs) that deal with triggering the airbag, measuring the tire pressure, braking, and fuel injection. The company's technology also recognizes malware attempting to manipulate the settings of those ECUs and returns them to factory settings. Google-Waymo took radical steps because of threats from hackers and keeps its test vehicles offline as much as possible.[176] The big manufacturers founded the Automotive Information Sharing and Analysis Center (Auto-ISAC), made up of companies cooperating against cybercrime. The threat is similar for everyone concerned, so for once the companies are not competing with each other in this area.

As a former software developer, I am only too aware of the fact that programs can have bugs. Debugging, test programs, test systems, and test data therefore can only be as good as the people writing them and the programs producing them. Test conditions furthermore never correspond to the complexity we find in real life. Errors in software are inevitable, even if you are using expensive ways of programming such as *extreme programming*, which always has at least two programmers generating the program code of the same line simultaneously instead of just one. This effort is mainly used for systems with extremely high levels of responsibility, for example, for control software used in nuclear power plants. And even this effort is not sufficient to anticipate all the possible traffic scenarios and translate them into software code. This is the reason why we use machine learning and also the reason why software regularly has to be updated remotely.

However, not everybody wants an "unhackable" car no matter what. Mechanics and private owners wishing to modify their vehicles want to continue to have an option of doing some work themselves. While father and son or daughter used to play around on the car, trying to get some extra horsepower from the engine, tuning will be more digital in the future. Initially, it was not clear whether vehicle owners would actually be allowed to interact with the digital interfaces and modify them. Automotive manufacturers referred to the Digital Millennium Copyright Act (DMCA), which regards software as subject to copyright just like any other text by an author. However, there are exceptions set out in the DMCA precisely for cases such as car tuning.[177]

Blockchain might prove to be an adequate helpful infrastructure to secure cars during use and communication. The underlying asymmetric cryptography in the security infrastructure system known as public key infrastructure (PKI) uses two keys with different functions. One is intended for encryption and the other for decryption. The characteristics of the blockchain, namely, the generation of a publicly visible transaction chain for each step of encryption and decryption, would document and save every step—basically, who uses the vehicle when and how—and would make manipulative operations much more difficult.

Potential aggressive scenarios can be categorized as follows:

- Attacks from a moving vehicle
- Attacks from the side of the road
- Hardware attached to the car itself

Signals from a moving vehicle following the vehicle under attack might maintain a malfunction for some time. Interference from the side of the road might affect a larger number of passing vehicles, and a relay of several interference devices at different points on the route can attack any car over a prolonged period of time. Alternatively, one could try to attach interference devices on a parking car once the owner has left. Every attack is a danger to the vehicle and to the passengers' safety. It is possible that other, unaffected sensors can compensate for the malfunction, but in an emergency, the vehicle has to stop. In this case, direct robberies are possible.[178]

But it may be very harmless. A few seagulls are enough to confuse sensors. For sensors, a flock of birds looks like a larger object. nuTonomy had such an "encounter of the third kind" when testing its autonomous taxi fleet in Boston. One more reason why autonomous vehicles must be tested comprehensively in different cities and landscapes.[179]

Chitty Chitty Bang Bang: The Dream of Flying Cars

In 1910, John Emory Harriman patented the first flying car, his Aerocar.[180] Since that time (and more so since Harry Potter), we are fascinated by this idea. Experts regularly predict that flying cars will soon take over traffic, and disappointment follows with similar regularity. Just look at what we have been promised: Aerocar, Aerobile, Airphibian, ConVairCar, Aircar, Aero-Car, AeroMobile, and even a Chitty Chitty Bang Bang. More than a hundred years have passed, and nothing has happened. Nowadays, the flying car is a synonym for failure. Investor Peter Thiel expressed his frustration with the newspaper hoaxes with the words:

We wanted flying cars; instead we got 140 characters.

Sometimes reality overtakes faster than we think. Even for me, despite the privilege I have had of already experiencing the future here in Silicon Valley, the vision that I wish to share with you in this book continues to be caught on the wrong foot at times. I actually planned to add this little section for fun only. And then we are told that Google founder Larry Page is a shareholder not in one but in two startup companies that want to build a flying car.[181] Zee. Aero and Kitty Hawk are the two companies in which he personally invested more than $100 million. Those two are joined by others such as Terrafugia, Volocopter, AeroMobil, Moller Skycar, Lilium Jet, and Joby. Airbus is also working on an autonomous flying taxi.[182] And Uber is publicly toying with the idea of vertical takeoff and landing (VTOL) and regards flying cars as the future of the transportation system.[183] Uber actually presented its prototype of the Bell Nexus at the CES in Las Vegas in 2019. We have a crowded field here. And will it bring us what has been promised for so long?

Part III

En Marche! Tools and Methodologies for Automotive Manufacturers and Suppliers

I always believed ironing boards are just surfboards that gave up on their dreams and got real jobs.

—UNKNOWN

One participant from a German delegation of premium manufacturer representatives asked what Silicon Valley thought of the German automobile industry. A good question. When I arrived in 2001, I was surprised at the number of German car brands I saw. The selection of brands and models—Mercedes, BMW, Porsche, and Volkswagen—was similar to what you see on German roads. Everybody who could afford it—and many in Silicon Valley can—drove a car "made in Germany." This is not really surprising because German engineering still sets the standard in terms of quality and design. Steve Jobs, for example, drove Mercedes, the "car with a star," for many years.

People admired German manufacturers. Even today, one can still get an idea of this appreciation for workmanship, although now it is mixed with nostalgia, the same emotion you always feel when something great—a brand, a nation, or a person—is past its prime, and from that point on it goes downhill. It's not as if Silicon Valley had not tried to warn German manufacturers. Every single industry expert I have met in recent years pointed out to German delegations the changes that were under way and emphasized the need for swift reactions and modifications in their own corporations. However, seven years after the introduction of the Tesla Model S, four years after the Model X, and one year after the Model 3 launch, German manufacturers (and those from other countries) still had nothing comparable in their portfolios, let alone something ready to launch on the market. Despite many announcements, their approach remained unstructured, and—pardon the pun—they could not seem to get into gear.

Driving a Tesla in Silicon Valley today shows that you "made it" and belong to the "in-group." Even the most hard-core fans of German brands see their loyalty put to the test. Driving a German car is starting to be a symbol of belonging to a dying species—like the dinosaur. This does not sit well in the sensitive global haven of innovation that claims to be a technology leader in all disciplines. The desire to help German carmakers slowly gives way to the realization that you can only help somebody who is ready to accept that help. The latest announcements from German manufacturers that they are planning to make billion-dollar investments in old combustion engine technologies mainly serve to prove one point: they have understood little of the signs of the times and do not want to be told what to do. One employee of a German manufacturer told of his frustration. He had already managed to survive the financial crisis in the American automobile metropolis of Detroit and now asked himself whether the future of Stuttgart and Wolfsburg—home to Daimler, Porsche, and Volkswagen—might be similar. He has not found an answer yet, but he may find one sooner rather than later.

In all of this, German and European car manufacturers and suppliers have the same technologies available as everyone else. They might even be able to leverage an advantage because they were the ones to first develop many of the technologies required for building a car and contributing to its functions. Many of the new enterprises that are racing ahead of traditional manufacturers with their concepts and ideas employ German car designers. They enticed entire teams away, in fact.

How do these newcomers manage to lead the automobile into a new era at such breathtaking speed, leaving the former prominent players speechless and close to panic? For example, in a single week, Mercedes not only announced it is retreating from fuel cell development and bringing the introduction of electric vehicles forward by an astounding three years—let me see, oh yes—to 2022 (this is a full nine years later than Tesla's Model S) but also announced it is collaborating with Bosch for the development of autonomous vehicles.[1] When EMC and Dell merged in late 2015, *Wired* magazine wrote: "Dell. EMC. HP. Cisco. These Tech Giants Are the Walking Dead."[2] This was exactly how I felt, too, when I heard the Mercedes announcement. German manufacturers, which were still making record profits in 2016 and 2017 but had to issue not one but multiple profit warnings for 2018 and 2019, are not immune to such headlines. Looking back on a long and successful history is apt to quickly become a ball and chain, stopping you from moving forward.

But what can you do when you see that a business model that worked well for more than 100 years is drawing to a close, that a third of the people employed will probably be made redundant, and that there are hardly any options for retraining them in new technologies? From piston rod expert to battery chemist? From engine designer to digital expert? Not likely. You cannot rise to the challenges posed by such changes by displaying a behavior that is focused on security and planning. How to ask the right questions, how to take a calculated risk, how to create a culture of innovation within a company—those are the prerequisites for competition with the United States and China.

Is it possible for traditional car manufacturers to move safely into the future by taking a couple of billion and going shopping, to thus obtain the necessary expertise and technology for their companies? Well, if we speak about German companies, they are not really good at that sort of thing. They hesitate. They buy when they are compelled to do so, but by then it is often too late. The Americans and the Chinese proceed more boldly. Germans put much more trust in their own developments. Intel buys Mobileye in one swift move. General Motors acquires Cruise Automation with a snap. Apple, Google, Microsoft, and others briskly buy companies and talents before others have even become aware of them. The deal is done before the competition hears of it. The acquisition of HERE by a consortium of German manufacturers went back and forth for weeks, a spectacle people followed in the news media. Will they buy or won't they? Do they even want to? Do they have a minimum degree of trust in each other? And why do they want to acquire HERE together? You could easily tell that HERE had to somehow justify itself to all members of the consortium. These are not the best conditions for confronting the nimble Silicon Valley league. Even Germany's top-of-the-class digital pioneer SAP often comes around late and has to take the crumbs that others have left behind. All the good and really exciting companies have left by that time.

It is more likely that we will see the opposite happening. Mercedes, BMW, and the VW Group will be bought by others—if they are lucky. Thus $47 billion for BMW, $77 billion for VW, and $56 billion for Daimler are the values that correspond to the market capitalization of these companies on the stock market in summer 2019: the well-filled war chests of Apple, Google, and Microsoft should have no problem with that. However, at the time of the possible purchase, the price will be much lower because the second automobile revolution will be in full swing by then, and nobody will spare the feelings of German manufacturers. By then, tens of thousands of jobs will have already been lost. Such radical changes have the potential of weakening the influence of workers' councils, as we saw in England during the 1980s when the crisis in the coal industry led to a total loss of union power. This is the only option to guarantee the survival of the companies.

So what can those car companies actually do to prepare for what is in store for them and to become more innovative? In addition to the right mindset and new ways of behavior, there are also tools and methodologies that can be learned and applied—but all of this must go together. As I am writing this, I am reminded of the example of the founder of a Hungarian startup company with whom I spent a week in Silicon Valley. Straightaway I sent him to a two-day design thinking workshop at Stanford University. He actually told me that he had participated in such a workshop in Budapest the previous year and that he had not been ready for it at the time. He had just set up a startup business and then had had to close it again shortly before his arrival, so he was looking for his next thing. Now he was open to new ideas and concepts, and suddenly, design thinking became a welcome tool to create something new.

In what follows, I present some concepts and approaches that illustrate the patterns of behavior in which we sometimes are stuck and how to break free. I will also briefly discuss some methods that I use for you to compare your own strong points with those of Silicon Valley.

Company Mission Statements

I will say it again: 2016 and 2017 were record-breaking years for the German automotive industry. All manufacturers were able to make record profits and realize huge growth. Porsche, for example, granted each employee a bonus of $10,256 (€9,111) in 2016 and even $10,7665 (€9,565.3) in 2017.[3] We are fine, so why should we change anything? If you take a closer look at the general sales figures, however, you can see that just about the only way to realize them was to offer very high discounts. The level of discounts was 35 percent higher in 2016 than it was in 2010. Diesel vehicles in particular are hard for dealers to move; in that year, diesel car registrations decreased by 2.8 percent. In 2018, the numbers plummeted. Over 40 percent fewer diesel cars were registered. Because diesel vehicles are mainly popular as company cars—only a quarter of all new registrations are from the private sector—the tendency of companies to hold back is especially tough on the manufacturers. One-third of all new registrations are actually registrations made by the manufacturers themselves, so-called day licenses so that they can then sell the vehicles at a high discount.[4] VW is among the biggest losers, and deservedly so. While the automobile market in total grew by 5.3 percent, the VW brand sold almost 2 percent less in the service car segment, 7.3 percent of them being diesel cars.[5] This is the effect of the uncertainty customers feel because of the diesel emissions scandal and the discussion of bans on certain vehicles. In addition, we now also have to take into account millions of diesel vehicles by Daimler and Audi.

And if that is not bad enough, there is the price-fixing scandal. Once again, we may expect billions of penalty payments from the German manufacturers involved, money they instead could have spent on developing new technologies. It is increasingly obvious, and nobody can justifiably deny it, that German car manufacturers stink to high heaven. Drastic changes are inevitable. Yet they still continue to stand in their own way.

German manufacturers strive for perfection, just like the ancient Egyptians, who had reached what they took for perfection and simply stopped. The Greeks, by contrast, wanted more. They always wanted to be the best.[6] This is one of the reasons why wealth and plenty often lead to stagnation. Everything seems perfect, and nothing needs to be improved. To be creative, however, you need a mix of wealth and poverty, beauty and ugliness, and things that work and things that break down frequently. This is partly the explanation for why Berlin is Germany's startup center rather than Munich, Hamburg, or Cologne.

If they want to get it right again, the car manufacturers have to take a step back and deal with the most fundamental issues anew. Why does this company exist? Why should it have a right to exist? What would happen if it didn't exist? If you look at the mission statements and corporate strategies of German car manufacturers, you understand why they were so successful and why they are so badly prepared for the coming changes.

- BMW Group, for example, has the mission statement of "becoming the global leader in quality products and services for individual mobility by 2020."[7] In addition to this, each brand in that group has its own brand identity. BMW is a symbol for the "Joy of driving. Sporty and dynamic performance with superb design and exclusive quality." The Mini, in turn, "Wins hearts and turns heads."[8]
- Audi follows the strategy of "we promise an edge" and continues, stating that it wants to "foster enthusiasm with sustainable, individual premium mobility. Our premium vehicles remain at the core."[9]
- Volkswagen's vision is, "We are a leading global provider of sustainable mobility." Its mission can be subdivided as follows:[10]
  - We inspire our customers with individual mobility solutions.
  - We meet our customers' various requirements with a portfolio of strong brands.
  - Every day, we take responsibility for the environment, safety, and society.
  - We act conscientiously and base our work on reliability, quality, and passion.
- Mercedes states, "As the inventors of the automobile, we believe it to be our mission and duty to shape the future of mobility in a safe and sustainable way—with groundbreaking technologies, superb products and personalized services."[11] Mercedes-Benz USA adds values under the motto, "What Drives US":

- The right to reject a compromise
- The instinct to protect what is important
- The obligation to preserve our heritage
- The vision to take every detail into account
- The foresight to accept responsibility
- The power to exceed expectations

Some of the statements, missions, and visions include general and sometimes empty objectives, whereas others are so detailed that they obscure the big picture. For example, what exactly is the value in being the world leader in an industry, that is, the biggest enterprise? Why is it important for a manufacturer that its products turn customers' heads? Other statements contradict the manufacturer's public actions. It seems almost perverse that Volkswagen claims to "take responsibility for the environment, safety, and society." Moreover, if all manufacturers thought and acted with sustainability in mind, Germany would be at the forefront of electric mobility and sustainable propulsion drives, not serve as the taillight. All the companies use the term *mobility* without explaining why this is important. Mobility is not an end in itself but instead enables other things, as electricity has already shown, for example. Today, what we see when looking at the German manufacturers are 40-year-old men making cars for 40-year-old men. No more, no less. For them, those are the best cars they can imagine. For the rest of society—that is, the majority—they are, in fact, not the best.

Just to complete the picture, here is the vision statement of a Chinese manufacturer of autonomous electric vehicles, NIO: "Give people time back—to be who they want to be."

Corporate Culture

We can do anything, except the future.

Culture (i.e., corporate culture) is a big word that gives us the impression of a prescribed, fixed value that we cannot modify ourselves. As I set out in my book, *The Silicon Valley Mindset*, culture is the end result of many small behavior patterns every one of us displays every day. When delegations visit Silicon Valley, I aim to enable the participants to experience this fact so that they can understand the degree to which their approach blocks that culture of innovation they so fervently hope for. Culture starts with every single one of us, regardless of whether we are at the head of our company or a regular employee, whether we are a workers' representative looking at matters from the inside or a journalist observing from the outside. Naturally, behavior at the top has more of an impact and tends to influence all the lower levels.

Several examples of companies that survived a crisis or, alternatively, became totally entrapped show just how massive the influence is of corporate culture on events. From 1976 to 1989, James E. Burke was chairman of the board of Johnson & Johnson, an American pharmaceutical company. Shortly after taking office, he called the management team into his office to discuss the internal company credo, which had been affixed to the walls in all company premises since 1943 for everyone to see. Burke suspected that the corporate philosophy was no longer taken seriously, that it was just "writing on the wall." He suggested taking down and destroying this credo, which included Johnson & Johnson's obligation to support mothers after giving birth. This suggestion was followed by an animated discussion about ethics in economy on the part of management, and in the end, they decided not only to maintain the philosophy but also to live it.

> With our credo, we commit to placing the health and wellbeing of humans at the center of our actions. We therefore work for the common good, the environment and our staff—and regularly report stories about people we were able to support.

Their chance to do this came soon enough. In 1982, it was discovered that bottles of Tylenol (a Johnson & Johnson brand) filled with capsules laced with fatal doses of potassium cyanide had been circulating in some Chicago pharmacies. The company reacted immediately. What followed was a nationwide recall action and a publicity campaign pointing out the danger and the situation to customers and pharmacists alike. In its entirety, it cost the company $100 million. The surprising element is that Burke was not informed of this at all because he was on a long-distance flight from the moment the matter came to light until the start and execution of the recall. While he was away, his staff had initiated all measures, acting in line with the company credo. Today this story is the perfect textbook example for excellent crisis management, with customers in mind and, eventually, serving company purposes.[12]

Then there is Volkswagen (and Daimler and BMW, because they were all caught in the diesel emissions scandal). If we compare the crisis management strategies of those companies, the Germans seem to be doing everything possible to do the opposite of what Johnson & Johnson did. They only admitted to the deception when threatened with severe consequences, and even then they tried to do everything they could to impede and prevent elucidation of the facts. In fact, they still do. The recent "diesel summits" (yes, there were multiple) with the German government brought nothing more than vague commitments to doing better. At Volkswagen, this kind of corporate culture had already started with the company patriarch and his successor, who ridiculed employees, slammed journalists, and blamed others. Pleasure trips with prostitutes to keep workers' representatives happy were seen as a small "irregularity" by the board.

Thus, what developed was the *normalization of deviance*. This is the term sociologist Diane Vaughan uses to describe behavior that would not have been acceptable under normal circumstances that suddenly morphs into being "completely okay."[13]

Another example of such deviant behavior becoming the standard involves events at Ford. There were problems with the Pinto model's fuel tank: in the case of a rear-impact crash, there was a danger that it could explode, thus burning the passengers. Management professor Dennis Gioia, who worked for Ford for some time, describes how much the corporate culture "turned him" around. Before and after his work for the car manufacturer, he thought it perfectly logical that in such cases the company had a moral obligation to withdraw the product from circulation. However, while he himself was "inside," he had an entirely different opinion. Difficult to believe, right?[14] Gioia blamed the "company scripts" for this. Because managers are bombarded with a huge wave of information in their work, some of it contradictory and incomplete, they use scripts for quick decision making, which makes their work easier and prevents cognitive overload. The problem is that those scripts may be flawed and tend to grow out of proportion over time while simultaneously interfering with critical inquiry. Scripts are surprisingly flexible, able to rationalize and include new, contradictory information. Sometimes you need a shock event before existing procedures are reassessed.

NASA engineers in charge of space shuttle starts had a similar problem. They had discovered unusual damage to some sealing rings during earlier starts, and their suspicion that the rings did not function correctly at low temperatures had even been confirmed in tests. Nevertheless, the internal scripts did not allow for further critical questions. Therefore, the absolutely necessary "no" for the next start became a fatal "yes"—and on January 28, 1986, all the astronauts on the *Challenger* perished.

The example of Volkswagen shows a similar *script culture* based on unrealistic requirements combined with fear so that instructions are carried out and not questioned. In the end, the installation of fraudulent software became the standard behavior and was considered okay. A little unethical behavior here, a little deviation there eventually brought forth a corporate culture that basically disregarded ethics. Such (erroneous) behavior seems to spread by itself to other areas as well, most dangerously to areas where the company's future is at stake. Corporate culture makes or breaks the culture of innovation.

The Silicon Valley Mindset

Dutch primatologist Frans de Waal studied the behavior of chimpanzees and capuchin monkeys while they tried to reach some tasty bits the scientist had hidden beforehand. Chimpanzees think before they act. After a few futile attempts, they sat down and thought the situation through until they found a solution. Capuchin monkeys, by contrast, are the ultimate trial-and-error machines. Being hyperactive, hypermanipulative, and totally fearless, they tested out a number of approaches to obtain the goodies. They did not mind failing a hundred times, and they didn't give up until they had reached their goal.[15]

Anyone who ever tried to deal with so-called lean and agile methods will immediately make the connection. Germans, Austrians, Swiss, and many others tend to first think everything through in detail, identify the problems, and write specifications and requirements before they start looking for a solution. In Silicon Valley, by contrast, rapid testing and, accordingly, frequent fast failures are permissible and a desired way of proceeding. The Germans act like chimpanzees; the Silicon Valley nerds act like capuchin monkeys.

Both ways have their advantages and disadvantages. The chimpanzee approach is great when the problem is largely known and you already have some expertise available. Efficiency innovation benefits from this approach. The capuchin monkey approach works if you are only vaguely aware of the problem, there are too many variables, and you have to include still unknown external expertise. This especially includes researching and testing, particularly helpful for disruptive innovation.

CHAPTER 12

Types of Innovation

Demanding innovation is like scheduling spontaneity.

—UNKNOWN

Innovation is what happens when a single discovery or invention is handed over to the general public and commercialized. Researchers at universities usually are limited to the state of discovery and invention. And the often-quoted inventor's lot frequently is simply incapacity and a lack of opportunity and appropriate framework for a profitable marketing of the discovery.

Innovation comes in two types: *incremental,* or *gradual, innovation* and *disruptive innovation.* Incremental innovation usually comes from experts and improves a technology or a process by some percentage points without replacing the technology. Over time, it can lead to huge increases in efficiency as well as cost reductions. Disruptive innovation, by contrast, is often driven by nonexperts and outsiders and overthrows the previous technology or a traditional process. It is violent and destructive because entire groups of professions are replaced by new ones during a transition period, and any investments already made are null and void.

For example, the longer a technology, a process, or a business model has been improved incrementally, the more forceful the possible effect of a disruptive innovation can be. Because nothing exists in a vacuum but is instead surrounded by other technologies and processes that are subject to the same kinds of innovative tendencies, it is only a question of time until a disruptive innovation concerns you.

In his book that has become something like the bible of innovation, titled *The Innovator's Dilemma*, Harvard professor Clayton Christensen in the 1990s first analyzed the reasons why established companies lost the innovative power and were blindsided by newcomers.[1] His latest research shifted the focus to economic crises and job losses. For his research, he investigated the 10 latest economic crises in the period from 1948 to 2008 to find out how long it takes until economic indicators and the number of employees reach the pre-crisis level again.

The job losses from the seven economic crises between 1948 and 1981 were compensated after only 6 months on average. This means that half a year later, the number of employees was similar to the number of employees before the economic crisis. However, this changed from 1990 onward. In that economic crisis, it took 15 months to return to the same number of employees. In 2001, the necessary time frame was 39 months. And when Christensen gave a lecture on this in 2013, almost 70 months had passed since the beginning of the economic crisis of 2008 without the number of employees reaching a level anywhere near the precrisis level (see Figure 12.1). In fact, it took until 2015—74 months later—to reach the precrisis employment numbers.[2] He referred to this phenomenon as *jobless recoveries* and asked himself what the

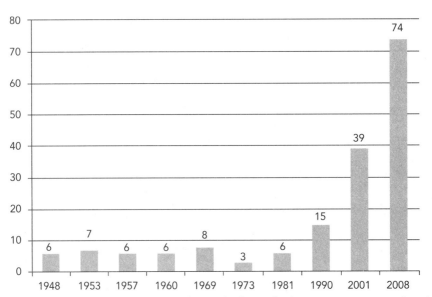

FIGURE 12.1 Duration in months until a loss of jobs was compensated and employment reached the same level as before the crisis

reasons for this might be. Why did the number of employees grow after every single crisis until 1981 but not after the more recent crises? What happened to those lost jobs?

During his research, he discovered that there are three types of innovation in general:

- **Empowering and market-creating innovation** creates jobs and markets by encouraging people to undertake new activities. His example of this is the introduction of the Ford Model T. Before this car, automobiles were regarded as a rich person's luxury toy. With the introduction of the Model T, however, more people were able to afford a car and use it for activities that allowed them to add value in other areas, thus accessing and creating new markets.
- **Sustaining innovation** supports existing markets and only creates a few new jobs. His example is the Toyota Camry (good innovation) and the Toyota Prius (better innovation). One was complemented by the other and then replaced; target groups and markets remained essentially the same.
- **Efficiency innovation** occurs when production processes become leaner, and you use less material and generate higher throughput, while keeping the number of employees at the same level or below. This kind of innovation destroys jobs.

If you compare those types of innovation with the effect on jobs and capital, you have the image depicted in Table 12.1.

TABLE 12.1   Types of Innovation and Their Effect on Jobs and Capital

	SUPPORTIVE INNOVATION	MAINTAINING INNOVATION	EFFICIENCY-INCREASING INNOVATION
Jobs	Creates many	Creates few	Destroys
Capital	Commits	Uses limits	Releases
Markets	Creates new markets	Maintains existing markets	Maintains existing markets

In the past, there was a kind of balance between these types of innovation. All three of them existed equally. The capital employed and gained in one area was also used for the other areas. From the 1980s onward, however, economics became a scientific discipline in its own right with its own language and methodology. Suddenly, ratios (indices) and similar instruments were introduced with the aim of tackling the scarcity of capital and using money only where it would make the most profit. The notorious Excel sheet was filled with economic figures, and suddenly seasoned CEOs had to answer to young students of economics in investment firms. The shareholder value became the new mantra. And the CEOs followed without protest.

Instead of investing money in building future-oriented and sustaining innovation projects, it was primarily put toward projects that increased efficiency. The *return on investment* (ROI)—another one of those key index figures—could then be reached more quickly and with a higher degree of success, although the amount of revenue was usually lower than for the other two types of innovation. The extended table (Table 12.2) then is as follows:

TABLE 12.2   Types of Innovation and Their Effects on Jobs, Capital, ROI Duration, the Chances of Success, and Revenue

	EMPOWERING INNOVATION	SUSTAINING INNOVATION	EFFICIENCY INNOVATION
Jobs	Creates many	Creates few	Destroys
Capital	Commits	Uses limits	Releases
ROI duration	Long term	Medium term	Short term
Chances of success	Very uncertain	Uncertain	Certain
Revenue	High	Medium	Low

Success-oriented quarterly reports that usually are drawn up in any publicly traded company "teach" the managers to think in short-term periods and behave accordingly. This tendency has been confirmed in practice and a number of studies. I already mentioned the micro- and macro-observations conducted by economic behavioral scientist Richard Thaler and Al Gore's survey among CEOs and CFOs on lower quarterly results compared with long-term profits at the beginning of this book.

The focus usually is on the financial figures that are used for bonus payments and incentives for managers. What do I care about an electric vehicle I can put on the market in five years' time if I can sell a diesel-powered vehicle today? Like Clayton Christensen, Susan Christopherson from Cornell University also emphasizes the increasing orientation toward numbers and results in economics since the 1980s. Companies that started out producing items soon found that creating financial constructs was more profitable than actual production, which thus became a secondary concern and was even considered to be an annoying element in the company structure.[3]

Add to this that there is no scarcity of capital as there was 50 years ago. Quite the contrary! Investment managers complain that there are not enough good opportunities for investment. What they mean is especially investments in efficiency-increasing innovations. In brief, too much money is looking for (too few available) investments. Innovators about to try something completely different and founders of startup enterprises, by contrast, complain that it is extremely difficult to obtain venture capital. Too much money leaps at efficiency-increasing innovations, while projects with empowering and sustaining innovation can barely get any support at all.

In other words, all of this has far-reaching consequences for our jobs. The number of new jobs created is insufficient. This makes companies much more liable to succumb to "attacks" when enterprises with lots of risk capital start introducing disruptive innovations. These innovations are coming especially often from Silicon Valley. Venture capitalists mainly put their money with the first two types of innovation. Efficiency-guided investors and capital providers weaken companies all over the world in such a way that they may become unresisting victims of Silicon Valley enterprises, a phenomenon that is found across all industries. You can tell the extent to which this problem is dispersed by taking a look at the wide range of delegations coming to visit.

According to Clayton Christensen, disruptors are more than just new competition: they create a completely new market and a very profitable one at that. The consulting firm McKinsey estimates that the entire data service sector in the automotive industry could make up to $750 billion by 2030.[4] We can predict—as my professor of mathematics used to put it—*with virtual certainty* that German companies will not play a big role in this. If they are lucky, they will be allowed to bend some metal for digital companies.

Car manufacturers regularly buy vehicles from the competition for testing and disassembly in order to analyze them. This happened to Tesla models, too. When a German manufacturer examined the USB port, the engineers found that Tesla did not use an industry standard for the port but had simply installed a consumer version. Thus, a defective USB stick would suffice to incapacitate all the vehicle systems. A German manufacturer would never have made such a mistake, and for a long time, this was the reason why they did not take this competitor seriously. And let's not even begin with the gap dimensions, the interior components, and the losses Tesla makes. By concentrating exclusively on the apparent weaknesses the opponent shows, you become blind to the disruption happening at the same time when a technology that is allegedly simple facilitates processes and allows faster progress. You simply ignore the danger and feel vastly superior. Basically, you feel that you are a Goliath—right until the moment when it is (almost) too late to get your act together and start running.

Clayton Christensen explained the dangers to the long-term opportunities of a company posed by a focus on short-term figures using an example from the steel industry.[5] In the past, there were two dominant types of steel manufacturers on the market: *integrated steel plants* and *compact steel plants*, or *minimills* as they call them. The first type nowadays costs about $10 billion and is able to produce a wide range of iron ore products for very diverse requirements. This range extends from cheap reinforcing steel to fairly expensive sheet metal for the automobile industry.

The cost-effective and comparatively small minimills, using an arc furnace, initially produced cheap construction steel from scrap metal such as the materials collected from vehicle recycling. They were usually built in areas where there was a lot of scrap metal to be had but no access to crude ore. You can expect to sell low-quality construction steel at a small price and the respective small margin. It still was about 20 percent cheaper than the steel the integrated plants produced. Still, none of the big steel works introduced compact steel plants, despite the fact that the price for this category of construction steel could have been reduced by one-fifth. As a matter of fact, operators of integrated steel mills were even happy to withdraw from this market and leave it to the smaller competition because the profit margin was only 7 percent. This withdrawal from the market increased their overall profitability, while the operators of compact steel mills were happy to find additional customers for simple construction steel. However, from the moment the integrated steel mills no longer were in this market as competitors, the price fell. The competition among them forced the compact steel mills to pass on monetary benefits to their customers. A strategy aimed at a low cost does, after all, only work if there is a competitor with a higher production cost in the market.

At first, the compact steel mills could still make a profit by producing more efficiently, but eventually that reached its limits, too, right up to the moment the first one of them managed to produce higher-quality construction steel while still maintaining a cost advantage of 20 percent compared with the integrated steel mills. Once again, the bigger companies retreated from this market segment. Why should they defend a product with a lower (12 percent) profit margin if the next product in line brought in 18 percent?

As they retreated, their total profit margin increased once again, and the small, compact steel mills also temporarily made a lot of money.

Are you beginning to see the pattern? This process repeated itself with increasingly better steel products, and today there is not a single integrated steel mill in the United States anymore. Based on the economic indicator figures for profit margin and profitability, the managers made perfectly rational decisions. The profit margins increased, but the market kept getting smaller.

Toyota started out at a similar low threshold as the compact steel mills did. The Japanese did not start with the luxury brand Lexus but in fact with a very bad and cheap car. In 1960, the Corona was first sold in the United States. It was then followed by the Tercel, Corolla, Camry, Avalon, Forerunner, and Sequoia, increasingly better cars at higher prices, until eventually the premium vehicle was launched on the market. Now, however, Toyota is under attack from companies from South Korea (Hyundai) and China that produce worse cars in product areas with low profit margins. And the game starts again.

American automobile manufacturers fared just like the integrated steel mills. Pickups guaranteed much higher profit margins than compact cars. We all have seen the result: General Motors and Chrysler filed for chapter 11 bankruptcy protection in 2009. Digital cameras also attacked the market from the bottom. The quality of the images was no match for the photographic paper Kodak offered. Still, you all know which company "disappeared" in the end in 2012.

Christensen calls this market phenomenon *competing against noncon-sumption*. In every market you find customers who do not participate in the consumption because the products are too expensive in their opinion. However, as soon as a competitor arrives with a similar product that is affordable for them, there is a new customer base. The alternative to having a cheap, scratchy radio is to have no radio. The alternative to owning a cheap car such as a Toyota Corona in 1960 was not owning a car at all. Nonconsumption therefore has the greatest potential for growth, but the dominant providers hardly ever recognize this opportunity.

In addition to the attack on the markets from the bottom, today we are witnessing a kind of pincer movement from the new automobile companies coming from the top and the bottom simultaneously. When you listen to employees of German manufacturers speaking about Tesla, they mainly point out the low manufacturing quality, larger gaps, use of cheap components, and "soft" steering. And all this for a price you would also pay for a high-class

Mercedes or BMW. At the same time, the Tesla vehicles offer a vast addition of digital services and acceleration that far exceeds that of any sports car, so we are actually talking about a partly much worse, partly much better quality than usual—all united in the same product. Uber has a similar process, although only its prices have a "bottom-up" process. Usually, the vehicles are superior to traditional taxis in terms of service, experience, and quality—at least in Europe.

Ockham's Razor: Is There an Austerity Principle for Innovation?

The technological progress in the automobile industry over the course of the last 100 years is impressive. Despite prophecies to the contrary and all the criticism encountered, engines today are more economical, efficient, and low-noise than ever before. This fact really sank in after I had spent time attending a language course in Russia in 1994. After one month in Saint Petersburg—the Soviet Union had just dissolved, and Western cars were still a rarity on the streets—I returned to Vienna, and the first thing I noticed was how quiet the cars were. I still remember sitting in the Mercedes taxi and pointing out the silence to my fellow passengers. No loud motor noises, no visible clouds of exhaust, no vibrations and humming, and last not least, no potholes in the roads. Innovative concepts such as the engine turning on and off automatically are based on such advanced technology that the driver often does not even notice whether the motor is actually running or not.

Incremental technological advantage, however, does not come without a price. The product becomes more complex. A modern combustion engine consists of about 100 to more than 1,000 parts depending on what you consider part of the engine and which model and year of construction you are considering. It won't surprise you to hear that the number of parts used increased over the years. Increasing complexity, however, does not automatically result in an increase in error frequency. Modern engines are much more reliable than their predecessors because materials and manufacturing quality levels have also been raised continually. Nonetheless, today's motors have already reached a degree of complexity and efficiency to which even extensions and restructuring can only make small improvements. The question is whether one might not apply Ockham's razor for innovations here, too.

Ockham's razor—also known as the *efficiency principle*—is a rule applied in science that demands accepting only one sufficient explanation for each object of investigation.[6] The simplest theory should be preferred to the more complex one. One example we might mention is the planetary model. As long as people regarded the earth as the center of the universe, the movements of the sun, the planets, and the stars appeared to be utterly complex. However, as soon as one gave this view up for a heliocentric world view with the sun in the center, the theory was much easier to understand, and the model made a lot more sense.

If we now apply Ockham's razor to our topic at hand, innovation should mainly start with the complexity of the current standard solutions. If you look at a horse as being a power for traction and add up all the parts of the skeleton and the internal organs, the first motors had much fewer movable parts than a horse. The step-by-step improvements of the engines added parts with time until the system became very complex and started to practically scream for innovative solutions.

This scream did not go unheard. The number of parts required for an EV is dramatically lower than that for a combustion engine vehicle. You do not need an engine, nor a transmission, nor an exhaust and tailpipe system. A self-driving vehicle does not need elements for switching gears or a steering wheel or side mirrors because it does not need a human driver. The computer takes over the functions of the steering wheel, the turn signals, and the accelerator pedal. However, the complexity is transferred from the analog, mechanical world to the digital sphere. At some point, each solution and innovation will cause new problems itself.

This also applies to other industries. A digital iPhone has fewer movable parts than a mechanical telegraph. However, it can still complete its tasks much better and faster than the "previous model" ever could. From drawing a picture of an object—this involves humans with their movable parts—to taking photos with a simple camera with plates, to using a complex mechanical camera with foldable mirror, zoom, shutter, and film roll, the path led to digital cameras, which often now even have zoom functions that are digital.

The Keeley Model of Innovation Types: The Secret Is the Combination

Here is another model of differentiation for innovation that I won't deny you. Innovation researcher Larry Keeley describes 10 types you will find listed in Figure 12.2.[7]

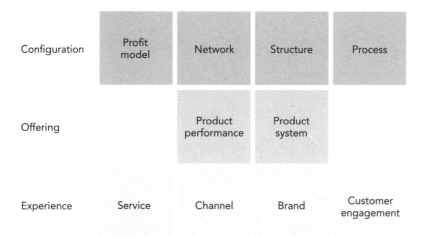

Configuration	Profit model	Network	Structure	Process
Offering		Product performance	Product system	
Experience	Service	Channel	Brand	Customer engagement

FIGURE 12.2   Types of innovation according to Larry Keeley

Keeley says: "Having analyzed and used the Ten Types [of innovation] for more than 15 years, we can now confidently generalize: you must look beyond products in order to innovate repeatedly and reliably. By combining multiple types of innovation, you will be more assured of bigger and more sustainable success." Especially disruptive innovators proceed in this way, like new forms of automobile manufacturing that were designed to revolutionize the production process. The idea for the production line surfaced when Henry Ford visited a slaughterhouse where every worker had to perform certain tasks at a conveyor belt—thus reaching high work speeds. He applied this principle to the production of the Model T, reducing the production time by one-eighth and therefore also allowing him to drastically reduce the price.

Frans Johansson, author of *The Medici Effect*, called the linking of several disciplines to form something new the *Medici effect*, after the Italian Medici family from the fifteenth and sixteenth centuries whose work and actions were the starting point for the Renaissance. The Medici joined creative people from various areas and disciplines, a step that gave crucial impetus to art, culture, architecture, science, and economics and brought them forward.[8]

Creativity in our Western society means creating something new that is original. And this is precisely where it becomes dangerous goings for the traditional automobile manufacturers looking at Tesla, Google, and Uber exclusively from a technological point of view. They believe that battery technology is well known and therefore there is no innovation.[9] And thinking this, they overlook the fact that much more is happening than a motor simply being replaced by an accumulator.

A Hindu believes that creativity can also be something that directs our attention to something that already exists, lighting it up as if with a flashlight and permitting us to actually see the creation behind it. For example, a room, even a dark one, may already exist, and the creative genius does not have to create or discover it. But the genius has to place it into the right light so that we become aware of its existence. Only then are we able to recognize the miracles hidden in it.

CHAPTER 13

A Psychologically Safe Environment

Fall Down, Get Up, Go On

The optimist proclaims that we live in the best of all possible worlds, and the pessimist fears this is true.

—James Branch Cabell

Nevertheless, innovation needs additional factors from the ones already mentioned in order to really become a success. One of those factors is a so-called psychologically safe environment that allows employees to take risks and to fail without having to fear repercussions. In this way, they are able to learn and become more innovative. Professor Amy Edmondson from Harvard Business School confirmed this fact in her study on the error frequency in hospitals. In places that had established a psychologically safe environment in case of failure, the number of errors documented in the protocols was higher. At first glance, this seems hardly desirable. Because people's lives are at stake here, the matter is literally deadly serious. When we look more carefully and consider independent data sources regarding errors in treatment, however, we can see that the hospitals with a lower number of documented errors in the protocols and a psychologically unsafe environment were much more dangerous for patients. To avoid punishment, the staff there did not report errors. This meant that, on the one hand, nobody could react to the implications of errors in treatment or they reacted too late, and on the other hand, nobody had the opportunity to learn from past mistakes and improve treatment.[1]

If we take a look at the Volkswagen corporate culture, we may say with reasonable certainty that the company does not provide a psychologically safe environment.[2] Some of the former CEOs appear to have been legendary in their ways of dealing with both staff and journalists. Even top executives were edgy days in advance of meetings with their CEO. And the former CEO of Porsche and VW, Matthias Müller, issued his view on the topic to the employees. In an interrview with the magazine *Auto Motor Sport*, Müller referred to autonomous vehicles as a hype by citing the "trolley problem" as an example for the unsolvability of this technological and ethical challenge.[3] First, this is the partly understandable reaction of a man who was responsible for making sports cars. The fun in such a racing machine is mainly that you control it yourself. Porsche sees its business go down the drain if people-controlled vehicles lose ground. Second, however, Müller's statement signals to his staff that their top boss smiles at innovation in disruptive technologies and does not take it seriously. Those who would dare to think in a way that did not agree with the boss's thinking and to pursue research in that direction are in danger of ruining their career. Consequently, nobody would dare to make any such suggestion that might change the company from the inside. And, voilà, there you are at a self-constructed dead end.

Other manufacturers are by no means immune to this danger. When Tesla stole the show with its Roadster in 2010, the BMW board of directors asked its Silicon Valley outpost for more details regarding this company. The suggestions on how to proceed further in the analysis provided by an external consultant—who also owned Tesla stocks and attended the Tesla shareholder telephone conferences—were cut out of the document because nobody at BMW dared to provide such proposals from the bottom up. They preferred waiting for instructions from the top. But from the top not much is coming either. At least not the encouragement for exploring new fields and taking on those new competitors. BMW's head of development Klaus Fröhlich exemplified that at a BMW internal event on electric mobility in summer 2019, when in front of the very teams developing BMW's answer to Tesla's electric vehicles, he called all the talk about electric vehicles electric vehicles a "hype" and blamed European customers for not buying BMW's not very competitive and overpriced EVs.

But how can an organization possibly learn if employees do not dare to make suggestions for improvements and innovative procedures? Innovation must be a joint effort by all the employees, as I stated earlier. Many traditional car manufacturers are still organized in a very hierarchical structure. Decision-making procedures are slow, and many sides must be heard and coordinated. And not all of the employees work with a focus on the best customer and company interests but may be more worried about their own careers.

Professor Robert Sutton from Stanford University describes the behavior of people you could well describe as assholes in his book, *The No Asshole Rule*.[4] But how do you recognize one? Remember two simple questions you should ask yourself after a conversation: do you feel oppressed, degraded, without energy, or like you have been given a roasting? Were you the person higher in rank at the meeting or not? If you can answer the first question with a yes and the second one with a no, then everything indicates that you were dealing with an asshole. To be absolutely sure, it is advisable to work through the following list—the so-called dirty dozen. Which of these applies to the conversation you had?

- Personal insults
- Intrusion into your personal space
- Unsolicited bodily contact
- Verbal and nonverbal threats and intimidations
- Sarcastic jokes and taunts
- Devastating e-mail flame wars
- Repeated invocation of status to humiliate the victim
- Shaming, public exposure, and status removal rituals
- Harsh interruptions
- Insidious attacks
- Dirty looks or ceaseless staring
- Snubbing people

If you can identify one or more of those elements in the meeting, it is obvious that this is definitely not a psychologically safe environment. Each of us has probably caught ourselves ridiculing a colleague's proposal behind that colleague's back. This is also an element conducive to a psychologically unsafe environment because we are simultaneously sending a signal to all staff to not make any "stupid" suggestions—their suggestions may be ridiculed as well.

I certainly do not want to give the impression that Silicon Valley entrepreneurs are choir boys (and girls). Quite the contrary. Steve Jobs met many of those asshole criteria, and Robert Sutton explicitly mentions him in his book. You have already heard about Uber's Travis Kalanick, and Elon Musk is no billboard boss either.[5] Nevertheless, you still have the feeling that Silicon

Valley entrepreneurs are mainly interested in advancing the cause, whereas traditional automobile manufacturers are mainly worried about market shares and retaining their power.

May I Ask a Question?

I would rather have questions that can't be answered than answers that can't be questioned.

—RICHARD FEYNMAN

A psychologically safe environment is the basis for asking the right questions. And that is not so easy, if not downright dangerous. Our society prepares us to find answers or have them ready. Asking a question can easily have negative effects. Perhaps you will appear ignorant or as though you are demonstrating a lack of respect. Despite that, our ability to ask questions is exactly what most distinguishes us from other primates. Children aged two to five years ask hundreds of questions every day and therefore about 40,000 questions every year. A question can be asked by words or just indicated by a gesture. Over time, the types of questions become more practical, from "What's that?" to "Why is that?" to "How does it work?"

We owe any scientific breakthroughs, innovations, discoveries, and organizations to a question. Reed Hastings, founder of Netflix, for example, forgot to return videos he had borrowed in time and found a hefty fine waiting for him. He asked himself, "Why are there any late charges in the first place?" and then "How about if we did not have them and you paid a monthly fee like at the fitness club?" Henry Ford asked himself, "How can I accelerate the production of the Model T?" Carl Benz wanted to know, "What happens if I place a motor on a carriage?"

Such sentences with *what, how,* and *why* are at the start of a journey of discovery and allow us to go to new limits and reach far-away shores. An apparently safe answer is likely to interrupt the discovery process prematurely, often without unearthing the full potential. This is precisely where the danger is for today's organizations. Many of them started out with a question and then found an answer to it, but then they asked fewer and fewer questions and eventually only wanted answers, regarding the asking of questions as a waste of time. As the head of Porsche, Matthias Müller might have done well to investigate a little further into the topic of "What should a sports car look like, and what kind of joy of driving could you experience in it if it was an autonomous car?" And then, as the CEO of Volkswagen, he should have asked himself, "What is environmentally friendly mobility of the future?" It would soon emerge that diesel engines definitely have no part.

The art of asking real questions will become more important than pulling full answers out of a hat—especially so as answers in the age of Google and friends are a readily available mass commodity.[6] Dan Rochstein, founder of the Right Question Institute, makes the participants in his workshops communicate in question form only. The questions must be proactive and not passive. Every question must be answered with a counterquestion. The effect is to expand the participants' thinking mode, and their power of imagination rises to unheard of heights. The participants show much more engagement and interest, and ideas just abound—all of them in question form.

What, Why, for What Reason: Question Storming Versus Brainstorming

> In the beginner's mind there are many possibilities, but in the expert's there are few.
>
> —Shunryu Suzuki

Eric Ries, author of *Lean Startup*, comments that the managers in the companies that receive most of the resources are those with the greatest amount of self-confidence and the best plan, so basically those people who seem to have an answer for everything or who—as far as everybody knows—failed with very few projects. Instead, resources and incentives should really go to those in the company who ask intelligent questions, conduct promising tests, and take calculated risks. Failures help us to learn, which is the only way on the path to innovation.

In contrast to brainstorming, "question storming" does not have the goal of generating a lot of ideas but only some few really good questions. When brainstorming, you hope that the participants exit the session with a solution—and if that does not happen, people are disappointed. At a question-storming session, this is not what is expected at all: the participants aim to generate at least 50 good questions about a problem they face. Subsequently, the questions are prioritized and subsumed under the top three questions. Usually it is easier to ask questions than to desperately try to drum up some ideas. Questions are like magnets; they draw the participants' attention. The degree of commitment is higher, too. And questions allow you to first highlight several different aspects. The critical point is to slowly filter out the best, that is, relevant, questions.[7] You obtain focus and momentum because the questions found by the group indicate the direction for further steps and research.

Another important aspect of question storming is the kinds of questions. Is it an open or a closed question? Questions you can answer with yes or no usually do not offer much in the way of progress. Open questions, by contrast, require some thinking from the person answering and allow space for explanations and new questions. Here are some examples of how open questions should be started:

- Why is . . . ?
- What if . . . ?
- How could . . . ?

You can easily stop people by simply asking why they think they know more than the experts. The answer is simple: they do not know more, rather less. This is effectively an advantage. You do not know all the potential problems by heart, do not have to take anybody into consideration owing to long-term and carefully maintained relationships, and can dare to begin a fresh approach. Asking questions helps to play a trick on the *certainty epidemic*, that is, the phenomenon of feeling too secure in our own knowledge and therefore neglecting to question or double-check our assumptions.[8] Neurologist Robert Burton suggests the following strategy to avoid that trap: pause for a moment and ask yourself, "Why did I think of that question?" "What are the assumptions behind that question?" "Is there another question I should be asking?"

- "What if our company did not exist?" allows for a new start that enables you to place yourself above your own industry and position.
- "What if money played no role? How would we then approach this project?" By abolishing limitations temporarily, participants are able to give their powers of imagination free rein.
- "What if we could not possibly fail? What would we do? How would we proceed?" These questions generate self-confidence and permit acting without fear.
- "How would IKEA solve this problem" or "How would a tough TV cop proceed?" Put yourself in the position and thoughts of someone else as if in a role play.
- "How can our company once more have the character of a startup?"

Let the so-called killer questions work their magic on you so that you can feel the difference. They are practical, reminiscent of an interrogation, and very popular, and they give the impression that the asker is particularly competent and important—and often looking for someone to blame.

- "How much is that going to cost us?"
- "Who is responsible for this problem?"
- "What do the figures say?"
- "What is our Tesla killer?"
- (Why didn't we come up with this idea?)

There is some justification for the killer questions, too, but they tend to shift the focus from the actual problem to details that may only emerge when working on a solution. Those kinds of questions push you into a defensive position. You assist in managing a company but are not helping to actually make it a success. The best managers ask open questions.

Appreciative positive questions motivate participants. You appreciate what has already been done and do not emphasize what has been omitted. Instead of just looking at the things that do not work, you should always and everywhere be conscious of what works. This starts with you personally and should be cultivated even for small things. Instead of being upset that the alarm clock is buzzing, be thankful that it works and that you won't be late for work. Thank your coffee machine for the early morning energy boost, the car or bicycle that got you to work without an accident, and the elevator that got you to the twentieth floor with no effort on your part and no hitches. Only then will you realize how many things in your surroundings simply work and that there are actually very few mishaps and inconveniences in your daily life.

The ultimate acid test for any solution is always, "Is this something that improves the lives of people?"

"Kill the Company," or How I Can Bring Down My Own Company

Warning words against imminent changes are often ignored. Successful companies in particular feel that they are immortal. Basically every project so far has been a success. You project your figures into the future, and the thumbs are definitely up. And yet, shit happens!

"Immortal" is exactly how Nokia and Polaroid felt. And Blockbuster and General Motors. Until they went bankrupt or had to be sold. In his book *How the Mighty Fall*, Jim Collins lists the individual phases of decline such companies may go through.[9] It is not so much their arrogance that brings them down—although that does play quite a large role in some of those companies—but it is their blind faith in tried-and-tested success factors and a reluctance to question them and leave the comfort zone to try something new, combined with the conviction that you already have all the expertise needed in your company.

"Kill the company" is an approach in which managers and employees in one company play the role of a competitor and try to find ways to destroy their own company. For example, if I were Paypal, how could I make my company's life difficult? Mercedes employees could take on Tesla's role and think about the technologies, business models, processes, or whichever kind of crazy innovation they can think of to attack the automobile giant.

Role plays are a tested means for allowing employees to break out of their mental cages. It is one thing to guard a bank if you are a police officer. Being permitted to connect with a robber's mentality, find gaps in the security system, and crack the safe is another. This role play can be a lot of fun and get the creative juices flowing. It can be boring to work your way through legal texts in order to comply with prescribed requirements for the security of financial transactions, but if you are allowed to think like Jack the Hacker or Sherlock Holmes for a while to find out how you could obtain valuable data and information from the company, then your own professional field suddenly adds a very different, exciting dimension. You only take your own weaknesses and inadequacies seriously from a competitor's point of view, and then you realize that action is urgently needed. If there are already 400,000 preorders for an as yet nonexistent vehicle by a competitor, then it is already five minutes past twelve.

Thinking at 180 Degrees: What Do You Do with an Electric Car That Does Not Run?

What purpose could there be for an oven that does not bake or roast, a fridge that does not keep things fresh, and a car that does not run? These are questions that counteract the actual purpose of the application, which is why we refer to this as *thinking at 180 degrees.*

Tom Monaghan, founder of Domino's Pizza, uses this technique to have a fresh view of things. The objective is not necessarily to find solutions but to open your mind for unusual questions and points of view. Questions like this allow you to turn a situation around. What if people did not have to pay for the food in your restaurant? How could you then make money? What if there were no tables and chairs and no menu either? If, in other words, exactly those things were missing that we regard as being essential elements of a matter, the restaurant might turn into a stand-up snack bar, or a novel evening event, or a once-a-week free buffet for people in need.

The question "What can you do with an electric car that does not run?" might get you to the solution of using the batteries as power storage for your household.

Are You Experiencing a Déjà-Vu or Rather Vujà-Dé?

We use the term *déjà-vu* to refer to something that is new for us but seems to be strangely familiar, as if we had seen it before. And that is exactly the translation from French: "seen before." The "vujà-dé," coined by American comedian George Carlin, aims for the opposite effect: suddenly we look at something that is very familiar to us from a new point of view.[10] Sometimes this happens to me when I look at a word and repeat and use it so many times that all at once it starts to look funny and seems to disappear into a fog, or it happens when a long-time friend suddenly tells me something or acts in a way that challenges everything I thought I knew about him.

Moore's Law: Interpreted in a Countersense by Auto Managers

Intel cofounder Gordon Moore wrote in a study as early as 1965 that the number of transistors on integrated circuits would double every 12 to 24 months, and consequently, the same would apply to the computing speed and memory volume of electronic components. For several decades, the development fit this observation with startling accuracy, a fact that is often mentioned as proof of the validity of what people had named *Moore's law*.

Gordon Moore himself, however, was not quite so sure. He believed that his "law" became a "self-fulfilling prophecy" because the development divisions in the semiconductor industry were guided in their product plans by his observations.[11] Companies used it as a guiding principle just because there was this "law" and started to plan development cycles accordingly in order not to be left behind and land on the scrapheap of their respective industry. An unexpected consequence of Moore's law was that the development in the semiconductor industry went ahead without pause.

The German automobile industry, by contrast, seems to almost ignore Moore's law or even turn it upside down. Manufacturers, management, and experts from the industry often assume that electric mobility is not wanted by customers and that autonomous driving hence is decades away from any kind of practical use. In this way, they confirm to each other that there is no need in this area to conform to the development speed in other countries. Ten years after Google started test drives with autonomous vehicles and seven years after Tesla launched its first electric sedan car, German manufacturers still have nothing solid to offer in comparison. While the U.S. government and the individual states outdo each other with regulations on autonomous vehicles, Germany is busy discussing a legislative text that the United States regards as obsolete even before it has become valid in Germany. Moore is highly valued in Silicon Valley: the development of sensors, algorithms, test drives, and expert training occurs at the speed he observed.

Open Sources: In-House Expertise Against the World

Whether you think you can or you think you can't—you're right!
—Henry Ford

Anyone who has lived in Silicon Valley for any length of time is astonished at how openly people here discuss questions and are ready to share information. Okay, you do not easily share the secret recipe of your own technology; but generally people are quick to help each other out with expertise or to link up with someone. In the early years of Silicon Valley, William Hewlett and Dave Packard took a week off from their own company to help their friends, the Varian brothers, with an issue in one of their medical technology devices. Whether at children's birthday parties or meet-ups, you constantly meet competitors and experts from other industries. You chat, exchange views, and learn from each other.

This readiness to exchange information results not only in a mutual assessment but also in accelerated development for everyone involved. The urgency is quickly recognized, and trends are included at a higher speed. You plan with the help of a lot of experts outside your own company. This includes the success story of InnoCentive, a platform on which companies post problems or questions; problem solvers from all over the world and any discipline are welcome to join in.[12] When NASA used this platform for the first time, everyone was very skeptical. For many years, experts had despaired over the question of how they could improve their predictions of solar flares. Such eruptions are a danger for astronauts and satellites in space, as well as for electronics on earth. Dubious at first, NASA eventually announced a prize of $30,000 for a comprehensive approach. More than 500 people participated, and the winner was a retired engineer from rural New Hampshire. He used his own equipment to provide forecasts of sun flares with 75 percent accuracy.[13]

NASA management was surprised to find its own experts outdone by others, and NASA engineers felt humiliated, as if they had let their company down and had failed. However, this feeling passed quickly. The NASA specialists soon discovered how they could use crowd intelligence in their favor. Today, it is understood that the mentality of "invent everything in-house" cannot be maintained. Why should you ignore experts from all disciplines "outside"? Surprising results and solutions abound. It is precisely this openness that many traditionally operating companies lack. The circle of experts is too small, and the industry is too secretive.

Innovation Outposts: The Music of the Future Is Played in the Valley

Automobile manufacturers and suppliers that want to at least maintain a small chance of keeping up with the most recent developments in the automobile industry have to set up branches in a place that is far away from the head offices and traditional automotive research centers: welcome to Silicon Valley. All renowned manufacturers have an "innovation outpost" in California or, rather, in a valley between San Francisco and San Jose. The idea is to recognize trends faster, come into contact with startups, and use the local infrastructure, legal framework, and available specialist expertise. Silicon Valley connects disciplines in such a range and such a depth that currently there is no other region on earth that can possibly keep up. Look for yourself in Table 13.1.

TABLE 13.1 Selected Innovation Outposts of Automobile Manufacturers and Suppliers in Silicon Valley

COMPANY	HEADQUARTERS
AIMotive (AdasWorks)	Hungary
Audi Innovation Research	Germany
BMW	Germany
Continental AG	Germany
Daimler	Germany
Delphi Automotive	USA
Denso	Japan
Efficient Drivetrains Silicon Valley Innovation Center	USA
Fiat Chrysler Automobiles	Italy
Ford Research and Innovation Center	USA
General Motors Advanced Technology SV	USA
Great Wall Motors	China
HERE	Finland
Honda Silicon Valley Lab	Japan
Hyundai	South Korea
Magna	Canada
Mazda	Japan
Mercedes Benz Research and Development	Germany
NextEV	China
Nissan/Nissan Research Center	Japan
Preh Car | TechniSat Automotive	Germany
Toyota Research Institute	Japan
VW Audi	Germany
Yamaha Motor Ventures and Laboratory Silicon Valley, Inc.	Japan
Zenrin	Japan

At least 25 companies from the automobile sector do research locally, with a number of employees ranging between just a few and several hundred. Not all of them work efficiently and make the best out of their branch office. Volkswagen, for example, runs the Electronics Research Lab (ERL) together with Audi Innovation Research (AIR). However, being physically present does not in the least automatically get you into the right mindset. When you enter the building, you are not looking for German, Japanese, South Korean, or Detroit culture but for the culture of the Bay Area. The approach of transferring a team from the head office to this place and then letting it act once more behind closed doors without creating connections and links to others is wrong. The right way would be to hire local staff with appropriate networks and a different mentality, rent coworking spaces between startups and other companies, participate at events, network, and, above all, keep your eyes open.

People who have never used an Uber or Lyft cannot comprehend the unique character of this service. People who do not pay attention while driving through Mountain View or the San Francisco financial district are likely to overlook the autonomous vehicles. Somewhat easier to spot is the large number of EVs that are almost everywhere and use the existing charging infrastructure. This proves that the head office's objections that certain things "cannot possibly work" or that "nobody needs that" are not true.

Innovation outposts furthermore offer managers from the head office a wonderful pretense for traveling to Silicon Valley to form their own opinions "on the spot." Although a prophet has no honor in his or her own country, and inside Cassandras are often looked on pityingly, they are actually often the ones to provide valuable information and context and thus complete the fragmentary puzzle in a way that makes sense. You have to reckon with Google, Uber, and Apple—especially in the automobile market. This is what the head office has to understand. It is not enough to open a shiny new outpost in California; you also need an "innovation inpost" at home so that acquired knowledge can be referred back to the company.

Nothing Going Without Training and Research: Who Currently Does What Where?

Where do they train the engineers who are to develop autonomous vehicles? There is no existing comprehensive training program for that yet. As mentioned earlier, there are first organizations aiming to fill this need. For example, there has been an offer of a so-called nanodegree as an engineering degree for programming self-driving vehicles on the online learning platform Udacity founded by Sebastian Thrun since late 2016.[14] And where did Sebastian Thrun do his research when he took part in the DARPA Grand Challenge? At the Stanford Artificial Intelligence Laboratory. It is hardly surprising that Stanford is also the home of the Stanford Center for Automotive Research, or CARS for short.[15] Before that, Thrun did research and taught at Carnegie Mellon's Robotics and AI Department, working on the basics of autonomous systems.[16]

In April 2017, a first group of students and teachers went independent and founded their own startup for self-driving technology, Voyage.[17] A&M University in Texas, in turn, has invested $150 million in a development center that is intended to allow students and companies to do research on autonomous technologies.[18]

In Germany, TU Braunschweig with its close ties to the Volkswagen Group, and the Free University of Berlin are among the most renowned universities doing research on autonomous driving systems. autoNOMOS Labs, dedicated to the further development of autonomous vehicles and currently testing two units in Berlin, started out as a project.[19]

Conclusion

"En Marche!" Politics
and Society on the Move

*Politics is the art of looking for trouble, finding it
everywhere, diagnosing it incorrectly, and applying
the wrong remedies.*

—GROUCHO MARX

I will stick my neck out and make a forecast: everything in the automobile
industry will change in the coming 10 to 15 years. The industry itself and
everything related to it will have to undergo a fundamental change. The
question is not so much if but how quickly this will occur. Politics and our
society are facing a huge challenge.

A few months ago, I was traveling in Silicon Valley with several EU par-
liamentarians. They wanted to understand what it is that makes Silicon Valley
so special in technology and entrepreneurship. During our conversations, one
of the conservative parliament members huffed about people expecting benefits without wanting to do anything for them. "If you don't want to work,
well, you won't get any money! So you won't have anything to eat!," he
stated. In his anger, the man overlooked the most important lesson from his
visit: the world has changed. There is a looming tsunami of machines and AI
systems that will replace human work and power. The figures and facts about

375

the automobile industry presented in this book alone show just how far the development has already progressed. Soon 540,000 truck drivers, 250,000 taxi drivers, and more than 300,000 workers in the German automobile industry itself are likely to lose their jobs. In the United States, we have 3.3 million truck drivers on the line and more than half a million taxi drivers. And this is just the beginning! New technologies probably will not be able to make up for this loss of jobs because we are facing a revolution in the working world that is unprecedented in human history. It will hit both low-skilled workers and highly skilled professional groups as well. Nobody can say that these people do not want to work. There just won't be any more work for them. So what will we do with them?

Some of the problems we have to deal with today can already be regarded as harbingers of this digital revolution. We have to lead completely different discussions based on the twenty-first century, detached from the ideological debates of the nineteenth century. It does not help to throw around terms such as *class warfare*, *machine tax*, *capitalism*, and *socialism* in a knee-jerk reaction. The twenty-first century requires solutions from the twenty-first century.

Silicon Valley giveth, and Silicon Valley taketh away. If we are not careful, if we fail to confront these issues, we will exacerbate our situation still further. In 2016, Switzerland took a vote on the issue of an unconditional basic income, which may be a possible solution to those fundamental changes. However, at that time, it was still rejected by a large majority. We should be debating this issue in other countries, too, seriously, honestly, and ready to let past ideologies be bygones, trying to find answers for the people of the future. The future is very close indeed.

Overcoming Cognitive Distortions

Our prejudices and human rationality or irrationality are not always the best guidance when trying to understand big upheavals and reacting appropriately. So-called cognitive distortions have very tangible effects:[1]

- **Fear of loss.** Potential overestimation of losses and underestimation of potential benefits
- **Endowment effect.** A higher estimation of the assets we already have
- **Status quo distortion.** Preferring the status quo to any change (These three types of cognitive distortion ensure that people still prefer their

own private cars that they steer personally to any of the future shared and/or autonomous vehicles.)

- **Faulty risk assessment.** Dramatic overestimation of unknown risks in comparison with existing risks even if the numbers contradict this assessment (This is the underlying reason why consumers think autonomous and shared cars are more dangerous than they actually are or will turn out to be.)
- **Optimistic distortion.** Overestimation of one's own abilities and underestimation of risks (Car owners believe that they themselves drive more safely and have better control than machines, overlooking the safety advantages that autonomous vehicles offer.)
- **Availability heuristics.** Application of rules of thumb when assessing facts without being able to draw on one's own experience or exact information (Human beings tend to focus on rare negative events such as accidents, the trolley problem, or cyberattacks that may come with the future of mobility.)

Put Unconditional Basic Income and Robot Taxes on the Agenda

According to the official definition, "unconditional basic income (UBI) is a social-political financial transfer concept according to which every citizen—independently of their economic situation—shall receive a legally fixed financial contribution that is the same for everyone and paid by the state, without having to provide any return service for it."[2] Assuming that every single one of us receives $500 or even $2,000 every month as a guaranteed income without having to work, how would we then spend our time? This is difficult to say. We must not forget that there have always been groups in our populations that were not necessarily included in the (paid) work cycle: children and senior citizens (depending on the country), rich heirs, religious groups, and aristocracy were and are among those groups. Some of them studied or honed their sports and arts talents; others took care of the home and children or turned to charitable causes.[3] And then there are those who just enjoy life.

We would be able to afford such benefit payments even today, although this is often vehemently denied. Actually, benefit payments already constitute the highest amounts in most countries' budgets. In the United States, 20 percent of all household income is benefit payments.[4] Tax rebates and subsidies for companies are nothing but benefit payments. The money the governments gave to failing banks during the financial crisis is benefit payments.

However, many people define themselves based on their jobs. Losing their job throws them into an ethical crisis of purpose, leaving them with the feeling that they are now useless or do not belong anywhere anymore. Our societies today look down on people without work.[5] Our entire life revolves around our occupation, the way we organize our daily tasks, the traffic system, mealtimes, holidays, even the times we use most of our electricity. Schools train us so that we can find a job. But what happens if jobs are no longer the solution but the problem?

Eighty-five percent of all jobs lost in the course of the last two decades were not moved to low-wage countries but were in fact made obsolete by technological progress.[6] Those jobs are not ever coming back. The combination of automation and AI will lead to the loss or complete redefinition of 47 percent of all jobs in the United States, 35 percent of all jobs in Great Britain, and a staggering 77 percent of all jobs in China.[7] The Organisation for Economic Co-operation and Development indicates that 57 percent of all jobs will be affected.

Bill Gates, philanthropist and founder of Microsoft, therefore suggests introducing a tax on robots and automation.[8] A human being doing a job pays income tax. A robot does not. This increases productivity and decreases the tax burden for companies. If you look at it closely, this shows that our tax systems are organized in a way to penalize human work and encourage people to replace humans by robots, a policy that inevitably leads entire countries on a downward spiral. Robots take jobs and decrease the tax income that would help us to retrain future unemployed workers, pay unemployment benefits, or simply pay an unconditional basic income to everyone. Taxes on robots would not, by contrast, really create a conflict of company interests. You do not have to keep provisions for expenses such as health insurance or retirement benefits for robots. They would still be cheaper than humans.

Whether the unconditional basic income or robot taxes or other measures are the solution, I have no better idea than you do. We should, however, be allowed to have an open discussion about this—overcoming traditional behavior and thinking patterns—without having to defend ourselves or be ridiculed. An unconditional basic income is just one of many tools we will have to examine for suitability. And we should do it soon.

Qualifying for the Future

Our future starts with our children. At present, we still educate children with the goal that they can hold a job later on and earn a living for themselves. In fact, many students at school and at university today still have a vision of working for Bosch, General Motors, Mercedes, General Electric, or IBM or of becoming a civil servant. There will be fewer jobs in those industries, however, and those that are available will require more rapidly changing profiles than ever before. In order to prepare our children for this, we have to provide them with an education that really enables them to create their future.

Gigi Read, an ex-colleague and a "mom on a mission" (as she frequently refers to herself), runs workshops for 8- to 14-year-olds to teach them the qualifications necessary for the twenty-first century. Our syllabus today teaches reading, mathematics, science, cultural and social values, and knowledge. According to the World Economic Forum, the curricula are lacking such competencies as critical thinking and problem-solving strategies, creativity, competencies in communication and cooperation, and the development of characteristics such as curiosity, initiative, persistence, adaptability, leadership skills, and social as well as cultural awareness. Read, for example, offers design thinking for children as a framework to identify and solve problems through stories. To accomplish this, the children combine digital tools with modeling clay and craft utensils. Prototypes are developed and modeled, transferred to a computer with a 3D scanner, and printed on a 3D printer. The children learn a programming language, Scratch, to program robots and immediately put the acquired knowledge into practical results. AI is also an issue. How do you create it, and how do you help it with machine learning?

If we indeed experience a wave of job losses, how will we then teach similar workshop strategies to those affected by unemployment? How can we help people start a wave of entrepreneurship and lifelong learning?

Show a Willingness to Change

I can talk until I am blue in the face and provide a million facts and data to prove that the second automotive revolution is already under way—all of which is to no avail if there is no willingness to embrace change. The example I included earlier, that of the Hungarian startup founder who only understood the value of design thinking when he had to close down his own startup and was on the lookout for new ideas, is a case in point.

Small exercises help you to prepare for mindset changes and to make crucial decisions. I have a few more suggestions for you: gain experience! Rent a Tesla or a BMW i3 and drive around in it for an hour, a day, or better still, a whole week. Consciously go to a place where self-driving cars and buses are being tested and observe them. Or get into a driverless subway or tram. This would be a good start for beginners.

Always remember how quickly changes can happen, even in the automobile industry. In 1900, Fifth Avenue in New York City was lined with horse-drawn carriages, but by 1913, the street was full of cars. Think of friends or people you know who have been involved in serious traffic accidents. Think about going to work. Would you still like to be "steamed" for your daily commute, or would you rather take an electric suburban train? So why do so many of you still insist on cars propelled by "dinosaur juice"?

Afterword

Germans invented the car. Americans created the automotive lifestyle. The Japanese perfected the production process of cars. Yet their reputations were of no use when it came to saving them; as an example, a sterling reputation didn't help the great coach builder Carl Marius in 1920. Kodak could no longer benefit from making the best film in the world when it went bankrupt in 2012. Nokia and RIM/Blackberry no longer drew any profit from producing the best cell phones with hardware keyboards. Ampex had the best audio and video recorders, and filed for bankruptcy in 2012. Eumig made the best super-8 film cameras and projectors and held 100 percent market shares—in a market that shrank to zero.

The signs of a turning point are obvious: the diesel emissions scandal, the bans on vehicles, great sales results for Tesla, sports cars left behind by electric vehicles (EVs), digital user interfaces that demonstrate how far behind we are compared with some tech companies, young adults who no longer buy cars and do not want a driver's license, traffic jams that worsen every day, automobile pioneers appearing from completely different industries, hundreds of companies developing technologies for autonomous cars, and billions of dollars redirected to new types of transportation by investors.

Max, Sofie, Julian, and my three sons won't have to sit behind the wheel themselves. Nor will they want to. They will be driven electrically, and perhaps they won't even have their own car. Being a father, I am relieved not only because in this way they are less likely to do something stupid but also because they are less likely to be put in danger by others. Who will provide the dominant

381

technologies for the next automobile generation is not quite so obvious. However, there are clear signals that traditional carmakers are probably not going to be among the leading global players. Based on the facts presented in this book, I would place my money on companies from Silicon Valley and Asia.

So are the automotive engineers who work in the development departments of automobile companies in Silicon Valley and in China the smart ones, whereas those who stayed at home the incapable ones without ambition? Of course not. The situation is due to the companies themselves. Their success made traditional manufacturers content and sated. A success story with millions of cars sold is not conducive to embracing a change, especially a change from a success model that is earning a lot of money to a mainly unknown, untested driving experience one needs to learn from scratch just like everyone else.

However, success should never be taken for granted. Just as Germany won the soccer World Cup in 2014 but failed miserably in the UEFA European Championship in 2016 and even more miserably in the World Cup in 2018, past heydays are no guarantee of future survival. Examples such as Nokia and Kodak should be warning enough. The staff at all companies should maintain a certain degree of paranoia. Remember that at some point—usually earlier than later—somebody might come along and blow you out of the water with a new technology, a new business model, or better execution. Some companies have had this kind of "near-death experience" and turned it into a fixed part of their corporate DNA, whereas others have rested on their laurels and disappeared. There is no need to look across the ocean to find examples of this.

These signals are not isolated, pale dots in the sky; they radiate like bright comets that could fall from the sky right onto our heads. Sticking our heads in the sand pretending that this has nothing to do with us or—worse even, telling everyone else that there is no danger if you simply ignore it—is negligent. Naturally, not all the signals I described will have the same impact or the same intensity or will strike at precisely the predicted moment. Yet there are more than enough of them that the effects—even if only a few of them come true—will be so significant that we will be torn and shaken.

The second automobile revolution is in full swing, and it requires courageous action with no ifs, ands, or buts. The discussion has long been over; now it is time to roll up our sleeves and get to work. Defaming electric mobility and refusing to accept autonomous driving and sharing models due to misconceived protectionism for the national economy are going to backfire and in the end only benefit the competition.

I will say it once more, loud and clear: the era of combustion motors for vehicles is over. End of story! Done! Finished! Politicians and car manufacturers claiming otherwise are doing a disservice to everyone and to themselves. We have to work on competencies for alternative propulsion drives now, hoping that it isn't already too late. Hundreds of thousands of jobs in those old professions will be lost. Not maybe lost, definitely lost. Digital competency, expertise in AI, and the ability to handle and master new technologies as well as create conducive framework conditions are part of the most important objectives of our time. We should build the structures to prepare people for jobs in new professions that will come out of those new technologies and support a new wave of startups now rather than later.

We cannot justify it to our children and grandchildren if we leave them a rundown nation with a ruined economy. Even today, we are not making their lives easier because it will be increasingly difficult for them to find permanent jobs and create financial security for themselves. This is why sharing models are so popular with the younger generation. With Brexit, the older generation flipped the younger one the bird. So where does that lead us?

Whichever way you twist and turn the numbers forging the change, we are looking at hundreds of thousands of jobs lost in a best-case scenario or possibly more than a million jobs. In one industry alone! This will look like a drop in the bucket compared with the changes in the past. In all this, countries have access to the same technologies and resources as the companies that are driving the second automobile revolution forward. Some countries actually often have better know-how than the companies that are disrupting have. But all that counts for nothing if managers in those companies continue with a mindset that is arrogant, afraid of risks, and without ambition and have a smart-aleck comment about everything. Then we will deserve what we get. The right mindset is not magic. Every country has countless examples of how its sons and daughters can be innovative and successful. All we need to do is follow that example.

There is no excuse for not doing anything: if we cannot make it, we have nobody to blame but ourselves. Tesla, Google, Apple, Facebook, and Uber are not "happening" to us; rather, it is the future that is putting pressure on us. Yet we allow this pressure to be put on us because we dream of past glory and are oblivious to what is fast approaching. We are no longer shaping our fate. To put it in soccer terms, you could say that our style is *catenaccio* compared with the competition: defensive, destructive, just trying to block the game. We want to increase our advantage, but the rules of the games are changing as we speak, and the audience is on its way to a different stadium. There is nothing left from the era when we had to catch up with Great Britain, the industrial powerhouse of the time.

To make things worse, our problem is not Chinese or American companies but German, English, French, and Japanese engineers in foreign companies who pull the rug from under the feet of German, English, French, and Japanese engineers in domestic enterprises. They have boarded the electric train, while we are still running to catch it. So let's change our mindset. The mindset of employees in the companies, of our politicians, of civil servants, and of our entire society! We need fewer administrators and apologists telling people what customers should want (big, showy cars), what doesn't work (self-driving vehicles), what nobody wants (electric cars), and who is to support them in this (politicians). People like this never solve problems; they only create them.

This work is the result of three years of research and more than two decades of exploring human behavior. I did not write this book to pick on Germany, France, or Japan or on the companies there and not even on those countries' societies and politics in general, ranting about their alleged ignorance and incapacity, although it may sometimes appear that way. I have lived in Silicon Valley for years, my sons are Americans, and I could simply say "I don't care!" But I don't say that, and I won't, because my parents, siblings, nieces and nephews, other relatives, and many, many friends live in some of these countries, and I am a European citizen, too. I cannot simply look on, especially when it is the automobile industry that is at stake, because it is just too important for many countries, and so are our children. Far too important to continue endangering their lives in potential traffic accidents and by exposing them to environmental pollution. I am presenting no secrets here: all the facts are public and available. The new "element" here is probably the intensity of the overview I provide.

That is because I want you to wake up and help shape the future! What are you waiting for?

Notes

Introduction

1. https://www.theglobaleconomy.com/rankings/Percent_urban_population/.
2. http://de.theglobaleconomy.com/rankings/Percent_urban_population/.
3. Ivan Arreguín-Toft, *How the Weak Win Wars: A Theory of Asymmetric Conflict* (Cambridge Studies in International Relations, Vol. 99). Cambridge University Press, 2005.
4. https://en.wikipedia.org/wiki/Samuel_Pierpont_Langley.
5. https://www.ted.com/talks/simon_sinek_how_great_leaders_inspire_action/transcript.
6. Lukas Bay, Thomas Tuma, "Elon Musk: All Charged Up in Berlin," Global. Handelsblatt.com, September 25, 2015, https://global.handelsblatt.com/edition/271/ressort/companies-markets/ article/ all-charged-up-in-berlin.
7. Al Gore, *The Future*. Pantheon, Munich, 2015.
8. Richard H. Thaler, *Misbehaving: The Making of Behavioral Economics*. Norton, New York, 2015.
9. http://www.mdr.de/nachrichten/politik/inland/steuerzahlerbund-absurde -foerderprojekte-100; html; https://schienestrasseluft.de/2016/03/21/update -foerdermillionen-fuer-einen-porsche/.
10. Amounts of national subsidies for German automotive manufacturers from 2010 to 2012, http://de.statista.com/statistik/daten/studie/197024/umfrage/subventionen -fuer-autohersteller-aus-dem-konjunkturpaket-ii/.
11. David McCullough, *The Wright Brothers*. Simon & Schuster, New York, 2015.
12. Eric Morris, *From Horse Power to Horsepower*, http://www.uctc.net/access/30/ Access%20 30%20-%2002%20-%20Horse%20Power.pdf.
13. Margaret Derry, *Horses in Society: A Story of Animal Breeding and Marketing, 1800–1920*. University of Toronto Press, Toronto, 2006, p. 131.
14. Jim Collins, *How the Mighty Fall: And Why Some Companies Never Give In*. Random House Business, New York, 2009.

15. "VW's U.S. Volume Tumbles 17% for Worst May Since 2010," https://
www.autonews.com/article/20160601/RETAIL01/306019995/vw-u-s-volume
-tumbles-17-for-worst-may-since-2010; and http://www.manager-magazin.de/
unternehmen/karriere/bmw-audi-vw-das-sind-die-beliebtesten-arbeitgeber
-deutschlands-a-1088279.html.

Part I

1. William F. Ogburn, Dorothy Thomas, "Are Inventions Inevitable? A Note on Social
Evolution," in *Political Science Quarterly* 37(1):83–98, March 1922.
2. Bill Aulet, *Disciplined Entrepreneurship: 24 Steps to a Successful Startup*. Wiley,
Hoboken, NJ, 2013.
3. Warren Berger, *A More Beautiful Question: The Power of Inquiry to Spark Breakthrough
Ideas*. Bloomsbury, London, 2014.
4. https://en.wikipedia.org/wiki/Electric_car.

Chapter 1

1. https://de.wikipedia.org/wiki/Hansa-Automobil.
2. https://www.lohner.at/about-us/?lang=en.
3. Clayton M. Christensen, Joseph L. Bower, "Customer Power, Strategic Investment
and the Failure of Leading Firms," in *Strategic Management Journal* 17:197–218,
1996.
4. Daniel Franklin, John Andrews, *Megachange: The World in 2050*. Economist Books,
2012.
5. "Charts of the Day: Creative Destruction in the S&P 500 Index," https://www.aei.org/
publication/charts-of-the-day-creative-destruction-in-the-sp500-index/.
6. https://techcrunch.com/2016/08/30/drive-ai-uses-deep-learning-to-teach-self-driving
-cars-and-to-give-them-a-voice/.
7. "The Unknown Start-up That Built Google's First Self-Driving Car," http://spectrum
.ieee.org/robotics/artificial-intelligence/the-unknown-startup-that-built-googles-first
-selfdriving-car.
8. Matt Warmann, "Can the Web Make the World Go Faster?," *The Telegraph* 18,
November 2010, http://www.telegraph.co.uk/technology/facebook/8140562/Can
-the-web-make-the-world-go-faster.html.
9. http://www.sueddeutsche.de/wirtschaft/lobbyismus-zum-wohle-des-deutschen-autos
-1.3035506.
10. Daniel Goleman, *Focus: The Hidden Driver of Excellence*. Bloomsbury, London, 2013.
11. John Micklethwait, Adrian Wooldridge, *The Fourth Revolution*. Allen Lane, London,
2014.
12. https://en.wikipedia.org/wiki/Seven_generation_sustainability.

Chapter 2

1. *Reclaim: The Magazine of Transportation Alternatives* 20:1, 2014, http://www.transalt
.org/sites/default/files/news/magazine/2014/Spring/Reclaim_2014-1_LQ.pdf.
2. "The Struggle for the Automobile," https://www.nzz.ch/schweiz/schweizer-geschichte/
sonderfall-graubuenden-der-kampf-ums-automobil-ld.103634.
3. *Reclaim: The Magazine of Transportation Alternatives* 20:1, 2014, http://www.transalt
.org/sites/default/files/news/magazine/2014/Spring/Reclaim_2014-1_LQ.pdf.

4. "The Invention of Jaywalking," http://www.citylab.com/commute/2012/04/ invention-jaywalking/1837/.

5. http://www-nrd.nhtsa.dot.gov/pubs/812115.pdf.

6. Christian Engel, "Jugendliche Automuffel: 'Führerschein? Unnötig!'" (Young Car Objectors: Driving License? No Need!"), Spiegel Online, September 21, 2015, http://www.spiegel.de/lebenundlernen/schule/auto-verweigerer-keine-lust-auf -fuehrerschein-a-1040493.html.

7. "Possession of Vehicles and Driving Licenses," https://www.bfs.admin.ch/bfs/de/ home/statistiken/mobilitaet-verkehr/personenverkehr/verkehrsverhalten/besitz -fahrzeuge-fahrausweise.html.

8. Jack Neff, "Is Digital Revolution Driving Decline in U.S. Car Culture?," http:// adage.com/article/digital/digital-revolution-driving-decline-u-s-car-culture/144155/.

9. "Jugend ohne Auto: Die Zweckmobilisten (Youth Without Automobile: Mobile for a Purpose)," http://www.tagesspiegel.de/politik/jugend-ohne-auto-die -zweckmobilisten/9752254.html.

10. http://www.goldmansachs.com/our-thinking/pages/millennials/.

11. https://de.statista.com/statistik/daten/studie/215576/umfrage/durchschnittsalter-von -neuwagenkaeufern/.

12. "Cruising Toward Oblivion: America's Fading Car Culture," http://www .washingtonpost.com/sf/style/2015/09/02/americas-fading-car-culture/?utm _term=.45ab2deae6ff.

13. "Record 10.8 Billion Trips Taken on U.S. Public Transportation in 2014," http:// www.apta.com/mediacenter/pressreleases/2015/pages/150309_ridership.aspx.

14. Intelligent Cities Initiative (poster), National Building Museum, Washington, DC, 2012 https://www.nbm.org/exhibitions/publications/intelligent-cities/.

15. BMW—das deutsche Apple, *Manager Magazine*, August 2015. https://www .manager-magazin.de/magazin/artikel/bmw-soll-unter-chef-harald-krueger-das -deutsche-apple-werden-a-1052278.html.

16. Amy Webb, *The Signals Are Talking: Why Today's Fringe Is Tomorrow's Mainstream*. Public Affairs Press, New York, 2016.

Chapter 3

1. http://www.spiegel.de/auto/aktuell/tesla-model-3-die-deutschen-hersteller-sind -weiter-als-viele-denken-a-1084896.html.

2. http://www.businessinsider.com/blackrock-topic-we-should-be-paying-attention -charts-2015-12/.

3. Tom Standage, *The Victorian Internet*. Weidenfeld & Nicholson, London, 1998.

4. John Freeman, *The Tyranny of E-mail*. Scribner, New York, 2009.

5. Stephen Kern, *The Culture of Time and Space, 1880–1918*. Harvard University Press, Cambridge, MA, 2003.

6. Larry Downes, Paul Nunes, *Big Bang Disruption*. Portfolio Press, New York, 2014.

7. Peter Thiel, Blake Masters, *Zero to One: Notes on Startups, or How to Build the Future*. Currency, 2014.

8. http://www.umweltbundesamt.at/umweltsituation/energie/energieszenarien/.

9. https://cleantechnica.com/2016/08/29/tesla-model-3-delivers-gas-vehicles-history -gasoline-automotive-services-dealers-america-exec-says/.

10. https://www.wsj.com/articles/why-electric-cars-will-be-here-sooner-than-you-think -1472402674.

11. Association of Automobile Importers, *Facts Instead of Prejudices: Clear Answers on Environment, Climate and the Automobile.* 2014, http://www.automobilimporteure.at/ wp-content/uploads/2015/06/Fakten-statt-Vorurteile.pdf.
12. http://www.marketwatch.com/investing/stock/gm/financials/cash-flow.
13. http://annualreport.ford.com/.
14. "Tesla Has Something Hotter Than Cars to Sell: Its Story," https://www.nytimes.com/ 2017/04/06/business/tesla-story-stocks.html.
15. http://fr.slideshare.net/capgemini/cars-online-2015.
16. Interview with Friedrich Indra, "Es gibt einen Hass gegen Verbrenner: Motoren-Papst rechnet mit Elektromobilität ab" (This Is Hatred Against Combustion Engines: The Motor Pope Squares Accounts with Electric Mobility), http://www.focus.de/auto/ elektroauto/interview-mit-friedrich-indra-es-gibt-einen-hass-gegen-verbrenner -motoren-papst-rechnet-mit-elektromobilitaet-ab_id_6512817.html.
17. "Disruptive Trends That Will Transform the Auto Industry," http://www.mckinsey .com/insights/high_tech_telecoms_internet/disruptive_trends_tha t_will_transform _the_auto_industry.
18. http://www.newyorker.com/magazine/2011/02/14/the-information.

Part II

1. https://en.wikipedia.org/wiki/Clarke%27s_three_laws.
2. Frances Anne Kemble, *Records of a Girlhood*, 1878.
3. Clifford Winston, "On the Performance of the U.S. Transportation System: Caution Ahead," *Journal of Economic Literature* 51(3):773–824, September 2013, http://pubs .aeaweb.org/doi/pdfplus/10.1257/jel.51.3.773.
4. "The Future Economic and Environmental Costs of Gridlock in 2030: An Assessment of the Direct and Indirect Economic and Environmental Costs of Idling in Road Traffic Congestion to Households in the UK, France, Germany and the USA," https://www.cebr.com/reports/the-future-economic-and-environmental-costs -of-gridlock/.
5. http://de.statista.com/statistik/daten/studie/30703/umfrage/beschaeftigtenzahl-in-der -automobilindustrie/.
6. https://www.vda.de/en/services/facts-and-figures/facts-and-figures-overview.
7. Matthias Breitinger, "Hört auf, die Autobranche zu hätscheln" (Stop Pampering the Automotive Industry), http://www.zeit.de/mobilitaet/2016-09/automobilindustrie -iaa-bundesregierung-abgasskandal-5vor8.
8. http://www.strategyand.pwc.com/innovation1000.

Chapter 4

1. http://www.statista.com/statistics/232958/ revenue-of-the-leading-car-manufacturers-worldwide/.
2. http://www.spiegel.de/auto/aktuell/zulieferer-die-heimlichen-autohersteller-a-1108529 .html.
3. http://www.cargroup.org/?module=Publications&event=View&pubID=103.
4. "Corporate Profits and Research and Development Spending," in *2010 Ward's Motor Vehicle Facts and Figures.* 2011.
5. https://www.nada.org/nadadata/.

6. http://de.statista.com/statistik/daten/studie/74642/umfrage/kfz-betriebe-in -deutschland-seit-2004/.
7. "Self-Driving Cars Might Need Standards, but Whose?," https://www.nytimes.com/ 2017/02/23/automobiles/wheels/self-driving-cars-standards.html.
8. Lawrence Livermore National Laboratory, "Estimated US Energy Use in 2014," https://flowcharts.llnl.gov/commodities/energy.
9. http://www.pumpthemovie.com/.

Chapter 5

1. Beijing Air Pollution: Real-Time Air Quality Index (AQI), http://aqicn.org/city/beijing/.
2. http://venturebeat.com/2016/12/10/israeli-startups-deliver-much-needed-tech-for -self-driving-cars/.
3. Georges Haour, Max von Zedtwitz, *Created in China: How China Is Becoming a Global Innovator*. Bloomsbury, London, 2016.
4. http://www.reuters.com/article/us-autos-china-leeco-idUSKCN0XH10D.
5. https://electrek.co/2016/08/11/faraday-futures-chinese-backer-leeco-electric -vehicle-factory/.
6. http://www.pcworld.com/article/2900452/foxconn-partners-with-chinas-tencent -on-smart-electric-cars.html.
7. http://fortune.com/2016/07/12/future-mobility-electric-car-2020/.
8. https://www.cbinsights.com/company/atieva-funding.
9. "Buffetts chinesische Wette auf den Erfolg der Elektro-Mobilität" (Buffett's Chinese Bet on the Success of Electric Mobility), http://www.wallstreet-online.de/ nachricht/8878929-warren-s-byd-buffetts-chinesische-wette-erfolg-elektro-mobilitaet.
10. "Electric Buses and Driverless Shuttles Are About to Solve Auckland's Traffic Woes," https://ecotricity.co.nz/electric-buses-and-driverless-shuttles-are-about-to -solve-aucklands-traffic-woes/.
11. "'Electric Buses Are Now Cheaper Than Diesel/CNG and Could Dominate the Market Within 10 Years,' Says Proterra CEO," https://electrek.co/2017/02/13/ electric-buses-proterra-ceo/.
12. "Cities Shop for $10 Billion of Electric Cars to Defy Trump," https://www .bloomberg.com/news/articles/2017-03-14/cities-shop-for-10-billion-of-electric -vehicles-to-defy-trump.
13. "So bremst China die E-Konkurrenz aus" (How China Thwarts the Electric Competition), http://www.n-tv.de/wirtschaft/So-bremst-China-die-E -Konkurrenz-aus-article18781721.html.
14. Greg Anderson, *Designated Drivers: How China Plans to Dominate the Global Auto Industry*. Wiley, Hoboken, NJ, 2012.
15. "Gas-to-Electric Cab Conversion in Beijing Brings Opportunity Worth 9 Bln Yuan," http://www.nbdpress.com/articles/2017-02-23/1613.html.
16. "China Takes Lead as Number One in Plug-in Vehicle Sales," http://www.hybridcars .com/china-takes-lead-as-number-one-in-plug-in-vehicle-sales/.
17. "Alternative Antriebe? Nicht mit den Deutschen" (Alternative Drive Trains? Not with the Germans), http://www.manager-magazin.de/unternehmen/autoindustrie/ elektroauto-boom-nicht-in-deutschland-a-1074200.html.
18. https://www.welt.de/motor/modelle/article154606460/Diese-Laender-planen-die -Abschaffung-des-Verbrennungsmotors.html.

19. http://deutsche-wirtschafts-nachrichten.de/2016/09/11/debatte-um-diesel-fahrverbot-in-deutschland-eroeffnet/.

20. "Warum in Deutschland kaum Elektroautos gebaut warden" (Why So Few Electric Cars Are Made in Germany), http://bauplan-elektroauto.de/kaum-elektroautos/.

21. http://www.faz.net/aktuell/wirtschaft/unternehmen/elektromobilitaet-bosch-steht-an-der-spitze-bei-subventionen-fuer-elektroautos-12190060.html.

22. https://www.spiegel.de/international/spiegel/germany-s-new-mercedes-museum-from-horsepower-to-the-popemobile-a-416896.html.

23. "Tesla Shows How the Model S Is Totally Disrupting the Large Luxury Car Market in the US," http://electrek.co/2016/02/10/tesla-shows-how-the-model-s-is-totally-disrupting-the-large-luxury-car-market-in-the-us/.

24. "Trial Illuminates Porsches' Rise to Power at Volkswagen," http://www.nytimes.com/2016/02/15/business/international/ex-porsche-executives-trial-sheds-light-on-a-familys-rise-at-volkswagen.html.

25. https://netzfrauen.org/2016/04/26/autokonzerne-wurden-mit-milliarden-subventionen-gespeist-und-verpennen-emobilitaet-emobilitaet-aus-china-erobert-die-welt-nun-soll-es-wieder-der-steuerzahler-richten/.

26. "Lucid (Formerly Known as Atieva) Will Be the Sole Battery-Pack Supplier for Formula E," http://blog.caranddriver.com/lucid-formerly-known-as-atieva-will-be-the-sole-battery-pack-supplier-for-formula-e/.

27. https://www.greencarcongress.com/2019/01/20190130-vwdiesel.html.

28. "Bericht der Untersuchungskommission 'Volkswagen': Untersuchungen und verwaltungsrechtliche Maßnahmen zu Volkswagen; Ergebnisse der Felduntersuchung des Kraftfahrt-Bundesamtes zu unzulässigen Abschalteinrichtungen bei Dieselfahrzeugen und Schlussfolgerungen" (Report of the "Volkswagen" Investigation Commission: Investigations and Administrative Measures Regarding Volkswagen; Findings of the Field Research Conducted by the German Federal Motor Vehicle Authority), April 22, 2016, https://www.bmvi.de/SharedDocs/DE/Anlage/VerkehrUndMobilitaet/Strasse/bericht-untersuchungskommission-volkswagen.pdf?__blob=publicationFile.

29. "Moderne Diesel-Pkw stoßen mehr Schadstoffe aus als Lastwagen" (Modern Diesel Vehicles Are More Polluting Than Trucks), http://www.zeit.de/mobilitaet/2017-01/icct-studie-diesel-pkw-stickoxide-ausstoss#a-03c3b9ea-62a7-476d-94ec-ed31a6d95b6b.

30. http://www.automobile-propre.com/marguerite-2-cv-electrique/.

31. http://www.bloomberg.com/news/articles/2016-05-12/why-tesla-s-mass-market-car-should-scare-mercedes-and-bmw.

32. http://www.goodcarbadcar.net/

33. "Ernstfall Schweiz – wie Tesla Mercedes, Audi und Co. das Geschäft verdirbt" (Switzerland Emergency: How Tesla Ruins Business for Mercedes, Audi and Co.), http://www.manager-magazin.de/unternehmen/autoindustrie/elektroautos-warum-die-schweizer-gern-mit-strom-fahren-a-1092086.html.

34. http://electrek.co/2016/05/19/the-math-and-evidence-all-around-you-that-shows-shared-autonomous-vehicles-powered-by-solar-power-and-batteries-are-inevitable/.

35. http://fortune.com/2015/11/17/electric-motors-crush-gas-engines/.

36. http://www.focus.de/auto/elektroauto/medienbericht-zu-elektroautos-geheimer-plan-ministerium-wollte-1000-euro-strafabgabe-fuer-autofahrer-mit-benzinfahrzeugen_id_5906218.html.

37. "All Charged Up in Berlin," https://global.handelsblatt.com/edition/271/ressort/companies-markets/article/all-charged-up-in-berlin.

38. "Tesla ist kein Vorbild: Interview mit Harald Kröger, Leiter Entwicklung Elektrik bei Mercedes-Benz" (Tesla Is No Model: Interview with Harald Kröger, Head of Electrics Development at Mercedes-Benz), http://mein-auto-blog.de/news/tesla-kein-vorbild.html.

39. "Battery Cell Production Begins at the Gigafactory," https://www.tesla.com/no_NO/blog/battery-cell-production-begins-gigafactory.

40. "Battery Material Could Reduce Electric Car Weight," https://www.kth.se/en/aktuellt/nyheter/de-gor-batterier-av-kolfiber-1.480780.

41. http://energyload.eu/elektromobilitaet/elektroauto/hanergy-solarauto/.

42. https://www.sonomotors.com/.

43. https://www.youtube.com/watch?v=9qi03QawZEk.

44. https://en.wikipedia.org/wiki/Lithium.

45. http://www.sueddeutsche.de/auto/elektromobilitaet-die-verspaetete-revolution-der-deutschen-autoindustrie-1.3046291-2.

46. http://www.pbqbatteries.com/media/datasheet/lithium-ferro-phosphate-batteries-vs-vrla-batteries.pdf.

47. http://www.kreiselelectric.com/.

48. http://www.pluginamerica.org/surveys/batteries/model-s/faq.php.

49. "Tesla Model S Battery Degradation Data," https://steinbuch.wordpress.com/2015/01/24/tesla-model-s-battery-degradation-data/.

50. "Battery Capacity Loss Warranty Chart for 2016: 30 kh Nissan LEAF," http://insideevs.com/battery-capacity-loss-chart-2016-30-kwh-nissan-leaf/.

51. http://www.greencarcongress.com/2013/03/vo2-20130315.html.

52. "Tesla CTO JB Straubel Talks Battery Technology, 'One-Stop Sustainable Lifestyle' Company and More," https://electrek.co/2016/11/14/tesla-cto-jb-straubel-battery-technology/.

53. McKinsey and Company, "Electrifying Insights: How Automakers Can Drive Electrified Vehicle Sales and Profitability," January 2017.

54. "Tesla Is Now Claiming 35% Battery Cost Reduction at 'Gigafactory 1': Hinting at Breakthrough Cost Below $125/kWh," https://electrek.co/2017/02/18/tesla-battery-cost-gigafactory-model-3/.

55. http://www.zeit.de/mobilitaet/2015-08/elektromobilitaet-batterie-recycling.

56. "Recycling of Lithium-Ion Batteries," http://www.elektroniknet.de/elektronik/power/recycling-von-lithium-ionen-akkus-106499.html.

57. Linda Gaines, "The Future of Automotive Lithium-Ion Battery Recycling: Charting a Sustainable Course," *Sustainable Materials and Technologies* 1–2:2–7, December 2014, http://www.sciencedirect.com/science/article/pii/S2214993714000037.

58. http://energyload.eu/elektromobilitaet/elektrofahrzeuge/hybrid-lkw-autobahn-oberleitung/.

59. http://www.charinev.org/.

60. "Japan Has More Car Chargers Than Gas Stations," http://www.japantimes.co.jp/news/2015/02/16/business/japan-has-more-car-chargers-than-gas-stations#.WJb3WBBOm5h.

61. "Japan Now Has More Electric Charging Points Than Petrol Stations," https://www.weforum.org/agenda/2016/05/japan-now-has-more-electric-charging-points-than-petrol-stations.

62. https://e-tankstellen-finder.com/.

63. "Auto-Weltmacht China" (China, the Automobile World Power), https://www.heise.de/tp/features/Auto-Weltmacht-China-3617797.html.

64. "Shell to Install Chargers for Electric Cars on European Forecourts," https://www
.ft.com/content/00d0f1ce-e22b-11e6-8405-9e5580d6e5fb.

65. "Why Electric Cars Will Be Here Sooner Than You Think," https://www.wsj.com/
articles/why-electric-cars-will-be-here-sooner-than-you-think-1472402674.

66. http://www.slam-projekt.de/.

67. "Die Stadt Wien errichtet bis zu 1000 E-Tankstellen – Ampeln sollen als Stromquelle
dienen" (City of Vienna to Install Up to 1000 e-Service Stations: Traffic Lights to
Be Used as Power Sources), http://diepresse.com/home/panorama/wien/4958347/
Ampeln-als-ETankstellen?_vl_backlink=%2Fhome%2Fpanorama%2Fwien%2Findex
.do.

68. http://www.openchargealliance.org/.

69. https://www.hubject.com/.

70. "Google Wants Its Driverless Cars to Be Wireless Too," http://spectrum.ieee.org/
cars-that-think/transportation/self-driving/google-wants-its-driverless-cars-to-be
-wireless-too.

71. "The U.K. Is Testing Roads That Recharge Your Electric Car as You Drive," http://
www.citylab.com/commute/2015/08/the-uk-is-testing-roads-that-recharge-your
-electric-car-as-you-drive/401276.

72. https://www.electreon.com/.

73. "Final Opinion on Potential Health Effects of Exposure to Electromagnetic Fields
(EMF)," http://ec.europa.eu/health/scientific_committees/consultations/public
_consultations/scenihr_consultation_19_en.htm.

74. Alexander Lerchl et al., "Tumor Promotion by Exposure to Radiofrequency
Electromagnetic Fields Below Exposure Limits for Humans," *Biochemical and
Biophysical Research Communications* 459(4):585–590, April 17, 2015, http://
www.sciencedirect.com/science/article/pii/S0006291X15003988.

75. "Segmenting the $10 Billion Battery Market for Plug-in Vehicles: Market Share
Projections for OEMs, Individual Models and Suppliers," https://portal
.luxresearchinc.com/research/report_excerpt/21944.

76. https://chargedevs.com/features/tom-gage-on-zev-mandates-teslas-early-days-bmws-ev
-commitment-and-v2g-tech/.

77. https://en.wikipedia.org/wiki/Vehicle-to-grid.

78. http://www.faz.net/aktuell/technik-motor/elektromobilitaet-lernimpuls-14504997.html.

79. Alfie Kohn, *Punished by Rewards*. Mariner Books, Boston, 1999.

80. http://www.reuters.com/article/2015/08/09/us-teslamotors-cash-insight.

81. http://germanaccelerator.com/.

82. https://en.wikipedia.org/wiki/Gigafactory_1.

83. https://global.handelsblatt.com/edition/271/ressort/companies-markets/article/
all-charged-up-in-berlin.

84. http://www.manager-magazin.de/unternehmen/autoindustrie/fahrbericht-bmw-i3
-rwe-eon-vattenfall-etc-verschlafen-e-mobilitaet-a-955489.html.

85. http://www.nfpa.org/safety-information/for-consumers/vehicles.

86. http://www.autozeitung.de/auto-news/auto-feuer-verhalten-fahrzeug-brand-statistik#.

87. http://ecomento.tv/2014/01/02/elektroautos-2014-die-elf-wichtigsten-fragen
-antworten-fuer-das-neue-jahr/.

88. https://www.thrillist.com/cars/the-beastly-car-collection-of-arnold-schwarzenegger.

89. https://www.facebook.com/notes/arnold-schwarzenegger/i-dont-give-a-if-we-agree
-about-climate-change/10153855713574658.

90. "Ökobilanz alternativer Antriebe" (Environmental Accounting of Alternative Propulsion Methods), http://www.umweltbundesamt.at/fileadmin/site/publikationen/REP0572.pdf.

91. http://www.ucsusa.org/clean-vehicles/electric-vehicles/life-cycle-ev-emissions#.V4_LPo5Om5g.

92. "Cleaner Cars from Cradle to Grave: How Electric Cars Beat Gasoline Cars on Lifetime Global Warming Emissions," http://www.ucsusa.org/sites/default/files/attach/2015/11/Cleaner-Cars-from-Cradle-to-Grave-full-report.pdf.

93. http://www.ucsusa.org/clean-vehicles/electric-vehicles/ev-emissions-tool#.WGVWj5JOm5g.

94. "Where in Europe Is Electric Car a Good Idea?," https://jakubmarian.com/where-in-europe-is-electric-car-a-good-idea/.

95. http://www.pri.org/stories/2012-11-02/energy-costs-oil-production.

96. http://www.eia.gov/Energyexplained/index.cfm?page=oil_refining.

97. National Research Council, *Hidden Costs of Energy: Unpriced Consequences of Energy Production and Use.* National Academies Press, Washington, DC, 2009, www.nap.edu/openbook.php?record_id=12794&page=1.

98. https://en.wikipedia.org/wiki/Gasoline.

99. http://www.oekonews.at/index.php?mdoc_id=1103262.

100. "Pump," https://www.pumpthemovie.com/.

101. Shuguang Ji, Christopher R. Cherry, Matthew J. Bechle, et al., "Electric Vehicles in China: Emissions and Health Impacts," *Environ. Sci. Technol.* 46(4):2018–2024, 2012, http://pubs.acs.org/doi/abs/10.1021/es202347q.

102. https://www.technologyreview.com/s/602458/planes-trains-and-automobiles-have-become-top-carbon-polluters/.

103. http://energycenter.org/sites/default/files/docs/nav/policy/research-and-reports/California%20Plug-in%20Electric%20Vehicle%20Owner%20Survey%20Report-July%202012.pdf.

104. http://energycenter.org/clean-vehicle-rebate-project/vehicle-owner-survey/feb-2014-survey.

105. Fraunhofer Institute for System and Innovation Research ISI, "Energiespeicher-Monitoring 2016: Deutschland auf dem Weg zum Leitmarkt und Leitanbieter?" (Energy Storage Monitoring 2016: Germany on Its Way to Leading Market and Leading Provider?), December 1, 2016, http://www.isi.fraunhofer.de/isi-de/t/publikationen/Energiespeicher-Monitoring-2016_Web.pdf.

106. http://www.handelsblatt.com/unternehmen/industrie/werk-kamenz-daimler-baut-produktionsverbund-fuer-batterien-auf/14729818.html.

107. http://nomadicpower.de/.

108. http://www.manager-magazin.de/unternehmen/autoindustrie/interview-wie-elektroautos-das-fahrzeugdesign-veraendern-koennten-a-1104660-2.html.

109. "Elektrische Motoren in Industrie und Gewerbe: Energieeffizienz und Ökodesign-Richtlinie" (Electric Motors in Industry and Commerce: Energy Efficiency and the Ecodesign Directive), https://web.archive.org/web/20111018090832 und http://www.industrie-energieeffizienz.de/fileadmin/InitiativeEnergieEffizienz/referenzprojekte/downloads/Leuchtturm/Ratgeber_Motoren_Energieeffizienz_OEkodesign.pdf.

110. http://www.auto-motor-und-sport.de/news/effizienz-wie-effizient-sind-elektromotoren
      -1322458.html.
111. "Wärmekraftwerke im energetischen Vergleich (in 2006, durchschnittliche
      Wirkungsgrade in Prozent)" [Thermal Power Plants in Energetic Comparison (in
      2006, Average Efficiency in Percent], http://kraftwerkforschung.info/quickinfo/
      energieversorgung/waermekraftwerke-im-energetischen-vergleich-in-2006
      -durchschnittliche-wirkungsgrade-in/.
112. http://ecomento.tv/2016/09/20/zf-bereitet-werk-saarbruecken-auf-elektromobilitaet-vor/.
113. https://www.wired.com/2016/05/hidden-battle-make-perfect-tires-electric-car-divas/.
114. Björn Nykvist, Måns Nilsson, "Rapidly Falling Costs of Battery Packs for Electric
      Vehicles," *Nature Climate Change* 5:329–332, 2015, http://www.nature.com/
      nclimate/journal/v5/n4/full/nclimate2564.html.
115. https://www.mckinsey.de/elektromobilitaet-mehrheit-der-deutschen-autokaeufer
      -vertraut-etablierten-herstellern.
116. "Tesla Is Now Claiming 35% Battery Cost Reduction at 'Gigafactory 1': Hinting at
      Breakthrough Cost Below $125/kWh," https://electrek.co/2017/02/18/tesla-battery
      -cost-gigafactory-model-3/.
117. http://journalistsresource.org/studies/environment/energy/electric-vehicles-battery
      -technology-renewable-energy-research-roundup.
118. https://www.db.com/cr/en/docs/solar_report_full_length.pdf.
119. http://www.afdc.energy.gov/calc/.
120. http://www.welt.de/motor/article157080589/Gebrauchte-Elektroautos-sind
      -echte-Restwertriesen.html.
121. "Garagisten geht wegen Tesla-Boom die Arbeit aus" (Garage Owners Out of Work
      Due to Tesla Boom), http://www.20min.ch/finance/news/story/25771189#videoid=
      524953?redirect=mobi&nocache=0.1541377262158543.
122. http://www.energietarife.com/index.php?lohnt-sich-ein-elektroauto.
123. "Tesla's Innovations Are Transforming the Auto Industry," http://www.forbes.com/
      sites/innovatorsdna/2016/08/24/
      teslas-innovations-are-transforming-the-auto-industry/#2be369e1578a.
124. Technical Museum Vienna (ed.), *Mobilitär: 30 Dinge, die bewegen* (*Mobility: 30
      Things That Move*). Czernin Publishers, Vienna, 2015.
125. *Monitoringbericht AustriaTech, Elektromobilität 2015* (*Monitoring Report AustriaTech:
      Electric Mobility 2015*). Vienna.
126. http://ecomento.tv/2016/09/01/elektroauto-transporter-streetscooter-vw-chef
      -mueller-sauer-auf-die-post/.
127. https://de.statista.com/statistik/daten/studie/200160/umfrage/neuzulassungen-von
      -fahrzeugen-in-deutschland/.
128. "Technik-Mythos: Wasserstoff revolutioniert die Energieversorgung" (Technical
      Myth: Hydrogen Revolutionizes Energy Supply), https://www.heise.de/newsticker/
      meldung/Technik-Mythos-Wasserstoff-revolutioniert-die-Energieversorgung-3638549
      .html.
129. "Brennstoffzelle Reloaded" (Fuel Cell Reloaded), http://www.wiwo.de/technologie/
      auto/wasserstoffautos-brennstoffzelle-reloaded/5666152-all.html.
130. "Global Automotive Executive Survey 2017," https://assets.kpmg.com/content/dam/
      kpmg/xx/pdf/2017/01/global-automotive-executive-survey-2017.pdf.

131. "Mercedes Says It Will Not Pursue Fuel Cell Development for Its Cars," http://gas2
    .org/2017/03/31/mercedes-will-not-pursue-fuel-cell-development/.
132. http://www.faz.net/aktuell/wirtschaft/neue-mobilitaet/warum-deutsche-gegenueber
    -elektroautos-skeptisch-sind-14603445.html.
133. https://www.mckinsey.de/elektromobilitaet.
134. https://de.statista.com/statistik/daten/studie/183003/umfrage/pkw---gefahrene
    -kilometer-pro-jahr/.
135. "U.S. Driving Tops 3.1 Trillion Miles in 2015, New Federal Data Show," https://
    www.fhwa.dot.gov/pressroom/fhwa1607.cfm.
136. "Der Kollaps bleibt aus" (The Collapse Fails to Materialize), http://www.spektrum.de/
    kolumne/der-kollaps-bleibt-aus/1444719.
137. "Bruttostromerzeugung in Deutschland für 2014 bis 2016" (Gross Energy Generation
    in Germany for 2014 to 2016), https://www.destatis.de/DE/ZahlenFakten/
    Wirtschaftsbereiche/Energie/Erzeugung/Tabellen/Bruttostromerzeugung.html.
138. National Research Council, *Hidden Costs of Energy: Unpriced Consequences of Energy
    Production and Use*. National Academies Press, Washington, DC, 2009, http://
    www.nap.edu/openbook.php?record_id=12794&page=1.
139. https://en.wikipedia.org/wiki/Gasoline.
140. Jaana I. Halonen, Anna L. Hansell, John Gulliver, et al., "Road Traffic Noise Is
    Associated with Increased Cardiovascular Morbidity and Mortality and All-Cause
    Mortality in London," *European Heart Journal*, DOI:10.1093/eurheartj/ehv216.
141. https://www.destatis.de/DE/ZahlenFakten/GesamtwirtschaftUmwelt/Umwelt/
    UmweltoekonomischeGesamtrechnungen/Umweltschutzmassnahmen/Aktuell.html.
142. https://www.bmf.gv.at/services/publikationen/Daten_und_Fakten_Steuer-_und_
    Zollverwaltung_2014.pdf.pdf?555a9o; http://www.ezv.admin.ch/zollinfo_firmen.
143. "BYD investiert in Produktion in Frankreich" (BYD Invests in Production in France),
    http://www.it-times.de/news/byd-investiert-in-produktion-in-frankreich-123323/.
144. "Auf China, nicht auf Tesla schauen" (Watch Out for China, Not for Tesla), https://
    www.nzz.ch/finanzen/elektromobilitaet-auf-china-nicht-auf-tesla-schauen-ld.1085290.
145. https://www.fhwa.dot.gov/policyinformation/statistics/2016/fe10.cfm.

Chapter 6

1. Burkhard Bilger, "Auto Correct," *New Yorker*, November 25, 2013, http://www
    .newyorker.com/magazine/2013/11/25/auto-correct.
2. http://www.nsc.org/NewsDocuments/2017/12-month-estimates.pdf.
3. Centers for Disease Control and Prevention, https://www.cdc.gov/injury/wisqars/pdf/
    leading_causes_of_death_by_age_group_2016-508.pdf.
4. National Safety Council, *Odds of Dying*, https://www.nsc.org/work-safety/tools
    -resources/injury-facts/chart.
5. *The Economic and Societal Impact of Motor Vehicle Crashes, 2010* (revised). National
    Highway Traffic Safety Administration, Washington, DC, May 2015.
6. U.S. Department of Transportation, NHTSA, "Critical Reasons for Crashes
    Investigated in the National Motor Vehicle Crash Causation Survey," February 2015,
    http://www-nrd.nhtsa.dot.gov/pubs/812115.pdf.
7. Bill Sanderson, "Epidemic of Fatal Crashes," *Wall Street Journal*, February 10, 2014,
    http://www.wsj.com/articles/SB10001424052702303465004579322441555410428.

8. Aimee Green, "Sober Drivers Rarely Prosecuted in Fatal Pedestrian Crashes in Oregon," OregonLive.com, November 15, 2011, http://www.oregonlive.com/portland/index.ssf/2011/11/sober_drivers_rarely_prosecute.html.

9. http://www.dvr.de/betriebe_bg/daten/unfallstatistik/eu_europa.htm.

10. http://www.lightningsafety.noaa.gov/odds.shtml.

11. Fred A. Manuele, *On the Practice of Safety*. Wiley Interscience, New York, 2013.

12. P. Chapman, D. Crundall, N. Phelps, G. Underwood, "The Effects of Driving Experience on Visual Search and Subsequent Memory for Hazardous Driving Situations," in *Behavioural Research in Road Safety*, Thirteenth Seminar. Department for Transport, 2003.

13. Stine Vogt, Svein Magnussen, "Expertise in Pictorial Perception: Eye-Movement, Patterns and Visual Memory in Artists and Laymen," *Perception* 36(1), 2007.

14. P. Lynn, C. R. Lockwood, "The Accident Liability of Company Car Drivers," *Transport Research Laboratory Report* 317, 1998.

15. "Distraction and Teen Crashes: Even Worse Than We Thought," AAA Foundation for Traffic Safety, March 25, 2015, http://newsroom.aaa.com/2015/03/distraction-teen-crashes-even-worse-thought/.

16. "Selfie Crash Death: Woman Dies in Head-on Collision Seconds After Uploading Pictures of Herself and 'HAPPY' Status to Facebook," http://www.independent.co.uk/news/world/americas/selfie-crash-death-woman-dies-in-head-on-collision-seconds-after-uploading-pictures-of-herself-and-9293694.html.

17. https://en.wikipedia.org/wiki/Yerkes%E2%80%93Dodson_law.

18. Andrea Glaze, James Ellis, "Pilot Study of Distracted Drivers," Center for Public Policy, Virginia Commonwealth University, January 2003.

19. Teck-Hua Hoa, Juin Kuan Chong, Xiaoyu Xia, "Yellow Taxis Have Fewer Accidents Than Blue Taxis Because Yellow Is More Visible Than Blue," *Proceedings of the National Academy of Sciences of the United States of America*, 2016, http://www.pnas.org/content/early/2017/02/28/1612551114.

20. Michelle J. White, "'The Arms Race' on American Roads: The Effect of Sport Utility Vehicles and Pickup Trucks on Traffic Safety," *Journal of Law and Economics*, October 2004.

21. Michael L. Anderson, Maximilian Auffhammer, "Pounds That Kill: The External Costs of Vehicle Weight," working paper, University of California, Berkeley.

22. "Google Cars Drive Themselves in Traffic," http://www.nytimes.com/2010/10/10/science/10google.html.

23. http://archive.darpa.mil/grandchallenge04/.

24. Sebastian Thrun, "Google's Driverless Car," March 2011, https://www.ted.com/talks/sebastian_thrun_google_s_driverless_car.

25. Burkhard Bilger, "AutoCorrect," *New Yorker*, November 25, 2013, http://www.newyorker.com/magazine/2013/11/25/auto-correct.

26. http://archive.darpa.mil/grandchallenge/.

27. "The Unknown Start-up That Built Google's First Self-Driving Car," http://spectrum.ieee.org/robotics/artificial-intelligence/the-unknown-startup-that-built-googles-first-selfdriving-car.

28. "Die Wiege des autonomen Fahrens steht in Neubiberg" (The Cradle of Autonomous Driving Is at Neubiberg), https://www.bundeswehrkarriere.de/it/autonomes-fahren.

29. http://motherboard.vice.com/read/carnegie-mellons-1986-self-driving-van-was
    -adorable.
30. "Wer hat das Roboterauto erfunden? Die Bundeswehr!" (Who Invented the Robot
    Car? The German Federal Armed Forces!), http://www.zeit.de/mobilitaet/2015-07/
    autonomes-fahren-geschichte.
31. "Levels of Driving Automation," https://www.sae.org/news/2019/01/sae-updates
    -j3016-automated-driving-graphic.
32. http://www.templetons.com/brad/robocars/levels.html.
33. "2016 Disengagement Reports," https://www.dmv.ca.gov/portal/dmv/detail/vr/
    autonomous/disengagement_report_2016.
34. Chris Urmson, "How a Driverless Car Sees the Road," https://www.youtube.com/
    watch?v=tiwVMrTLUWg.
35. "Autonomous Vehicles in California," https://www.dmv.ca.gov/portal/dmv/detail/vr/
    autonomous/testing.
36. "Ford's Dozing Engineers Side with Google in Full Autonomy Push," https://www
    .bloomberg.com/news/articles/2017-02-17/ford-s-dozing-engineers-side-with
    -google-in-full-autonomy-push.
37. "Autonomous Car Companies WITHOUT a Test License, But Still Testing in
    California," https://thelastdriverlicenseholder.com/2017/03/03/autonomous-car
    -companies-without-a-test-license-but-still-testing-in-california/.
38. https://www.dmv.ca.gov/portal/dmv/detail/vr/autonomous/auto.
39. "Autonomous Vehicle Regulations in Nevada," https://scoe.transportation.org/,
40. http://www.dmvnv.com/autonomous.htm.
41. "Who's Who in the Rise of Autonomous Driving Startups," https://www.cbinsights
    .com/blog/early-stage-autonomous-driving-startups/.
42. https://techcrunch.com/2017/01/13/nissans-first-european-self-driving-car-trials-begin
    -on-london-roads-next-month/.
43. http://www.theverge.com/2016/8/1/12337516/delphi-self-driving-car-service
    -singapore.
44. https://www.washingtonpost.com/business/economy/why-uber-is-going-to-test-its-new
    -self-driving-cars-in-pittsburgh/2016/08/24/ab48c3be-696f-11e6-99bf-f0cf3a6449a6_
    story.html.
45. http://qz.com/688003/ubers-self-driving-cars-are-on-the-road/.
46. https://www.wired.com/2015/12/baidus-self-driving-car-has-hit-the-road/.
47. http://www.businessinsider.com/r-bmw-seeks-to-be-coolest-ride-hailing-firm-with
    -autonomous-car-2016-12.
48. http://www.detroitnews.com/story/business/autos/2016/08/23/opposite-strategie
    s-fuel-driverless-car-development/89239658/.
49. http://www.businessinsider.com/how-otto-defied-nevada-scored-a-680-million
    -payout-from-uber-2016-11.
50. https://www.dmv.ca.gov/portal/dmv/detail/vr/autonomous/testing.
51. "Uber's Autonomous Cars Drove 20,354 Miles and Had to Be Taken Over at Every
    Mile, According to Documents," http://www.recode.net/2017/3/16/14938116/
    uber-travis-kalanick-self-driving-internal-metrics-slow-progress.
52. http://www.reuters.com/article/us-tech-ces-autos-idUSKBN0UJ1UD20160105.
53. http://www.consumerwatchdog.org/resources/cadmvdisengagereport-dec.2015.pdf.

54. Michael Sivak, Brandon Schoettle, "Road Safety with Self-Driving Vehicles: General Limitations and Road Sharing with Conventional Vehicles," http://www.umich.edu/%7Eumtriswt/PDF/UMTRI-2015-2_Abstract_English.pdf.

55. World Economic Forum and Boston Consulting Group, "Self-Driving Vehicles in an Urban Context," http://www3.weforum.org/docs/WEF_Press%20release.pdf.

56. http://www.economist.com/blogs/economist-explains/2015/07/economist-explains.

57. http://www.alphr.com/cars/7038/how-do-googles-self-driving-cars-work.

58. "Clever AI Turns a World of Lasers into Maps for Self-Driving Cars," https://www.wired.com/2016/07/civil-maps-self-driving-car-autonomous-mapping-lidar/.

59. "Lower-Cost LiDAR Is Key to Self-Driving Future," http://articles.sae.org/13899/.

60. "Quanergy Announces $250 Solid-State LiDAR for Cars, Robots, and More," http://spectrum.ieee.org/cars-that-think/transportation/sensors/quanergy-solid-state-lidar.

61. "The Race to Affordable LiDAR," https://www.allaboutcircuits.com/news/the-race-to-afforable-lidar/.

62. http://spectrum.ieee.org/transportation/advanced-cars/cheap-lidar-the-key-to-making-selfdriving-cars-affordable.

63. "Ford and Baidu Invest $150 Million into Major Supplier of Self-Driving Car Tech," http://fortune.com/2016/08/16/ford-baidu-invest-velodyne-lidar/.

64. "The 22-Year-Old at the Center of the Self-Driving Car Craze," https://www.bloomberg.com/news/articles/2017-03-30/the-22-year-old-at-the-center-of-the-self-driving-car-craze.

65. Keynote by Waymo CEO John Krafcik at the Detroit Auto Show, https://derletztefuehrerscheinneuling.com/2017/01/09/keynote-von-waymo-ceo-john-krafcik-auf-der-detroit-autoshow/.

66. http://qz.com/637509/driverless-cars-have-a-new-way-to-navigate-in-rain-or-snow/.

67. https://archive.ll.mit.edu/publications/technotes/LGPR.html.

68. "Why Better Paint Coatings Are Critical for Autonomous Cars," https://www.caranddriver.com/news/a15342871/why-better-paint-coatings-are-critical-for-autonomous-cars/.

69. "SensL Solid State LiDAR Design Consideration," https://youtu.be/npnAr1BlQhw.

70. http://media.nxp.com/phoenix.zhtml?c=254228&p=irol-newsArticle&ID=2125903.

71. "Camera-Based Technology Tracks People in Car Interiors," http://www.fraunhofer.de/en/press/research-news/2016/august/camera-based-technology-tracks-people-in-car-interiors.html.

72. "Tesla Motors Club Connect 2016 in Reno, NV, July 29, 2016," https://www.youtube.com/watch?v=E-qqRTugknI.

73. http://blogs.nvidia.com/blog/2016/01/05/eyes-on-the-road-how-autonomous-cars-understand-what-theyre-seeing/.

74. http://jacobsschool.ucsd.edu/news/news_releases/release.sfe?id=1883.

Chapter 7

1. https://www.udacity.com/course/artificial-intelligence-for-robotics--cs373.

2. Pranav Rajpurkar, Toki Migimatsu, Jeff Kiske, et al., "Driverseat: Crowdstrapping Learning Tasks for Autonomous Driving," http://arxiv.org/pdf/1512.01872v1.pdf.

3. "The Moral Life of Babies," http://www.nytimes.com/2010/05/09/magazine/09babies-t.html.

4. *Biology of Fun*, 25th Anniversary Special Issue, http://www.cell.com/current-biology/issue?pii=S0960-9822(14)X0025-4.

5. Ashesh Jain, Hema S. Koppula, Shane Soh, et al., "Brain4Cars: Car That Knows Before You Do via Sensory-Fusion Deep Learning Architecture," http://arxiv.org/pdf/1601.00740v1.pdf.

6. http://www.cnet.com/news/nvidias-computer-for-self-driving-cars-as-powerful-as-150-macbook-pros/.

7. http://www.forbes.com/sites/aarontilley/2016/04/05/nvidia-redoubles-focus-on-artificial-intelligence-and-autonomous-cars/#69e57456e2b3.

8. "Self-Driving Cars Rattle Supply Chain," http://semiengineering.com/self-driving-cars-rattle-supply-chain/.

9. http://www.nytimes.com/2016/11/29/business/intel-to-team-with-delphi-and-mobileye-for-self-driving-cars.html.

10. http://mi.eng.cam.ac.uk/projects/segnet/.

11. "Google's Former Self-Driving Car Guru Raises Cash for His Own Startup," https://www.axios.com/the-former-cto-of-google-self-driving-car-has-raised-money-for-his-own-2344944616.html.

12. "Wir brauchen keine Regulierung für Künstliche Intelligenz, sondern mehr Förderung" (We Don't Need Regulations for Artificial Intelligence, We Need More Financing), http://bootstrapping.me/politik-kuenstliche-intelligenz-2017/.

13. "Nissan's Path to Self-Driving Cars? Humans in Call Centers," https://www.wired.com/2017/01/nissans-self-driving-teleoperation/.

14. "What the AI Behind AlphaGo Can Teach Us About Being Human," http://www.wired.com/2016/05/google-alpha-go-ai/.

15. "A Conversation with Koko the Gorilla," http://www.theatlantic.com/technology/archive/2015/08/koko-the-talking-gorilla-sign-language-francine-patterson/402307/.

16. Christoph Keese, *Silicon Germany: Wie wir die digitale Transformation schaffen* (*Silicon Germany: How We Can Manage Digital Transformation*). Knaus, Munich, 2016.

17. "Elite: Dangerous' Latest Expansion Caused AI Spaceships to Unintentionally Create Super Weapons," https://www.eurogamer.net/articles/2016-06-03-elite-dangerous-latest-expansion-caused-ai-spaceships-to-unintentionally-create-super-weapons.

18. https://www.bloomberg.com/news/articles/2018-11-28/tesla-customers-rack-up-1-billion-miles-driven-on-autopilot.

19. http://electrek.co/2016/06/03/tesla-share-autopilot-data-department-of-transport/.

20. https://www.tesla.com/blog/master-plan-part-deux.

21. "Tesla Driver Dies in First Fatal Crash While Using Autopilot Mode," https://www.theguardian.com/technology/2016/jun/30/tesla-autopilot-death-self-driving-car-elon-musk.

22. https://static.nhtsa.gov/odi/inv/2016/INCLA-PE16007-7876.pdf.

23. http://www.nytimes.com/2016/09/02/automobiles/big-carmakers-merge-cautiously-into-the-self-driving-lane.html.

24. http://fortune.com/2016/03/11/gm-buying-self-driving-tech-startup-for-more-than-1-billion/.

25. http://research.comma.ai/; Eder Santana, George Hotz, "Learning a Driving Simulator," https://www.scribd.com/document/320095885/.

26. http://www.golem.de/news/mercedes-entwickler-warum-autonome-autos-nicht-selbst
    -lernen-duerfen-1606-121003.html.
27. "Matthias Müller kritisiert selbstfahrende Autos: 'Ein Hype, der durch nichts zu
    rechtfertigen ist'" (Matthias Müller Criticizes Self-Driving Vehicles: "A Hype That
    Cannot Be Justified"), http://www.manager-magazin.de/unternehmen/autoindustrie/
    porsche-chef-nennt-autonomes-fahren-hype-a-1052709.html.
28. https://www.wired.com/2015/12/baidus-self-driving-car-has-hit-the-road/.
29. http://blog.caranddriver.com/nhtsa-chief-autonomous-cars-should-cut-death-rate
    -in-half/.
30. http://blog.caranddriver.com/nhtsa-chief-autonomous-cars-should-cut-death-rate
    -in-half/.
31. https://thelastdriverlicenseholder.com/2017/01/08/keynote-by-waymo-ceo-john
    -krafcik-at-the-detroit-auto-show/.
32. https://www.technologyreview.com/s/602317/self-driving-cars-can-learn-a-lot-by
    -playing-grand-theft-auto/.
33. http://www.wsj.com/articles/drivers-ed-startup-uses-videogames-to-teach-cars-to
    -drive-themselves-1480933804.
34. http://www.gizmag.com/synthia-dataset-self-driving-cars/43895/.
35. http://newatlas.com/synthia-dataset-self-driving-cars/43895/.
36. http://www.wsj.com/articles/is-uber-a-friend-or-foe-of-carnegie-mellon-in
    -robotics-1433084582.
37. "Autonomous Car Race Creates $400k Engineering Jobs for Top Silicon Valley
    Talent," https://www.forbes.com/sites/alanohnsman/2017/03/27/autonomous
    -car-race-creates-400k-engineering-jobs-for-top-silicon-valley-talent/#28102a914a37.
38. https://www.udacity.com/course/self-driving-car-engineer-nanodegree--nd013.
39. http://www.wired.com/2016/01/gm-and-lyft-are-building-a-network-of
    -self-driving-cars/.
40. http://www.bloomberg.com/news/articles/2016-05-03/fiat-google-said-to-plan
    -partnership-on-self-driving-minivans.
41. http://www.bloomberg.com/graphics/2016-merging-tech-and-cars/.
42. https://www.brookings.edu/research/gauging-investment-in-self-driving-cars/.
43. https://magazin.spiegel.de/SP/2016/4/141826740/index.html.
44. https://en.wikipedia.org/wiki/Trolley_problem.
45. http://www.vox.com/2016/6/13/11896166/self-driving-cars-ethics.
46. https://www.youtube.com/watch?v=Uj-rK8V-rik.
47. http://fortune.com/self-driving-cars-silicon-valley-detroit/.
48. http://www.vtti.vt.edu/featured/?p=422.
49. Vinand M. Nantulya, Michael R. Reich, "The Neglected Epidemic: Road Traffic
    Injuries in Developing Countries," *British Medical Journal*, May 2002.
50. "400 Road Deaths per Day in India; Up 5% to 1.46 lakh in 2015," http://
    timesofindia.indiatimes.com/india/400-road-deaths-per-day-in-India-up-5-to-1-46
    -lakh-in-2015/articleshow/51919213.cms.
51. "Road Accidents Due to Speed Breakers." https://www.financialexpress.com/
    india-news/speed-breakers-in-india-kill-more-people-than-accident-do-in-uk
    -australia/728537/.

52. *Traffic Safety Facts 2004*. National Highway Traffic Safety Administration, Washington, DC, 2005

53. "Fools and Bad Roads," *The Economist*, May 22, 2007, http://www.economist.com/node/8896844.

54. Anand Swamy, Stephen Knack, Young Lee, Omar Azfar, "Gender and Corruption," working paper, Center for Development Economics, Department of Economics, Williams College, 2000.

55. B. G. Simons-Morton, N. Lerner, J. Singer, "The Observed Effects of Teenage Passengers on Risky Driving Behavior of Teenage Drivers," *Accident Analysis & Prevention* 37, 2005.

56. http://www.popsci.com/volvo-on-self-driven-car-liability-i-volunteer.

57. "When Driverless Cars Crash, Who Gets the Blame and Pays the Damages?," https://www.washingtonpost.com/local/trafficandcommuting/when-driverless-cars-crash-who-gets-the-blame-and-pays-the-damages/2017/02/25/3909d946-f97a-11e6-9845-576c69081518_story.html.

58. http://ideas.4brad.com/enough-trolley-problem-already.

59. http://www.theatlantic.com/technology/archive/2013/10/the-ethics-of-autonomous-cars/280360/; http://www.theatlantic.com/technology/archive/2016/03/google-self-driving-car-crash/471678/.

60. Dan Ariely, *Predictably Irrational: The Hidden Forces That Shape Our Decisions*. HarperCollins, New York, 2008.

61. http://www.europarl.europa.eu/sides/getDoc.do?pubRef=-//EP//NONSGML%2BCOMPARL%2BPE-582.443%2B01%2BDOC%2BPDF%2BV0//EN.

62. https://www.facebook.com/Beipackzettelpresse/.

63. http://www.spiegel.de/netzwelt/gadgets/juergen-schmidhuber-der-weltraum-ist-fuer-roboter-gemacht-a-1074759.html.

64. Teresa M. Amabile, "Brilliant but Cruel: Perceptions of Negative Evaluators," *Journal of Experimental Social Psychology*, March 1983.

65. Khaled Saleh, Mohammed Hossny, Saeid Nahavandi, "Kangaroo Vehicle Collision Detection Using Deep Semantic Segmentation Convolutional Neural Network," International Conference on Digital Image Computing: Techniques and Applications (DICTA), 2016, http://ieeexplore.ieee.org/abstract/document/7797057/.

66. http://www.telegraph.co.uk/news/2017/01/06/driverless-cars-will-cause-congestion-britains-roads-worsen/.

67. Brett Stern, *Inventors at Work*. Apress, New York, 2012.

68. P. W. Singer, *Wired for War: The Robotics Revolution and Conflict in the 21st Century*. Penguin Press, New York, 2009.

69. http://www.bloomberg.com/news/articles/2015-12-18/humans-are-slamming-into-driverless-cars-and-exposing-a-key-flaw.

70. http://www.dailymail.co.uk/sciencetech/article-3592567/The-self-driving-car-behaves-like-person-Audi-s-robotic-vehicle-taught-human-manners.html.

71. "Why Google's Self-Driving Cars Are Considered 'Too Polite,'" http://bigthink.com/ideafeed/googles-self-driving-cars-are-too-polite.

72. https://www.technologyreview.com/s/602292/top-safety-official-doesnt-trust-automakers-to-teach-ethics-to-self-driving-cars/.

73. https://en.wikipedia.org/wiki/Skeuomorph.

74. "Driving Is Social. Autonomous Cars Aren't, Argues Computer Scientist," https:// motherboard.vice.com/en_us/article/driving-is-social-autonomous-cars-arent-argues -computer-scientist.

75. Barry Brown, Eric Lautier, "The Trouble with Autopilots: Assisted and Autonomous Driving on the Social Road," http://www.ericlaurier.co.uk/resources/Writings/Brown -2017-Car-Autopilots.pdf.

76. Tom Vanderbilt, *Traffic: Why We Drive the Way We Do and What It Says About Us.* Vintage Books, New York, 2008.

77. "The Secret UX Issues That Will Make (or Break) Self-Driving Cars," http://www. fastcodesign.com/3054330/innovation-by-design/the-secret-ux-issues-that-will-make -or-break-autonomous-cars.

78. "This Self-Driving Car Smiles at Pedestrians to Let Them Know It's Safe to Cross," https://www.fastcoexist.com/3063717/this-self-driving-car-smiles-at-pedestrians-to-let -them-know-its-safe-to-cross.

79. "Drive.ai Uses Deep Learning to Teach Self-Driving Cars—and to Give Them a Voice," https://techcrunch.com/2016/08/30/drive-ai-uses-deep-learning-to-teach -self-driving-cars-and-to-give-them-a-voice/.

80. https://www.iflscience.com/technology/google-self-driving-car-now-knows-when -honk-horn/.

81. https://patents.google.com/patent/US9014905B1/en.

82. https://www.humanisingautonomy.com/.

83. http://www.emercedesbenz.com/autos/mercedes-benz/concept-vehicles/mercedes -benz-looks-to-the-future/attachment/mercedes-benz-14c634_029/.

84. http://venturebeat.com/2015/12/07/chinese-researchers-unveil-brain-powered-car/.

85. http://www.wired.com/2016/02/googles-self-driving-car-may-caused-first-crash/.

86. "Nissan Anthropologist: 'We Need a Universal Language for Autonomous Cars,'" https://www.2025ad.com/in-the-news/blog/nissan-melissa-cefkin-driverless-cars/?WT .tsrc=Newsletter&WT.mc_id=07/2017.

87. http://www.jdpower.com/press-releases/2016-us-tech-choice-study.

88. https://www.wpi.edu/Pubs/E-project/Available/E-project-043013-155601/ unrestricted/A_Study_of_Public_Acceptance_of_Autonomous_Cars.pdf.

89. https://newsroom.cisco.com/press-release-content?articleId=1184392.

90. http://www.fastcodesign.com/3054330/innovation-by-design/the-secret-ux-issues -that-will-make-or-break-autonomous-cars.

91. http://spectrum.ieee.org/automaton/robotics/artificial-intelligence/children-beating -up-robot.

92. http://blogs.wsj.com/digits/2016/01/21/human-driver-taking-over-from-computer -crashes-autonomous-car/.

93. http://thenextweb.com/insider/2015/11/25/these-defiant-robots-are-learning-to -reject-human-orders/.

94. Don Norman, *Emotional Design: Why We Love (or Hate) Everyday Things.* Basic Books, New York, 2004.

95. https://en.wikipedia.org/wiki/Three_Laws_of_Robotics.

96. Nick Bostrom, *Superintelligence: Paths, Dangers, Strategies.* Oxford University Press, Oxford, 2014.

97. Bryant Walker Smith, "Automated Vehicles Are Probably Legal in the United States," *Texas A&M Law Review* 1(411), 2014.

98. https://en.wikipedia.org/wiki/Locomotive_Acts.

99. https://www.dmv.ca.gov/portal/wcm/connect/dbcf0f21-4085-47a1-889f -3b8a64eaa1ff/AVRegulationsSummary.pdf?MOD=AJPERES.

100. "Next Milestone on the Road to Autonomous Driving: 'One More Christmas Present': Mercedes-Benz Receives Approval from Regional Council for the Next Generation Of Autonomous Vehicles," https://media.daimler.com/marsMediaSite/en/ instance/ko.xhtml?oid=15142248.

101. http://ideas.4brad.com/alternative-specific-regulations-robocars-liability-doubling.

102. https://www.bloomberg.com/news/articles/2016-12-22/uber-pulls-self-driving -cars-from-california-for-arizona.

103. https://cyberlaw.stanford.edu/wiki/index.php/Automated_Driving:_Legislative_ and_Regulatory_Action.

104. https://www.transportation.gov/sites/dot.gov/files/docs/AV%20policy%20 guidance%20PDF.pdf.

105. "Can Automated Driving Make People Love the EU?," https://www.2025ad.com/ in-the-news/blog/automated-driving-conference-brussels/.

106. https://www.faa.gov/uas/.

107. http://diepresse.com/home/recht/rechtallgemein/5042568/Wenn-Vertraege -automatisiert-werden.

108. http://www.reuters.com/article/us-germany-autos-idUSKCN0ZY1LT.

109. Ferdinand Dudenhöffer, *Wer kriegt die Kurve? Zeitenwende in der Autoindustrie* (*Who Gets Their Act Together? A Turning Point in the Automotive Industry*). Campus, Frankfurt a. M., 2016.

110. "Battery Material Could Reduce Electric Car Weight," https://phys.org/news/2014 -06-battery-electric-car-weight.html.

111. Sherry Turkle, *Alone Together: Why We Expect More from Technology and Less from Each Other*. Basic Books, New York, 2011.

112. P. W. Singer, *Wired for War: The Robotics Revolution and Conflict in the 21st Century*. Penguin, New York, 2009.

113. http://www.businessinsider.com/bmw-reveals-concept-interior-for-driverless-car -pictures-2017-1.

114. "Monetizing Car Data," http://www.mckinsey.com/industries/automotive-and -assembly/our-insights/monetizing-car-data.

115. "Waymo Could Be a $250 Billion Win for Alphabet, Jefferies Says (GOOGL)," https://markets.businessinsider.com/news/stocks/alphabet-stock-waymo-could-be-a -250-billion-deal-jefferies-says-2018-12-1027823079.

116. Fabio Caiazzo, Akshay Ashok, Ian A. Waitz, et al., "Air Pollution and Early Deaths in the United States," *Atmospheric Environment* 79, November 2013.

117. "Autonomous Taxis Could Greatly Reduce Greenhouse-Gas Emissions of US Light-Duty Vehicles," http://www.nature.com/articles/nclimate2685.epdf.

118. "Cost and Weight Added by the Federal Motor Vehicle Safety Standards for Model Years 1968–2001 in Passenger Cars and Light Trucks," https://icsw.nhtsa.gov/cars/ rules/regrev/evaluate/809834.html.

119. Michael Anderson, Maximilian Auffhammer, "Pounds That Kill: The External Costs of Vehicle Weight," *Review of Economic Studies* 81(Suppl. 2), 535–571, 2014, http://www.nber.org/papers/w17170.

120. https://www.google.com/selfdrivingcar/reports/.

121. "Uber's Autonomous Cars Drove 20,354 Miles and Had to Be Taken Over at Every Mile, According to Documents," http://www.recode.net/2017/3/16/14938116/uber-travis-kalanick-self-driving-internal-metrics-slow-progress.

122. "Waymo One: The Next Step on Our Self-Driving Journey," https://medium.com/waymo/waymo-one-the-next-step-on-our-self-driving-journey-6d0c075b0e9b.

123. "10 Cities at the Forefront of Automated Driving," https://www.2025ad.com/in-the-news/blog/driverless-cities/.

124. http://www.gizmag.com/google-reveals-lessons-learned-from-self-driving-car-program/37481/.

125. http://www.theverge.com/2016/4/27/11517926/googles-self-driving-car-graduating-alphabet-x.

126. https://www.technologyreview.com/s/601297/a-simple-way-to-hasten-the-arrival-of-self-driving-cars/.

127. "2015 Urban Mobility Scorecard," Texas Transportation Institute, http://tti.tamu.edu/documents/mobility-scorecard-2015-wappx.pdf.

128. European Commission, http://ec.europa.eu/transport/themes/urban/urban_mobility/.

129. "CEBR: 50% Rise in Gridlock Costs by 2030," https://www.cebr.com/reports/the-future-economic-and-environmental-costs-of-gridlock/.

130. Tom Vanderbilt, *Traffic*. Allen Lane, London, 2008.

131. Daniel Sperling, Deborah Gordon, "Two Billion Cars," *Transportation Research News*, December 2008, http://onlinepubs.trb.org/onlinepubs/trnews/trnews259billioncars.pdf.

132. http://www.tomtom.com/en_gb/trafficindex/.

133. "Record Number of Miles Driven in U.S. Last Year," http://www.npr.org/sections/thetwo-way/2017/02/21/516512439/record-number-of-miles-driven-in-u-s-last-year.

134. Ferdinand Dudenhöffer, *Wer kriegt die Kurve? Zeitenwende in der Autoindustrie* (*Who Gets Their Act Together? A Turning Point in the Automotive Industry*). Campus, Frankfurt a. M., 2016.

135. "Ford: Skip Level 3 Autonomous Cars—Even Engineers Supervising Self-Driving Vehicle Testing Lose 'Situational Awareness,'" https://cleantechnica.com/2017/02/20/ford-skip-level-3-autonomous-cars-even-engineers-supervising-self-driving-vehicle-testing-lose-situational-awareness/.

136. M. Jeon, A. Riener, J. Sterkenburg, et al. *An International Survey on Autonomous and Electric Vehicles; Austria, Germany, South Korea and USA*. ACM, 2016.

137. Don Tapscott, Alex Tapscott, *Die Blockchain Revolution* (*The Blockchain Revolution*). Plassen, Kulmbach, 2016.

138. http://www.umich.edu/%7Eumtriswt/PDF/UMTRI-2015-12_Abstract_English.pdf.

139. https://techcrunch.com/2016/07/13/land-rovers-lead-engineer-explains-autonomous-off-road-driving/.

140. https://autoweek.com/article/car-news/watch-audis-autonomous-rs-7-fly-around-hockenheim-circuit.

141. http://www.theverge.com/2016/6/15/11944112/self-racing-cars-george-hotz-polysync-autonomoustuff-thunderhill.

142. http://selfracingcars.com/.
143. http://roborace.com/.
144. http://robogames.net.
145. https://www.nature.com/articles/nclimate2685.epdf.
146. http://link.springer.com/chapter/10.1007%2F978-3-319-05990-7_13.
147. http://papers.sae.org/2012-01-0494/.
148. "Scania Takes Lead with Full-Scale Autonomous Truck Platoon," https://www.scania .com/group/en/scania-takes-lead-with-full-scale-autonomous-truck-platoon/.
149. https://nacfe.org/technology/two-truck-platooning/.
150. https://www.whitehouse.gov/the-press-office/2014/02/18/fact-sheet-opportunity -all-improving-fuel-efficiency-american-trucks-bol.
151. https://www.epa.gov/ghgemissions/sources-greenhouse-gas-emissions.
152. "Help or Hindrance? The Travel, Energy and Carbon Impacts of Highly Automated Vehicles," http://www.sciencedirect.com/science/article/pii/S0965856415002694.
153. http://www.recycle-steel.org/steel-markets/automotive.aspx.
154. https://www.daimler.com/karriere/jobsuche/standorte/detailseiten/standort -detailseite-18184.html.
155. http://gomentumstation.net/.
156. https://www.engadget.com/2016/09/30/cali-unmanned-autonomous-trials/.
157. http://www.reuters.com/article/us-usa-selfdriving-idUSKBN1A41UK.
158. http://www.mtc.umich.edu/test-facility.
159. "Michigan Lets Autonomous Cars on Roads Without Human Driver," http:// fox17online.com/2016/12/09/michigan-lets-autonomous-cars-on-roads-without -human-driver/.
160. https://techcrunch.com/2016/11/22/michigans-335-acre-willow-run-autonomous -car-test-facility-breaks-ground/.
161. https://www.acmwillowrun.org/.
162. https://news.kettering.edu/news/kettering-university-gm-mobility-research-center-will -position-flint-and-michigan-forefront; http://www.vtti.vt.edu/.
163. http://www.mynews13.com/content/news/cfnews13/news/article.html/content/news/ articles/bn9/2016/9/26/construction_of_polk.html.
164. https://backchannel.com/license-to-not-drive-6dbea84b9c45#.dw3t23da7.
165. "Ford's Dozing Engineers Side with Google in Full Autonomy Push." https://www .bloomberg.com/news/articles/2017-02-17/ford-s-dozing-engineers-side-with-google -in-full-autonomy-push.
166. https://www.transportation.gov/briefing-room/dot1717.
167. http://www.iwkoeln.de/presse/pressemitteilungen/beitrag/autonomes-fahren-deutsche -starten-von-guter-basis-286200.
168. "In München fahren bald Geister-BMWs" (Ghost BMWs Soon to Drive Around Munich), https://www.welt.de/wirtschaft/article159973041/In-Muenchen-fahren -bald-Geister-BMWs.html.
169. "Next Milestone on the Road to Autonomous Driving: 'One More Christmas Present': Mercedes-Benz Receives Approval from Regional Council for the Next Generation of Autonomous Vehicles," https://media.daimler.com/marsMediaSite/en/ instance/ko.xhtml?oid=15142248.

170. "Please Get on Board, Today Without a Driver," http://www.spiegel.de/auto/aktuell/autonomes-fahren-pilotprojekte-in-hamburg-kassel-und-berlin-a-1126368.html.

171. "A Test Track for the Traffic of the Future Is Under Construction in Berlin," https://www.wired.de/collection/tech/digitale-teststrecke-diginet-ps-selbstfahrende-autos-berlin-tu-strasse-17-juni.

172. http://www.govtech.com/fs/Will-US-83-Become-the-First-Driverless-Highway.html.

173. "Google, Ford, Uber Launch Coalition to Further Self-Driving Cars," http://www.reuters.com/article/us-autos-selfdriving-idUSKCN0XN1F1.

174. http://bbj.hu/business/pm-announces-plans-to-build-test-track-for-self-driving-cars-_116326.

175. Press Release: "AVL Testing a Self-Driving Car on Austrian Highways for the First Time," https://www.avl.com/press-releases-2016/-/asset_publisher/AFDAj3gOfDFk/content/press-release-avl-testet-erstmals-selbstfahrendes-auto-auf-osterreichischer-autobahn; "Self-Driving Cars: Test Tracks in Salzburg," http://salzburg.orf.at/news/stories/2815254/.

176. "Autonomous Shuttles in the Center of Sion," https://actu.epfl.ch/news/autonomous-shuttles-in-the-center-of-sion/.

177. "Ford Will Begin Testing Self-Driving Cars in Europe in 2017," https://techcrunch.com/2016/11/29/ford-will-begin-testing-self-driving-cars-in-europe-in-2017/.

178. "Nissan Hopes to Test Driverless Cars on London Roads Next Month," https://arstechnica.com/cars/2017/01/nissan-test-driverless-cars-london-roads/.

179. "Automated Vehicles Coming to Ontario Roads," https://news.ontario.ca/mto/en/2016/11/automated-vehicles-coming-to-ontario-roads.html.

180. "Finally, There's a Company with the Courage to Test Driverless Cars on Indian Roads," https://qz.com/887754/tata-elxsi-finally-theres-a-company-with-the-courage-to-test-driverless-cars-on-indian-roads/.

181. "Russia's Self-Driving Car Company Is Coming for the World," https://www.inverse.com/article/29452-cognitive-pilot-russian-autonomous-car-system.

182. "Meet Zoox, the Robo-Taxi Startup Taking on Google and Uber," http://spectrum.ieee.org/transportation/advanced-cars/meet-zoox-the-robotaxi-startup-taking-on-google-and-uber.

183. http://www.internationaltransportforum.org/Pub/pdf/15CPB_Self-drivingcars.pdf.

184. "80% of Driverless Car Users Would 'Relax and Enjoy the Scenery,' Ford Survey Says," http://www.connectedcar-news.com/news/2016/nov/30/80-people-using-driverless-cars-would-relax-and-enjoy-scenery-ford-survey-says/.

185. "Driverless Cars Set to Save World Economies Billions—World Study," http://www.gps.com.au/fleet-management-solutions/driverless-cars-set-to-save-world-economies-billions-world-study.

186. "Autonomous Drive Vehicles to Contribute €17 Trillion to European Economy by 2050," https://newsroom.nissan-global.com/releases/autonomous-drive-vehicles-to-contribute-17-trillion-to-european-economy-by-2050.

187. "Autonomous Cars: The Future Is Now," http://www.morganstanley.com/articles/autonomous-cars-the-future-is-now.

188. "Could Self-Driving Cars Spell the End of Ownership?," https://www.wsj.com/articles/could-self-driving-cars-spell-the-end-of-ownership-1448986572.

189. "Why Alphabet Thinks Minivans Make Perfect Self-Driving Taxis," https://www.technologyreview.com/s/602240/why-alphabet-thinks-minivans-make-perfect-self-driving-taxis/.

190. "Female Crash Dummy Upends Safety Ratings for Some Top-Selling Cars," http://bangordailynews.com/2012/03/26/health/female-crash-dummy-upends-safety-ratings-for-some-top-selling-cars/.

191. "When Bias in Product Design Means Life or Death," https://techcrunch.com/2016/11/16/when-bias-in-product-design-means-life-or-death/.

192. "Autos der Zukunft: Forscher stellen erst mal die richtigen Fragen" (Cars of the Future: Researchers First Asking the Right Questions), http://www.zeit.de/mobilitaet/2016-12/auto-zukunft-renault-nissan-forschung-autonomes-fahren.

193. "Why Self-Driving Cars 'Can't Even' with Construction Zones," https://www.wired.com/2017/02/self-driving-cars-cant-even-construction-zones/.

194. https://techcrunch.com/2016/06/11/investment-opportunities-in-the-autonomous-vehicle-space/.

195. "Udacity Self-Driving Car Software," https://github.com/udacity/self-driving-car; "Udacity Self-Driving Car Simulator," https://github.com/udacity/self-driving-car-sim.

196. "Open Pilot," https://github.com/commaai/openpilot.

197. http://oscc.io/.

198. http://opensourcesdc.com/.

199. http://www.cvlibs.net/datasets/kitti/.

200. http://mscoco.org/dataset/#download.

201. "How Far Are We from Solving Pedestrian Detection?," https://www.mpi-inf.mpg.de/departments/computer-vision-and-multimodal-computing/research/people-detection-pose-estimation-and-tracking/how-far-are-we-from-solving-pedestrian-detection/.

202. http://host.robots.ox.ac.uk/pascal/VOC/index.html.

203. https://www.cityscapes-dataset.com/downloads/.

204. http://spectrum.ieee.org/cars-that-think/transportation/self-driving/why-ai-makes-selfdriving-cars-hard-to-prove-safe.

Chapter 8

1. "In Japan, Priuses Can Talk to Other Priuses," https://techcrunch.com/2016/08/16/in-japan-priuses-can-talk-to-other-priuses/.

2. Audi Crosslinks Cars with Traffic Lights in Las Vegas," http://www.golem.de/news/verkehrssteuerung-audi-vernetzt-autos-mit-ampeln-in-las-vegas-1612-124937.html.

3. http://techcrunch.com/2016/01/28/security-and-privacy-standards-are-critical-to-the-success-of-connected-cars/.

4. https://www.wired.com/2015/07/hackers-remotely-kill-jeep-highway/.

5. Crag Smith, *The Car Hacker's Handbook: A Guide for the Penetration Tester*, http://opengarages.org/index.php/Car_Hacker%27s_Handbook.

6. "Tesla's Car Data Network Is Down in the US, It's a 'Top Priority' and 'Currently Being Fixed,'" https://electrek.co/2016/08/15/teslas-car-data-network-down-in-the-us-its-a-top-priority-currently-being-fixed/.

7. http://www.openautoalliance.net/.

8. http://www.autosar.org/.
9. https://www.weforum.org/agenda/2016/03/this-chinese-city-plans-to-track-all-cars -electronically/.
10. https://incardelivery.volvocars.com.
11. "Microsoft Launches a New Cloud Platform for Connected Cars," https://techcrunch .com/2017/01/05/microsoft-launches-a-new-cloud-platform-for-connected-cars/.
12. "Gartner Says by 2020, a Quarter Billion Connected Vehicles Will Enable New In-Vehicle Services and Automated Driving Capabilities," https://www.gartner.com/ en/newsroom/press-releases/2015-01-26-gartner-says-by-2020-a-quarter-billion -connected-vehicles-will-enable-new-in-vehicle-services-and-automated-driving -capabilities.
13. "How Connected Cars Are Turning into Revenue-Generating Machines," https:// techcrunch.com/2016/08/28/how-connected-cars-are-turning-into-revenue -generating-machines/.
14. "5G Will Help Autonomous Cars Cruise Streets Safely," http://www.itworld.com/ article/3173850/consumer-electronics/5g-will-help-autonomous-cars-cruise-streets -safely.html.
15. "Average Speed of Internet Connections in the Leading Countries Worldwide in the 3rd Quarter 2016 (in Mbit/s)," https://de.statista.com/statistik/daten/studie/224924/ umfrage/internet-verbindungsgeschwindigkeit-in-ausgewaehlten-weltweiten-laendern/.
16. "Elon Musk's Sleight of Hand," https://medium.com/@gavinsblog/elon-musk-s -sleight-of-hand-ea2b078ed8e6.
17. "Why Auto Designs Take So Long," http://semiengineering.com/designing-for-safety/.
18. "Uber's Big China Rival: 'The Market Will Pick the Best,'" http://money.cnn.com/ 2016/05/19/technology/jean-liu-didi-chuxing/.
19. "How a Global Alliance Against Uber Could Topple Its Monopoly," http://www.inc .com/alex-moazed/how-a-global-alliance-against-uber-could-topple-its-monopoly .html.
20. "Uber Sells China Operations to Didi Chuxing," http://www.wsj.com/articles/ china-s-didi-chuxing-to-acquire-rival-uber-s-chinese-operations-1470024403.
21. "Where Do All the Cabs Go in the Late Afternoon?," http://www.nytimes.com/2011/ 01/12/nyregion/12taxi.html?_r=0.
22. "Taxi Owners, Lenders Sue New York City over Uber," http://www.reuters.com/ article/us-newyorkcity-taxis-uber-idUSKCN0T700J20151118.
23. http://www.schipholtaxi.nl/en/.
24. http://www.handelsblatt.com/unternehmen/industrie/verordnung-bremst -elektroautos-meine-teslas-kann-ich-einstampfen/19292188.html.
25. "Update: Percentage of Young Persons with a Driver's License Continues to Drop," http://www.tandfonline.com/doi/abs/10.1080/15389588.2012.696755#.VnIFCcp325g.
26. http://www.kbb.com/car-news/all-the-latest/uber-wont-kill-car-sales-but-ride_sharing -may-affect-what-we-buy/2000010954/#survey.
27. http://www.zipcar.com/; Jeremy Rifkin, *The Zero Marginal Cost Society*. Palgrave Macmillan, New York, 2014.
28. https://www.bcgperspectives.com/content/articles/automotive-whats-ahead-car -sharing-new-mobility-its-impact-vehicle-sales/?chapter=8#chapter8.

29. "No Parking Here," http://www.motherjones.com/environment/2016/01/future
    -parking-self-driving-cars.
30. https://boostbybenz.com/aboutus.
31. https://flightcar.com/.
32. https://techcrunch.com/2016/12/15/mercedes-launches-car-sharing-service-croove/.
33. "Renault-Nissan Alliance and Transdev to Jointly Develop Driverless Vehicle Fleet
    System for Future Public and On-Demand Transportation," http://media.renault.com/
    global/en-gb/Media/PressRelease.aspx?mediaid=87743.
34. http://www.car2come.com/.
35. Panel Mobility Innovators Forum, Stanford, August 5, 2016.
36. "Why New Yorkers Can't Find a Taxi When It Rains," http://www.citylab.com/
    weather/2014/10/why-new-yorkers-cant-find-a-taxi-when-it-rains/381652/.
37. "Taxi Drivers and Beauty Contests," http://people.hss.caltech.edu/~camerer/Camerer
    %20Feature.pdf.
38. Morgan Stanley Research, Amnon Shashua CVPR 2016 Keynote: "Autonomous
    Driving, Computer Vision and Machine Learning," https://youtu.be/n8T7A3wqH3Q.
39. Margaret Derry, *Horses in Society: A Story of Animal Breeding and Marketing,
    1800–1920*. University of Toronto Press, Toronto, p. 131.
40. http://data.worldbank.org/indicator/SP.URB.TOTL.IN.ZS.
41. United Nations, "A World of Cities," August 2014, http://www.un.org/en/
    development/desa/population/publications/pdf/popfacts/PopFacts_2014-2.pdf.
42. https://en.wikipedia.org/wiki/List_of_cities_in_China_by_population_and_
    built-up_area.
43. McKinsey Global Institute, "Preparing for China's Urban Billion," February 2009,
    http://www.mckinsey.com/global-themes/urbanization/preparing-for-chinas-urban
    -billion.
44. http://www.fastcompany.com/3060860/what-saudi-women-really-think-about-their
    -countrys-investment-in-uber.
45. https://newsroom.uber.com/us-illinois/dui-rates-decline-in-uber-cities/.
46. "Impacts of Car2Go on Vehicle Ownership, Modal Shift, Vehicle Miles Traveled,
    and Greenhouse Gas Emissions: An Analysis of Five North American Cities,"
    http://innovativemobility.org/wp-content/uploads/2016/07/Impactsofcar2go_
    FiveCities_2016.pdf.
47. "Welcome to Uberville: Uber Wants to Take Over Public Transit, One Small Town
    at a Time," http://www.theverge.com/2016/9/1/12735666/uber-altamonte-springs
    -fl-public-transportation-taxi-system.
48. https://kurier.at/chronik/wien/wien-verdacht-auf-steuerbetrug-bei-taxiunternehmen/
    232.621.667.
49. https://techcrunch.com/2016/12/21/new-regulations-could-limit-didis-taxi-on
    -demand-service-in-chinas-top-cities.
50. Tom Slee, *What's Yours Is Mine: Against the Sharing Economy*. OR Books, 2017.
51. Don Tapscott, Alex Tapscott, *Blockchain Revolution: How the Technology Behind
    Bitcoin Is Changing Money, Business, and the World*. Portfolio, 2016.
52. https://en.wikipedia.org/wiki/Communications_Decency_Act.
53. Mike Hearn, "Future of Money," Turing Festival, Edinburgh, Scotland, August 23,
    2013, http://www.slideshare.net/mikehearn/future-of-money-26663148.

54. Don Tapscott, Alex Tapscott, *Blockchain Revolution: How the Technology Behind Bitcoin Is Changing Money, Business, and the World*. Portfolio, 2016.
55. "La'Zooz: The Decentralized, Crypto-Alternative to Uber," http://www.shareable.net/blog/lazooz-the-decentralized-crypto-alternative-to-uber.

Chapter 9

1. http://www.strategyand.pwc.com/innovation1000.
2. http://www.faz.net/aktuell/wirtschaft/wirtschaft-in-zahlen/grafik-des-tages-tesla-forscht-und-forscht-14488476.html.
3. "Europe's Innovation Deficit Isn't Disappearing Any Time Soon," https://www.washingtonpost.com/news/innovations/wp/2015/06/08/europes-innovation-deficit-isnt-disappearing-any-time-soon/.
4. Mary Meeker, "Internet Trends 2015 = Code Conference," https://www.kleinerperkins.com/perspectives/2015-internet-trends.
5. http://www.strategyand.pwc.cobayrische mottorenwerkem/innovation1000.
6. Fred Block, Matthew R. Keller, "Where Do Innovations Come From? Transformations in the U.S. National Innovation System 1970–2006," report issued by the Information Technology and Innovation Foundation, July 2008, http://www.itif.org/files/Where_do_innovations_come_from.pdf.
7. Sadao Nagaoka, John P. Walsh, "The R&D Process in the U.S. and Japan: Major Findings from the RIETI–Georgia Tech Inventor Survey," working paper from the Research Institute of Economy, Trade and Industry, July 5, 2009, http://www.rieti.go.jp/jp/publications/dp/09e010.pdf.
8. Mary Tripsas, Giovanni Gavetti, "Capabilities, Cognition, and Inertia: Evidence from Digital Imaging," *Strategic Management Journal* 21:1147–1161, 2000.
9. http://www.reuters.com/article/us-audi-strategy-idUSKCN1030HW.
10. http://www.manager-magazin.de/unternehmen/autoindustrie/porsche-1500-jobs-fuer-mission-e-und-gegen-tesla-a-1104843.html.
11. "Toyota Loses Sales Crown to VW as U.S. Trade Barriers Loom," https://www.bloomberg.com/news/articles/2017-01-30/toyota-loses-sales-crown-to-vw-as-threat-of-trade-barriers-looms?xing_share=news.
12. Jonathan Tepperman, *The Fix: How Nations Survive and Thrive in a World in Decline*. Tim Duggan Books, New York, 2016.

Chapter 10

1. http://www.pri.org/stories/2012-11-02/energy-costs-oil-production.
2. http://www.eia.gov/Energyexplained/index.cfm?page=oil_refining.
3. https://en.wikipedia.org/wiki/Internal_combustion_engine.
4. http://techcrunch.com/2016/06/01/an-open-letter-to-tesla-and-google-on-driverless-cars/.
5. http://www.kurzweilai.net/autonomous-vehicles-might-have-to-be-test-driven-tens-or-hundreds-of-years-to-demonstrate-their-safety; http://electrek.co/2016/06/03/tesla-share-autopilot-data-department-of-transport/.
6. http://www.abc.net.au/news/2015-10-18/rio-tinto-opens-worlds-first-automated-mine/6863814.

7. http://www.businessinsider.com/interview-gett-ceo-shahar-waiser-uber-automation -plans-future-self-driving-taxis-vw-2016-12.

8. http://qz.com/781113/how-florida-became-the-most-important-state-in-the-race -to-legalize-self-driving-cars/.

9. http://www.ncsl.org/research/transportation/autonomous-vehicles-self-driving -vehicles-enacted-legislation.aspx.

10. http://www.usinenouvelle.com/article/la-france-autorise-les-tests-de-voitures -autonomes-sur-ses-routes.N422102.

11. https://electrek.co/2016/08/16/ ford-fully-autonomous-cars-high-volume-available-2021/.

12. "Who Will Build the Next Great Car Company?," http://fortune.com/self-driving -cars-silicon-valley-detroit/.

13. https://de.statista.com/statistik/daten/studie/183003/umfrage/pkw---gefahrene -kilometer-pro-jahr/.

14. *Your Driving Cuts: How Much Are You Really Paying to Drive?* American Automobile Association, Washington, DC, 2015.

15. "Number of U.S. Aircraft, Vehicles, Vessels, and Other Conveyances," http://www .rita.dot.gov/bts/sites/rita.dot.gov.bts/files/publications/national_transportation_ statistics/html/table_01_11.html.

16. "3.2 Trillion Miles Driven on U.S. Roads in 2016," https://www.fhwa.dot.gov/ pressroom/fhwa1704.cfm.

17. "The Coming Nightmare for the Car Industry," http://robohub.org/the-coming -nightmare-for-the-car-industry/.

18. Alison Chaiken, http://she-devel.com/.

19. http://www.digitaltrends.com/cars/tesla-fremont-factory-drives-bay-area -manufacturing-growth/.

20. http://www.spiegel.de/auto/aktuell/zulieferer-die-heimlichen-autohersteller-a -1108529.html.

21. http://www.bain.com/publications/articles/winning-in-europe-truck-strategies-for-the -next-decade.aspx.

22. American Trucking Association, http://www.trucking.org/_layouts/ATARedesign/ News_and_Information_Reports_Industry_Data.aspx.

23. "Number of U.S. Aircraft, Vehicles, Vessels, and Other Conveyances," http://www .rita.dot.gov/bts/sites/rita.dot.gov.bts/files/publications/national_transportation_ statistics/html/table_01_11.html; "Large Truck and Bus Crash Facts 2014," https:// www.fmcsa.dot.gov/safety/data-and-statistics/large-truck-and-bus-crash-facts-2014; "Fatality Analysis Reporting System (FARS)," https://www-fars.nhtsa.dot.gov/Main/ index.aspx.

24. "Tractor-Trailers Without a Human at the Wheel Will Soon Barrel onto Highways Near You. What Will This Mean for the Nation's 1.7 Million Truck Drivers?," https://www.technologyreview.com/s/603493/10-breakthrough-technologies-2017 -self-driving-trucks/.

25. http://www.strategyand.pwc.com/media/file/The-era-of-digitized-trucking.pdf.

26. "Otto and Budweiser: First Shipment by Self-Driving Truck," https://youtu.be/ Qb0Kzb3haK8.

27. http://embarkdrive.com/.

28. Http://peloton-tech.com/.
29. http://starsky.io/.
30. "Baidu Unveils Self-Driving Truck with Foton," http://usa.chinadaily.com.cn/business/2016-11/16/content_27395804.htm.
31. http://www.tusimple.com/.
32. "First Self-Driving 'Pod' Unleashed on Britain's Streets," http://www.telegraph.co.uk/technology/news/11866132/First-self-driving-pod-unleashed-on-Britains-roads.html.
33. "CITY eTAXI—Complete ShowCar at the CeBIT," https://www.electrive.net/2017/01/30/city-etaxi-fertiges-leichtbaufahrzeug-auf-der-cebit/.
34. http://litmotors.com/.
35. "BMW's Self-Balancing Motorcycle of Tomorrow," http://money.cnn.com/2016/10/11/technology/bmw-next100-motorrad-motorcycle/.
36. http://de.statista.com/statistik/daten/studie/37088/umfrage/anteile-der-wirtschaftssektoren-am-bip-ausgewaehlter-laender/.
37. "Freelancing in America: A National Survey of the New Workforce," https://www.slideshare.net/oDesk/global-freelancer-surveyresearch-38467323.
38. http://de.statista.com/statistik/daten/studie/1376/umfrage/anzahl-der-erwerbstaetigen-mit-wohnort-in-deutschland/.
39. http://de.statista.com/statistik/daten/studie/158665/umfrage/freie-berufe---selbststaendige-seit-1992/.
40. http://www.nachhaltig-selbstaendig.at/ein-personen-unternehmen-in-oesterreich/.
41. http://de.statista.com/statistik/daten/studie/30703/umfrage/beschaeftigtenzahl-in-der-automobilindustrie/.
42. http://www.npr.org/sections/money/2015/05/21/408234543/will-your-job-be-done-by-a-machine.
43. "A To-Do List for the Tech Industry," *Wired Magazine*, November 2016, https://www.wired.com/2016/10/obama-six-tech-challenges/.
44. http://de.statista.com/statistik/daten/studie/294128/umfrage/anzahl-der-berufskraftfahrer-im-gueterverkehr/.
45. http://de.statista.com/statistik/daten/studie/294138/umfrage/anzahl-der-berufskraftfahrer-in-den-usa/.
46. "Map: The Most Common Job in Every State," http://www.npr.org/sections/money/2015/02/05/382664837/map-the-most-common-job-in-every-state.
47. "Trucking Industry: One Out of Three Trucks to Be Semi-autonomous by 2025," https://www.mckinsey.de/deliveringchange.
48. http://taxipedia.info/zahlen-und-fakten/.
49. "Study on Passenger Transport by Taxi, Hire Car with Driver and Ridesharing in the EU," http://www.astrid-online.it/static/upload/2016/2016-09-26-country-reports.pdf.
50. "Elektro-Autos: Wie viele Jobs fallen weg?" (Electric Vehicles: How Many Jobs Are Lost?), http://www.daserste.de/information/wirtschaft-boerse/plusminus/sendung/elektro-auto-mobilitaet-arbeit100.html.
51. "Daimler-Betriebsrat fürchtet Jobschwund durch E-Autos" (Daimler Worker's Council Fears Loss of Jobs Because of Electric Vehicles), http://derstandard.at/2000044550931/Daimler-Betriebsrat-fuerchtet-Jobschwund-durch-E-Autos.
52. http://www.spiegel.de/auto/aktuell/ig-metall-fordert-rasche-abkehr-von-benzin-und-dieselautos-a-1119779.html.

53. "Car Suppliers Vie for Major Role in Self-Driving Boom," https://www.wsj.com/articles/car-suppliers-vie-for-major-role-in-self-driving-boom-1483980527; "For Suppliers, Self-Driving Payday Nears: Sensor, Software Boom Expected by 2020," http://www.autonews.com/article/20160704/RETAIL01/307049984/for-suppliers-self-driving-payday-nears.

54. "Statistik zeigt: Anzahl der Fahrlehrer sinkt weiter, Trend zu angestellten Fahrlehrern" (Statistics Show Decreasing Numbers of Driving Instructors, Trend to Employed Driving Instructors), http://www.moving-roadsafety.com/wp-content/uploads/2016/04/2016-04-29-PM-Fahrlehrerstatistik-2016-Presse.pdf.

55. "Stifterverband für die Deutsche Wissenschaft e.V.: Zahlen und Fakten aus der Wissenschaftsstatistik" (Figures and Facts from the Science Statistics), January 2011, http://www.stifterverband.de/pdf/fue_facts_2011-01.pdf.

56. "Garagisten geht wegen Tesla-Boom die Arbeit aus" (Garage Owners Out of Work Due to Tesla Boom), http://www.20min.ch/finance/news/story/25771189#videoid=524953?redirect=mobi&nocache=0.15413772621585423.

57. Automotive Service Technicians and Mechanics, https://www.bls.gov/ooh/installation-maintenance-and-repair/automotive-service-technicians-and-mechanics.htm.

58. "Number of Employees in the U.S. Motor Vehicle and Parts Dealer Industry from 2007 to 2018 (in 1,000s)," https://www.statista.com/statistics/276514/automotive-dealer-industry-employees-in-the-united-states/.

59. Terry Tamminen, *Lives per Gallon: The True Cost of Our Oil Addiction*. Island Press, Washington, DC, 2006.

60. Catherine Lutz, Anne Lutz Fernandez, *Carjacked: The Culture of the Automobile and Its Effect on Our Lives*. St. Martin's Press, New York, 2010.

61. http://www.sourcewatch.org/index.php/Coal_and_jobs_in_the_United_States; https://de.statista.com/statistik/daten/studie/185209/umfrage/belegschaft-im-steinkohlebergbau-in-deutschland-seit-1950/.

62. "Number of Service Stations in Germany from 1950 to 2016," https://de.statista.com/statistik/daten/studie/2621/umfrage/anzahl-der-tankstellen-in-deutschland-zeitreihe/.

63. "Cars and Second-Order Consequences," http://ben-evans.com/benedictevans/2017/3/20/cars-and-second-order-consequences.

64. "Convenience Stores Hit Record In-Store Sales in 2015," http://www.nacsonline.com/Media/Press_Releases/2016/Pages/PR041216-2.aspx#.WOMZE461vUL.

65. "Self-Driving Cars to Cut U.S. Insurance Premiums 40%, Aon Says," http://www.chicagotribune.com/business/ct-self-driving-cars-insurance-premiums-20160912-story.html.

66. "The Future of Motor Insurance: How Car Connectivity and ADAS Are Impacting the Market," http://media.swissre.com/documents/HERE_Swiss%20Re_white%20paper_final.pdf.

67. "Tesla Enters Car Insurance Business as Self-Driving Cars Prepare to Disrupt the Industry," https://electrek.co/2016/08/30/tesla-enters-car-insurance-business-self-driving-cars-prepare-disrupt-industry/.

68. "Tesla Wants to Sell Future Cars with Insurance and Maintenance Included in the Price," http://www.businessinsider.com/tesla-cars-could-come-with-insurance-maintenance-included-2017-2.

69. https://optn.transplant.hrsa.gov/media/1161/ddps_03-2015.pdf.
70. http://fortune.com/2014/08/15/if-driverless-cars-save-lives-where-will-we-get-organs/.
71. http://docplayer.net/38025619-How-airbnb-combats-middle-class-income-stagnation-by-gene-sperling.html.
72. "Flying Cars Are Closer Than You Think," http://www.theverge.com/a/verge-2021/marc-andreessen-horowitz-verge-interview.

Chapter 11

1. https://www.youtube.com/watch?v=Uj-rK8V-rik.
2. https://en.wikipedia.org/wiki/Lost_time.
3. "Light Traffic | MIT Senseable City Lab," https://www.youtube.com/watch?v=4CZc3erc_l4.
4. "Rush Hour Intersection Traffic Condensed into One Minute," https://youtu.be/HFrrdhbC6pg.
5. T. Nagatani, "Traffic Jam Induced by Fluctuation of a Leading Car," *Physical Review E* 61, 2000.
6. https://youtu.be/7wm-pZp_mi0.
7. "The Present and Future of Trucking, Our Country's Broken, Inefficient Economic Backbone," https://techcrunch.com/2016/11/02/the-present-and-future-of-trucking-our-countrys-broken-inefficient-economic-backbone/.
8. "China's Driverless Trucks Are Revving Their Engines," https://www.technologyreview.com/s/602854/chinas-driverless-trucks-are-revving-their-engines/.
9. https://freight.uber.com/.
10. "Amazon Is Secretly Building an 'Uber for Trucking' App, Setting Its Sights on a Massive $800 Billion Market," http://www.businessinsider.com/amazon-building-uber-for-trucking-app-2016-12.
11. "Intel Announces $250 Million for Autonomous Driving Tech," https://techcrunch.com/2016/11/15/intel-announces-250-million-for-autonomous-driving-tech/.
12. "Bumps in the Road to Self-Driving Car Storage," http://itknowledgeexchange.techtarget.com/storage-disaster-recovery/bumps-in-the-road-to-self-driving-car-storage/.
13. "Data Storage Issues Grow for Cars," http://semiengineering.com/data-issues-grow-for-cars/.
14. "Data to Become New Profit Centre for Car Makers," http://www.telegraph.co.uk/technology/news/12033458/Data-to-become-new-profit-centre-for-car-makers.html.
15. "Joint Statement from the Conference of German Federal and State Independent Data Protection Authorities and the German Association of the Automotive Industry (VDA)," https://www.vda.de/de/themen/innovation-und-technik/vernetzung/gemeinsame-erklaerung-vda-und-datenschutzbehoerden-2016.html.
16. "Three Sneaky Ways Google Wins with Android Auto," http://www.wired.com/2014/06/android-auto-2/.
17. "How Ford Has Slammed the Door on Silicon Valley's Autonomous Vehicles Drive," https://www.theregister.co.uk/2017/03/27/keep_out_how_ford_is_keeping_silicon_valley_out_of_autonomous_vehicles/.
18. https://www.accenture.com/us-en/service-connected-vehicle.
19. https://caruma.tech/.

20. "Monetizing Car Data," http://www.mckinsey.com/industries/automotive-and -assembly/our-insights/monetizing-car-data.

21. https://www.transportation.gov/sites/dot.gov/files/docs/AV%20policy%20guidance %20PDF.pdf.

22. https://techcrunch.com/2016/09/20/federal-policy-for-self-driving-cars-pushes -data-sharing/.

23. http://www.bloomberg.com/news/articles/2016-05-03/fiat-google-said-to-plan -partnership-on-self-driving-minivans.

24. http://www.bbc.com/news/technology-36912700.

25. http://techcrunch.com/2016/01/26/lyft-cabify-99taxis-others-to-integrate-wazes -routing-software-in-their-own-apps/.

26. https://maps.apple.com/vehicles/.

27. http://techcrunch.com/2016/01/05/here-launches-cloud-based-maps-for -automated-driving/.

28. http://www.usatoday.com/story/tech/news/2015/12/28/heres-3d-maps-connected -cars-ces/77766922/.

29. http://www.bloomberg.com/news/articles/2016-03-13/race-to-guide-self-driving -cars-is-getting-another-competitor.

30. https://techcrunch.com/2015/06/29/uber-acquires-part-of-bings-mapping-assets-will -absorb-around-100-microsoft-employees/; http://www.theverge.com/2016/7/31/ 12338268/uber-maps-investment-500-million.

31. Amnon Shashua CVPR 2016 keynote: "Autonomous Driving, Computer Vision and Machine Learning," https://youtu.be/n8T7A3wqH3Q.

32. https://www.wired.com/2016/07/civil-maps-self-driving-car-autonomous -mapping-lidar/.

33. http://techcrunch.com/2016/06/01/mapbox-enters-the-autonomous-vehicle-market -with-mapbox-drive-an-sdk-for-cars/.

34. http://www.cultofmac.com/435571/mystery-vans-likely-making-3-d-road-maps-for -apples-self-driving-car/.

35. https://youtu.be/qu3ZuNjQMcQ.

36. http://www.autoblog.com/2016/03/28/volkswagen-egolf-recall-battery-software/.

37. http://www.wiwo.de/unternehmen/auto/funk-updates-tuev-fordert -nachpruefungen-fuer-frisierte-tesla-autos/13483414.html.

38. "Three Sneaky Ways Google Wins with Android Auto," https://www.wired.com/ 2014/06/android-auto-2/.

39. "Ford Is Adding Support for Apple CarPlay and Android Auto to Its Vehicles," http://techcrunch.com/2016/01/03/ford-is-adding-support-for-apple-carplay-and -android-auto-to-its-vehicles/.

40. "Warum Daimler sein Taxi-Geschäft riskiert?" (Why Does Daimler Risk Its Taxi Business?), http://www.spiegel.de/wirtschaft/unternehmen/daimler-legt-sich-wegen -mytaxi-und-car2go-mit-taxibranche-an-a-1074271.html.

41. "Uber's No-Holds-Barred Expansion Strategy Fizzles in Germany," https://www .nytimes.com/2016/01/04/technology/ubers-no-holds-barred-expansion-strategy -fizzles-in-germany.html?mabReward=A7&_r=0.

42. http://www.gallup.com/poll/1654/honesty-ethics-professions.aspx.

43. https://www.nada.org/WorkArea/DownloadAsset.aspx?id=21474839497.

44. "Multi-state Study of the Electric Vehicle Shopping Experience," https://www.scribd .com/document/321167667/1371-Rev-Up-EVs-Report-09-web-FINAL.

45. https://www.facebook.com/groups/50339366788/permalink/ 10154947776166789/?comment_id=10154948270041789&notif_t=group_ comment_follow&notif_id=1483114784037106.

46. https://en.wikipedia.org/wiki/Tesla_US_dealership_disputes.

47. "Economic Effects of State Bans on Direct Manufacturer Sales to Car Buyers," May 2009, http://www.justice.gov/atr/economic-effects-state-bans-direct-manufacturer -sales-car-buyers.

48. http://app.handelsblatt.com/unternehmen/industrie/autohaendler-hilferufe-aus-dem -industriegebiet/13689420.html.

49. Kim Hill, Debra Maranger Menk, Joshua Cregger, "Assessment of Tax Revenue Generated by the Automotive Sector for the Year 2013," Center for Automotive Research, January 2015, http://www.cargroup.org.

50. http://www.autoalliance.org/files/dmfile/2015-Auto-Industry-Jobs-Report.pdf.

51. https://www.destatis.de/DE/Themen/Gesellschaft-Umwelt/Umwelt/Publikationen/ Umweltnutzung-Wirtschaft/umweltnutzung-und-wirtschaft-bericht-5850001147004 .html.

52. https://www.bmf.gv.at/services/publikationen/Daten_und_Fakten_Steuer-_und_ Zollverwaltung_2014.pdf.pdf?555a9o.

53. https://www.ezv.admin.ch/ezv/de/home/information-firmen/steuern-und-abgaben/ einfuhr-in-die-schweiz/mineraloelsteuer.html.

54. https://www.caranddriver.com/news/a15347274/nhtsa-chief-autonomous-cars-should -cut-death-rate-in-half/.

55. http://fortune.com/2015/10/07/volvo-liability-self-driving-cars/.

56. http://www.autonews.com/article/20160329/OEM11/160329864.

57. https://www.joinroot.com/.

58. http://www.trefis.com/stock/hig/articles/218036/an-analysis-of-the-u-s-personal -automobile-insurance-market-part-1/2013-12-05.

59. https://www.metromile.com/.

60. https://qz.com/124721/the-secret-financial-market-only-robots-can-see/.

61. Department for Transport, Centre for Connected and Autonomous Vehicles, "Pathway to Driverless Cars: Consultation on Proposals to Support Advanced Driver Assistance Systems and Automated Vehicles," January 2017, https://assets.publishing .service.gov.uk/government/uploads/system/uploads/attachment_data/file/581577/ pathway-to-driverless-cars-consultation-response.pdf.

62. "State of the Automotive Finance Market: A Look at Loans and Leases in Q2 2016," https://www.experian.com/assets/automotive/quarterly-webinars/2016-Q2-SAFM.pdf.

63. https://de.statista.com/statistik/faktenbuch/225/a/services-leistungen/finanzen/ autokredit/.

64. "Uber Is Trying to Lure New Drivers by Offering Bank Accounts," https:// qz.com/533492/exclusive-heres-how-uber-is-planning-using-banking-to-keep-drivers -from-leaving/.

65. Brett King, *Augmented: Life in the Smart Lane*. Marshall Cavendish International, London, 2016.

66. "ZF, UBS and Innogy Innovation Hub Announce the Jointly Developed Blockchain Car eWallet," https://press.zf.com/press/en/releases/release_2638.html.

67. "The Death of Bank Products Has Been Greatly Under-Exaggerated," https:// medium.com/@brettking/the-death-of-bank-products-has-been-greatly-under -exaggerated-153cdb21a5d4#.uo025qbh0.

68. Jeff Speck, *Walkable City: How Downtown Can Save America, One Step at a Time.* North Point Press, San Francisco, 2012.

69. WSP/Parsons Brinckerhoff, Farrells, "Making Better Places: Autonomous Vehicles and Future Opportunities," http://www.wsp-pb.com/Globaln/UK/WSPPB-Farrells -AV-whitepaper.pdf.

70. "8 Cities That Show You What the Future Will Look Like," http://www.wired.com/ 2015/09/design-issue-future-of-cities/.

71. "Smart City Company Telensa Lights Up $18M in Funding," https://techcrunch.com/ 2016/01/19/telensa/.

72. William Whyte, *City: Rediscovering the Center.* University of Pennsylvania Press, Philadelphia, 2009.

73. Chuck Kooshian, Steve Winkelman, "Growing Wealthier: Smart Growth, Climate Change and Prosperity," 2011, http://growingwealthier.info/docs/growing_wealthier .pdf.

74. https://en.wikipedia.org/wiki/Marchetti%27s_constant.

75. Tom Vanderbilt, *Traffic.* Allen Lane, London, 2008.

76. Chris McCahill, Norman Garrick, Carol Atkinson-Palombo, "Visualizing Urban Parking Supply Ratios," Congress for the New Urbanism 22nd Annual Meeting, Buffalo, NY, June 4–7, 2014, https://www.cnu.org/sites/default/files/cnu22_ visualizing_urban_parking_supply_ratios.pdf.

77. "Chinese City Wuhu Embraces Driverless Vehicles," http://www.bbc.com/news/ technology-36301911.

78. National League of Cities, "City of the Future," http://www.nlc.org/sites/default/ files/2016-12/City%20of%20the%20Future%20FINAL%20WEB.pdf.

79. https://www.whitehouse.gov/blog/2015/12/07/american-innovation-autonomous -and-connected-vehicles.

80. https://www.transportation.gov/smartcity.

81. "This New Super-Sustainable Town Will Run on Solar Power and Use Driverless Cars for Public Transit," http://www.fastcoexist.com/3058874/this-planned-super -sustainable-town-will-run-on-solar-power-and-use-only-driverless-cars.

82. "LA's Big Plan to Change the Way We Move," http://la.curbed.com/2016/9/9/ 12824240/self-driving-cars-plan-los-angeles.

83. Gov. Scott Walker, "Wisconsin Road Projects May Be Scaled Back to Save Money," http://www.jsonline.com/story/news/politics/2017/03/01/gov-scott-walker -wisconsin-road-projects-may-scaled-back-save-money/98605250/.

84. "Road to Zero: New Partnership Aims to End Traffic Fatalities Within 30 Years," http://www.nsc.org/learn/NSC-Initiatives/Pages/The-Road-to-Zero.aspx.

85. "Right of Way for 'Vision Zero,'" http://www.dvr.de/presse/informationen/873.htm; "Time for Zero Traffic Fatalities," https://www.vcd.org/themen/verkehrssicherheit/ vision-zero/.

86. "Say Hello to Waymo," https://www.youtube.com/watch?v=uHbMt6WDhQ8.

87. "Flying Cars Are Closer Than You Think," http://www.theverge.com/a/verge-2021/ marc-andreessen-horowitz-verge-interview.

88. "Paris Mayor Unveils Plan to Restrict Traffic and Pedestrianise City Centre," https:// www.theguardian.com/world/2017/jan/08/paris-mayor-anne-hidalgo-plan-restrict -traffic-pedestrianise-city-centre-france.

89. "Diesel Car Sales Pie Halves to 26% in Four Years," http://economictimes.indiatimes
    .com/industry/auto/news/passenger-vehicle/cars/diesel-car-sales-pie-halves-to-26-in
    -four-years/articleshow/53056370.cms.

90. "Fahrverbote in Oslo: Diesel müssen draußen bleiben" (Driving Bans in Oslo: Diesels
    Must Stay Out), http://www.spiegel.de/auto/aktuell/diesel-fahrverbote-in-oslo-smog
    -erfordert-drastische-massnahmen-a-1130242.html.

91. "Results from the Reinforced ADAC EcoTest," https://www.adac.de/infotestrat/
    adac-im-einsatz/motorwelt/ecotest_feinstaub.aspx.

92. "Buy Up All the Street Cars," https://medium.com/@rynmcmns/buy-up-all-the-street
    -cars-d5c48db6039d#.1tk6j07ab.

93. "Here's How Self-Driving Cars Will Transform Your City," https://www.wired.com/
    2016/10/heres-self-driving-cars-will-transform-city/.

94. "Autonomous Tractor at Work," https://www.youtube.com/watch?v=Ybxhvlyw-X0.

95. Timothy J. Gates, Robert E. Maki, "Converting Old Traffic Circles to Modern
    Roundabouts," Michigan State University Case Study, ITE Annual Meeting
    Compendium, 2000.

96. "Roundabout Benefits," https://www.wsdot.wa.gov/Safety/roundabouts/benefits.htm.

97. Neal E. Wood, "Shoulder Rumble Strips: A Method to Alert 'Drifting' Drivers,"
    Pennsylvania Turnpike Commission, Harrisburg, PA, January 1994.

98. Heidi Garrett-Peltier, "Pedestrian and Bicycle Infrastructure: A National Study of
    Employment Impacts," Baltimore, 2011.

99. "New Urban Network Study: Transit Outperforms Green Building," http://
    newurbannetwork.com/study-transit-outperforms-green-buildings/.

100. Martin Wachs, "Fighting Traffic Congestion with Information Technology," *Issues in
     Science and Technology* 19, 2002.

101. https://en.wikipedia.org/wiki/Braess%27s_paradox.

102. Gilles Duranton, Matthew A. Turner, "The Fundamental Law of Road Congestion:
     Evidence from US Cities," *American Economic Review* 101:2616–2652, October
     2011, http://pubs.aeaweb.org/doi/pdfplus/10.1257/aer.101.6.2616.

103. "Yes, Sometimes I Drive Around Town to Get My Kids to Sleep," http://www
     .huffingtonpost.com/jennie-sutherland/yes-sometimes-i-drive-aroyes-sometimes-i
     -drive-around-town-to-get-my-kids-to-sleep_b_8124776.html.

104. https://en.wikipedia.org/wiki/Interstate_405_(California)#.22Carmageddon.22.

105. "Ewig lockt die Schnellstraße" (And God Created the Highway), http://www
     .sueddeutsche.de/wissen/ewig-lockt-die-schnellstrasse-1.913440.

106. "Would Standing on the Left Get You Through Tube Stations Quicker?," https://
     www.uk.capgemini.com/blog/business-analytics-blog/2016/04/would-standing-on
     -the-left-get-you-through-tube-stations#about-the-author-anchor.

107. German Institute for Urbanistics (DIFU), *Der kommunale Investitionsbedarf von 2006
     bis 2020* (*Municipal Investment Requirements from 2006 to 2020*). Berlin, 2008.

108. https://de.statista.com/themen/1199/strassen-in-deutschland/.

109. "Percentage of Settlement Areas and Traffic Areas in the Territorial Area on a German
     Regional Level," https://www.ioer-monitor.de/en.

110. "Maintenance Management for Municipal Roads," https://www.adac.de/_mmm/pdf/
     fi_erhaltungsmanagement_0412_238773.pdf.

111. "Expenses and Revenue for German Highways from 1992 to 2010 (in Million Euro),"
     https://de.statista.com/statistik/daten/studie/7002/umfrage/ausgaben-und-einnahmen
     -im-deutschen-strassenwesen-seit-dem-jahr-1992/.

112. "U.S. Driving Tops 3.1 Trillion Miles in 2015, New Federal Data Show," https://www.fhwa.dot.gov/pressroom/fhwa1607.cfm.

113. Interview conducted with employees of Aachener-Grund in Palo Alto.

114. Donald C. Shoup, *The High Cost of Free Parking*. American Planning Association, Chicago, 2005.

115. "Who Pays for Parking? The Hidden Costs of Housing," http://www.sightline.org/research_item/who-pays-for-parking/.

116. Donald C. Shoup, *The High Cost of Free Parking*. American Planning Association, Chicago, 2005.

117. Donald C. Shoup, "Cruising for Parking," *Transport Policy* 13, 2006.

118. Donald C. Shoup, *The High Cost of Free Parking*. American Planning Association, Chicago, 2005.

119. "ParkWhiz Acquires BestParking, Announces $24M Raise," https://techcrunch.com/2016/01/26/parkwhiz-acquires-bestparking-announces-24m-raise/.

120. SFPark.org.

121. Paul C. Box, "Curb Parking Findings Revisited," Transportation Research Circular 501, 2000.

122. A. J. Velkey, C. Laboda, S. Parada, et al. "Sex Differences in the Estimation of Foot Travel Time," presented at the Annual Meeting of the Eastern Psychological Association, Boston, 2002.

123. http://www.itf-oecd.org/sites/default/files/docs/15cpb_self-drivingcars.pdf.

124. Kevin Spieser, Kyle Ballantyne Treleaven, Rick Zhang, et al., "Toward a Systematic Approach to the Design and Evaluation of Automated Mobility- on-Demand Systems: A Case Study in Singapore," in Gereon Meyer, Sven Beiker (eds.), *Road Vehicle Automation* (Lecture Notes in Mobility). Springer, Berlin, 2014.

125. https://www.csail.mit.edu/ridesharing_reduces_traffic_300_percent.

126. Lawrence D. Burns, William C. Jordan, Bonnie A. Scarborough, "Transforming Personal Mobility," 2013, http://wordpress.ei.columbia.edu/mobility/files/2012/12/Transforming-Personal-Mobility-Aug-10-2012.pdf.

127. "Munich: 18,000 Electric Self-Driving Taxis May Replace 200,000 Private Vehicles," https://ecomento.tv/2017/04/11/muenchen-18-000-elektrische-selbstfahr-taxis-koennten-200-000-privat-pkw-ersetzen/.

128. Brandon Schoettle, Michael Sivak, "Potential Impact of Self-Driving Vehicles on Household Vehicle Demand and Usage," http://www.umich.edu/%7Eumtriswt/PDF/UMTRI-2015-3_Abstract_English.pdf.

129. Lawrence D. Burns, William C. Jordan, Bonnie A. Scarborough, "Transforming Personal Mobility," http://wordpress.ei.columbia.edu/mobility/files/2012/12/Transforming-Personal-Mobility-Aug-10-2012.pdf.

130. Emilio Frazzoli, "Can We Put a Price on Autonomous Driving?," *MIT Technology Review* 18, March 2014, https://www.technologyreview.com/s/525591/can-we-put-a-price-on-autonomous-driving/.

131. "The Future of the $100 Billion Parking Industry," https://pando.com/2014/01/30/the-future-of-the-100-billion-parking-industry/.

132. http://sf.streetsblog.org/2014/08/21/personal-garages-become-cafes-in-the-castro-thanks-to-smarter-zoning/.

133. http://dip21.bundestag.de/dip21/btd/15/033/1503378.pdf.

134. Leslie George Norman, "Road Traffic Accidents: Epidemiology, Control and Prevention," World Health Organization Public Health Papers 12, 1962.

135. http://www.srf.ch/konsum/themen/umwelt-und-verkehr/unnoetige-verkehrsschilder-teuer-und-gefaehrlich.
136. https://en.wikipedia.org/wiki/Drachten.
137. https://en.wikipedia.org/wiki/Shared_space.
138. https://scienceblog.com/489337/pedestrians-may-run-rampant-world-self-driving-cars/.
139. https://web.de/magazine/auto/100-jahre-ampel-kuriose-fakten-lichtzeichenanlage-32675132.
140. http://www.tagesspiegel.de/wirtschaft/teures-gruen-was-kostet-eigentlich-eine-ampel/11625462.html.
141. "Revealed: How Long You Really Spend Waiting at Traffic Lights," http://www.telegraph.co.uk/cars/news/revealed-how-long-you-really-spend-waiting-at-traffic-lights/.
142. "Jeder steht zwei Wochen seines Lebens vor roten Ampeln" (Everyone Spends Two Weeks of Their Lives Waiting at Red Traffic Lights), http://www.augsburger-allgemeine.de/wirtschaft/Jeder-steht-zwei-Wochen-seines-Lebens-vor-roten-Ampeln-id30907737.html.
143. "Audi schafft das Fließband ab" (Audi Does Away with Assembly Lines), https://www.welt.de/wirtschaft/article159622953/Audi-schafft-das-Fliessband-ab.html.
144. "Elon Musk Goes on a 'Machines Building Machines' Rant About the Future of Manufacturing," https://electrek.co/2016/06/01/elon-musk-machines-making-machines-rant-about-tesla-manufacturing/.
145. "AI Software Learns to Make AI Software," https://www.technologyreview.com/s/603381/ai-software-learns-to-make-ai-software/.
146. "The Alien Style of Deep Learning Generative Design," https://medium.com/intuitionmachine/the-alien-look-of-deep-learning-generative-design-5c5f871f7d10?linkId=35086660.
147. "Auto- und Internetfirmen erobern den Energiesektor" (Automobile and Internet Companies Conquer the Energy Sector), http://www.spiegel.de/wirtschaft/unternehmen/energiewende-branchenfremde-konzerne-erobern-den-stromsektor-a-1061546.html.
148. "Ford Wants to Develop Its Own Battery Chemistries for Hybrids, Electric Cars, but Why?," http://www.greencarreports.com/news/1101606_ford-wants-to-develop-its-own-battery-chemistries-for-hybrids-electric-cars-but-why.
149. "Roadway and Environment: Urban/Rural Comparison," http://www.iihs.org/iihs/topics/t/roadway-and-environment/fatalityfacts/roadway-and-environment.
150. "The Case for an Emphasis on Traffic Management," https://techcrunch.com/2015/12/14/the-case-for-an-emphasis-on-traffic-management/.
151. "Mapping the Self-Driving Car with Traffic Analytics," http://digitally.cognizant.com/mapping-the-self-driving-car-with-traffic-analytics.
152. http://www.anoukwipprecht.nl/.
153. http://safecarnews.com/unece-updates-vienna-convention-on-road-traffic-to-allow-automated-vehicles-ma7237/.
154. https://en.wikipedia.org/wiki/World_Forum_for_Harmonization_of_Vehicle_Regulations.
155. https://assets.documentcloud.org/documents/3111057/Federal-Automated-Vehicles-Policy.pdf.

156. http://gizmodo.com/5985682/jony-ive-chats-lunchbox-design-on-a-british-kids-tv-show.

157. https://en.wikipedia.org/wiki/Road_train.

158. Christoph Keese, *Silicon Germany: Wie wir die digitale Transformation schaffen* (*Silicon Germany: How We Can Manage Digital Transformation*). Knaus, Munich, 2016.

159. "Will Self-Driving Cars Kill Transit as We Know It? It Could Be Charlotte's $6 Billion Bet," http://www.charlotteobserver.com/news/politics-government/article134742964.html.

160. Armin Kaltenegger (ed.), *Unterwegs in die Zukunft: Visionen zum Straßenverkehr* (*On the Way to the Future: Visions of Road Traffic*). Manz, Vienna, 2016.

161. "No Parking Here," http://www.motherjones.com/environment/2016/01/future-parking-self-driving-cars.

162. "Forget Tesla, It's China's E-Buses That Are Denting Oil Demand," https://www.bloomberg.com/news/articles/2019-03-19/forget-tesla-it-s-china-s-e-buses-that-are-denting-oil-demand.

163. http://www.bloomberg.com/news/articles/2016-05-05/oil-isn-t-the-only-commodity-threatened-by-tesla-s-rise.

164. "China's Rare-Earths Bust," http://www.wsj.com/articles/chinas-rare-earths-bust-1468860856.

165. "New York City Says Electric Cars Are Now the Cheapest Option for Its Fleet," https://qz.com/1571956/new-york-city-says-electric-cars-cheapest-option-for-its-fleet/.

166. "Wind in Power: 2016 European Statistics," https://windeurope.org/wp-content/uploads/files/about-wind/statistics/WindEurope-Annual-Statistics-2016.pdf.

167. Firmin DeBrabander, "What If Green Products Make Us Pollute More?," http://articles.baltimoresun.com/2011-06-02/news/bs-ed-consumers-20110602_1_green-cars-green-products-greenhouse-gas-emissions.

168. David Owen, *Green Metropolis: Why Living Smaller, Living Closer, and Driving Less Are the Keys to Sustainability*. Riverhead Books, New York, 2010.

169. Michael Mehaffy, "The Urban Dimensions of Climate Change," https://www.planetizen.com/node/41801.

170. http://www.tomsguide.com/us/self-driving-car-crash-dc2016,news-23145.html.

171. Jonathan Petit, "Self-Driving and Connected Cars: Fooling Sensors and Tracking Drivers," https://www.blackhat.com/docs/eu-15/materials/eu-15-Petit-Self-Driving-And-Connected-Cars-Fooling-Sensors-And-Tracking-Drivers.pdf; Jonathan Petit, Bas Stottelaar, Michael Feiri, Frank Kargl, "Remote Attacks on Automated Vehicles Sensors: Experiments on Camera and LiDAR."

172. "Securing Driverless Cars from Hackers Is Hard. Ask the Ex-Uber Guy Who Protects Them," https://www.wired.com/2017/04/ubers-former-top-hacker-securing-autonomous-cars-really-hard-problem.

173. http://www.forbes.com/2008/11/21/data-breaches-cybertheft-identity08-tech-cx_ag_1121breaches.html.

174. "Justice Dept. Group Studying National Security Threats of Internet-Linked Devices," https://www.yahoo.com/news/justice-dept-group-studying-national-security-threats-internet-172351993--finance.html?ref=gs.

175. "Karamba Security Raises $2.5 Million to Keep Self-Driving Cars Safe from Hackers," https://techcrunch.com/2016/09/29/karamba-security-raises-2-5-million-to-keep-self-driving-cars-safe-from-hackers/.

176. "Google Keeps Self-Driving Cars Offline to Hinder Hackers," https://www.ft.com/content/8eff8fbe-d6f0-11e6-944b-e7eb37a6aa8e.
177. "The Car Hacking Village Brings Car Knowledge to the CES Masses," http://readwrite.com/2017/01/04/the-car-hacking-village-brings-car-knowledge-to-the-ces-masses-tl1/.
178. "Will Autonomous Cars Leave Us Vulnerable to Gangs of Armed Teens? Study Says Maybe," http://jalopnik.com/will-autonomous-cars-leave-us-vulnerable-to-gangs-of-ar-1792042072.
179. "Biggest Challenge for Self-Driving Cars in Boston? Sea Gulls," http://www.bostonglobe.com/business/2017/02/06/the-biggest-challenge-for-self-driving-cars-boston-sea-gulls/N5UHSUIyXlar4r60TXupdN/story.html.
180. "1890–1968 Flying Cars," http://mashable.com/2015/08/03/flying-car-evolution/.
181. http://www.bloomberg.com/news/articles/2016-06-09/welcome-to-larry-page-s-secret-flying-car-factories.
182. "Airbus Wants to Make Self-Flying Airborne Taxis a Real Thing," https://techcrunch.com/2016/08/17/airbus-wants-to-make-self-flying-airborne-taxis-a-real-thing.
183 "Fast-Forwarding to a Future of On-Demand Urban Air Transportation," https://medium.com/@UberPubPolicy/fast-forwarding-to-a-future-of-on-demand-urban-air-transportation-f6ad36950ffa#.52t6yxbia.

Part III

1. "Mercedes Targets Silicon Valley Rivals with Robo-Taxis by 2023," https://www.bloomberg.com/news/articles/2017-04-04/mercedes-bosch-join-forces-to-accelerate-rollout-of-robo-taxis.
2. "Dell. EMC. HP. Cisco. These Tech Giants Are the Walking Dead," https://www.wired.com/2015/10/meet-walking-dead-hp-cisco-dell-emc-ibm-oracle/.
3. "Porsche-Mitarbeiter bekommen Riesen-Bonus" (Porsche Employees Receive Huge Bonus), https://www.welt.de/wirtschaft/article163062200/Porsche-Mitarbeiter-bekommen-Riesen-Bonus.html; "9,656 Euro Bonus and Jubilee Payment for Porsche Employees," https://newsroom.porsche.com/fallback/en/company/porsche-bonus-payment-2018-employees-record-year-70-years-sports-car-15093.html.
4. "Autohersteller geben Rabatte in Rekordhöhe" (Automotive Manufacturers Make Record Discounts), http://www.wiwo.de/unternehmen/auto/neuwagen-autohersteller-geben-rabatte-in-rekordhoehe/19633862.html.
5. "Abgasskandal kostet VW Marktanteile bei Firmenwagen" (VW Loses Market Share for Company Cars in Emission Scandal), https://www.welt.de/wirtschaft/article163534898/Abgasskandal-kostet-VW-Marktanteile-bei-Firmenwagen.html.
6. Eric Weiner, *The Geography of Genius: A Search for the World's Most Creative Places, from Ancient Athens to Silicon Valley.* Simon & Schuster, New York, 2016.
7. https://www.bmwgroup.com/en/company.html.
8. "Mission Statements of Auto Manufacturers," https://www.thebalance.com/auto-industry-mission-statements-4068550.
9. "Company Strategy," https://www.audi.com/en/company/strategy.html.
10. http://together.volkswagenag.com/.
11. "Our Strategy," https://www.daimler.com/company/strategy/.
12. "What Was Volkswagen Thinking?," http://www.theatlantic.com/magazine/archive/2016/01/what-was-volkswagen-thinking/419127/.

13. Diane Vaughan, *The Challenger Launch Decision: Risky Technology, Culture, and Deviance at NASA.* University of Chicago Press, Chicago, 1997.
14. Dennis A. Gioia, "Pinto Fires and Personal Ethics: A Script Analysis of Missed Opportunities," *Journal of Business Ethics* 11(5–6), May 1992.
15. Frans de Waal, *Are We Smart Enough to Know How Smart Animals Are?* W. W. Norton, New York, 2016.

Chapter 12

1. Christensen, von den Eichen, Matzler, *The Innovator's Dilemma.* Vahlen, Berlin, 2015.
2. "Number of Full-Time Employees in the United States from 1990 to 2018 (in Millions)," https://www.statista.com/statistics/192356/number-of-full-time-employees-in-the-usa-since-1990/.
3. Susan Christoperson, "Short-Term Profit Seeking Risks the Future of Manufacturing; The Conversation," September 24, 2013, http://theconversation.com/short-term-profit-seeking-risks-the-future-of-manufacturing-18573.
4. http://www.mckinsey.com/industries/automotive-and-assembly/our-insights/monetizing-car-data.
5. Clayton Christensen, "Principles of Innovation and Measuring Success," https://www.youtube.com/watch?v=MpEmjwrOuxI.
6. https://en.wikipedia.org/wiki/Occam%27s_razor.
7. Larry Keeley, *Ten Types of Innovation: The Discipline of Building Breakthroughs.* Wiley, Hoboken, NJ, 2013.
8. Frans Johansson, *The Medici Effect: Breakthrough Insights at the Intersection of Ideas, Concepts and Cultures.* Harvard Business School Press, Cambridge, MA, 2004.
9. "Tesla ist kein Vorbild: Interview mit Harald Kröger, Leiter Entwicklung Elektrik bei Mercedes-Benz" (Tesla Is No Model: Interview with Harald Kröger, Head of Electrics Development at Mercedes-Benz), http://mein-auto-blog.de/news/tesla-kein-vorbild.html.

Chapter 13

1. A. L. Tucker and A. C. Edmondson, "Why Hospitals Don't Learn from Failures: Organizational and Psychological Dynamics That Inhibit System Change," *California Management Review* 45(2):55–72, 2003.
2. http://www.focus.de/finanzen/news/tid-5538/ferdinand-piech_aid_53701.html.
3. "O. Verf.: Zukunftstechnik: Porsche-Chef bezeichnet selbstfahrende Autos als 'Hype'" (Porsche Boss Calls Self-Driving Cars a "Hype"), www.spiegel.de, September 13, 2015, http://www.spiegel.de/auto/aktuell/porsche-chef-mat-thias-mueller-bezeichnet-autonomes-fahren-als-hype-a-1052688.html.
4. Robert I. Sutton, *Der Arschloch-Faktor (The Asshole Factor).* Heyne, Munich, 2008.
5. Jack Linshi, "Peter Thiel: Uber Is the 'Most Ethically-Challenged Company in Silicon Valley,'" Time.com, November 19, 2014, http://time.com/3593701/peter-thiel-uber/.
6. Warren Berger, *A More Beautiful Question: The Power of Inquiry to Spark Breakthrough Ideas.* Bloomsbury, London, 2014.
7. Ibid.
8. Ibid.

9. Jim Collins, *How the Mighty Fall: And Why Some Companies Never Give In*. Random House Business, New York, 2009.

10. Warren Berger, *A More Beautiful Question: The Power of Inquiry to Spark Breakthrough Ideas*. Bloomsbury, London, 2014.

11. "1965: Moore's Law Predicts the Future of Integrated Circuits," http://www .computerhistory.org/siliconengine/moores-law-predicts-the-future-of-integrated -circuits/.

12. https://www.innocentive.com/.

13. Hila Lifshitz-Assaf, "Dismantling Knowledge Boundaries at NASA: From Problem Solvers to Solution Seekers," May 14, 2016; https://ssrn.com/abstract=2431717.

14. https://www.udacity.com/course/self-driving-car-engineer-nanodegree--nd013.

15. https://cars.stanford.edu/.

16. https://www.ri.cmu.edu/.

17. "A New Self-Driving Car Startup Just Spun Out of Udacity to Challenge Uber with Its Own Autonomous Taxi Service," http://www.businessinsider.com/voyage -autonomous-taxi-udacity-2017-4.

18. https://unmanned.tamu.edu/.

19. http://autonomos.inf.fu-berlin.de/.

Conclusion

1. "Framing the Future of Mobility," https://www2.deloitte.com/content/dam/Deloitte/ de/Documents/human-capital/DR20_Framin_the_future_of_mobility.pdf.

2. "Unconditional Basic Income," https://en.wikipedia.org/wiki/Basic_income.

3. "The Robot Revolution Will Be the Quietest One," https://www.nytimes.com/ 2016/12/07/opinion/the-robot-revolution-will-be-the-quietest-one.html.

4. "Fuck Work," https://aeon.co/essays/what-if-jobs-are-not-the-solution-but -the-problem.

5. "Slush 2016—Universal Basic Income 'Has to Happen,'" http://diginomica.com/ 2016/12/02/slush-2016-universal-basic-income-happen/.

6. "Analysis: Between 2000 and 2010, 85% of Manufacturing Jobs Were Lost to Technology, Not Globalization," https://theintellectualist.co/analysis-between-2000 -and-2010-85-of-manufacturing-jobs-were-lost-to-technology-not-globalization/.

7. "Robots Will Steal Your Job: How AI Could Increase Unemployment and Inequality," http://www.businessinsider.com/robots-will-steal-your-job-citi-ai-increase -unemployment-inequality-2016-2?r=UK&IR=T.

8. "'Tax the Robots,' Says Bill Gates," https://www.forbes.com/sites/ianmorris/2017/02/ 17/tax-the-robots-says-bill-gates/ #72e7f80d1096.

Index

accidents
  90 percent drop with driverless cars in, 210
  caused by human drivers, 198
  driver assistance systems preventing, 191
  examples of what could go wrong, 177–181
  fatal crash of Uber, 123, 176–177
  frequency of, 111–114, 311–313
  lightweight design of self-drivers and, 191
  lost jobs in healthcare due to less EV, 275–276
  question of who pays for, 301–303
  reducing number of, 161–163
  statistics on, 194, 208–209
adapter plugs, charging EV vehicles, 76–81
agile (lean) technologies, rapid testing, 346
agricultural sector, impact of mechanization, 266
AI. *See* artificial intelligence (AI)
AIR (Audi Innovation Research), 372
Airbnb, 39–40, 229–230
Alibaba, 39–40
Allmers, Robert, 23–24
AlphaGo, 145–146, 161, 163
aluminum, EV batteries, 72
Amabile, Teresa, 165
American Automobile Association, 322
American Union of Concerned Scientists (UCS), 88
Andreesen, Marc, 309
Android Auto, 294
Apollo program, Baidu, 213

Apple
  battle for big data, 283–288
  Carplay, 294–295
  changes in electronic goods, 257–258
  dashboard battle, 294–295
  GPS accuracy issues, 288–290
  iCar, 283
  potential consumers and trust of, 43
  signs of change, 383
  skeumorphism in design of, 170
  TV market, 39–40
  vehicle map and navigation solutions, 138
Argo.ai, 156
Ars Electronic Future Lab, 174
artificial intelligence (AI)
  autonomous driving simulations, 153–155
  changing world and, 375–376
  dangers, 164–167
  developing own behavior, 145–147, 160–161
  digital updates, 148–149
  ethical issues, 157–166
  evolution of production, 319–320
  human brains vs., 137
  human expectations of, 163
  identifying intent of other participants, 155–156
  job loss from, 263–265, 378
  machine learning, 144–145
  new startups, 151–152
  processing power, 141–142
  programming basics for, 139–140

artificial intelligence (AI) (*continued*)
reducing number of accidents, 162–163
remote control, 145
safety, 147–148, 152–153
self-driving engineers, 156–157
semantic pixel labeling, 143
sensor malfunction and, 142–143
specialists in, 143–144
talent auctions for engineers, 156
testing self-driving vehicles, 150–151
training, 140–141
transgression of rules, 167–169
vehicle map/navigation, 137–139
Asimov, Isaac, 181–182
assembly-line production, 47, 319–320
Audi
driver assistance systems, 150
fast driving with RS7 prototype, 201
interior camera to alert driver, 150–151
mission statement, 342–343
record profits based on discounts, 341
ridesharing services, 225–226
sales models, 298–299
work on self-driving vehicles, 242
Audi Innovation Research (AIR), 372
Aurora.ai, 143
austerity, principle of innovation, 356–357
automation, tax on, 378
automobile clubs, 322
automobile industry, traditional
consumers switching EV instead of, 43
data and facts, 53–56
first automobile revolution, 19–22
future is being planned without, 3–4
lagging behind in technology, 5–7
lobbyists playing to, 28
love for cars, 29–35
pioneers of, 23–28
public subsidies for, 12
reaching "peak car," 15, 66, 253
ridesharing services, 225
second automobile revolution, 45–51
Tesla Model 3 may be final call for, 37–44
vertical integration systems, 149
Automotive Open System Architecture group (AUTOSAR), 218–219
autoNOMOS Labs, 374

autonomous and self-driving vehicles
autonomous vs. self-driving, 129
business models, 296
cameras, lasers, and LiDAR systems, 130–135
car sickness and, 199–200
classification of, 118–119
cybercrime, 331–335
drastic change coming from, 6
first experiments from other teams, 117–118
impact on revenues, 300–301
legal framework for, 323–325
manufacturer disengagement reports, 124–127
novice vs. experienced drivers, 112–114
patent applications, 128
on public streets now, 2
racing, 200–201
reservations about, 195–199
sleeping cars, 328
smart traffic management, 321–322
standards for technology comparisons, 127
status of automotive technology, 339
test licenses, 120–122
testing, 203–204
testing in the *GoMentum Station* area, 128
traffic accidents with, 114–115, 128
traffic activities/other side effects, 209–213
training and research, 373–374
trips to countryside, 311
understanding, 118–119
wave effects, 280–282
Waymo as leader, 127–128
when we can purchase, 192–194
who pays for accidents, 301–303
autopilot
continuous improvement in Tesla, 150, 232–233
fatal accidents in Tesla, 150, 153
in many current transportation modes, 196
problems of network failures, 218
aviation, pioneers of, 9–10

Baidu, 123, 154, 213, 226, 259, 261, 307
banks, financing options for cars, 303–304

batteries and accumulator
   argument against EVs, 106
   battery industry, 91–92
   battery market, and power grid, 81–82
   for clean future, 87–91
   cost of, 74
   cost of charging, 96
   creating higher EV purchasing cost,
      95–96, 98
   dangers of EVs, 86
   German industry of, 91
   high-performance battery cells, 320–321
   lithium, graphite, nickel, cobalt, and
      aluminum, 71–72
   memory effect/coffee filter effect, 72–74
   overview of, 69–71
   plugs, standards, charging obstacles,
      76–81
   polypropylene and ethyl carbonate, 72
   recycling, 75–76
   regulations and emergency solutions,
      82–86
behavior, machines have to understand/
   accept human, 175
benefit payments, unconditional basic
   income, 377–378
Benkler, Yochai, 229
Benz, Bertha, 19–21, 45, 85
Benz, Carl, 19–24, 85, 364
black swan events, 253
blame, killer questions, 367
Blockbuster, 25, 39–40
blockchain technology, 199, 230–231
BMW EV
   charging station test failure, 85
   competing against nonconsumption,
      355–356
   corporate culture, 345, 362
   mission statement, 342–343
   pricing, 98–99
   ridesharing services, 225
   sales models, 297–300
   self-driving vehicles project with Baidu, 123
   test area for self-driving, 206
   vulnerable to buy out, 340
   willingness for change, 380
Braess, Dietrich, 312
Braess paradox, 312–313

brainstorming, question storming vs.,
   364–365
Brexit, 383
Brin, Sergey, 25
Brown, Joshua, 150, 153
bugs/debugging, cybersecurity, 334
Burke, James E., 344
Burton, Robert, 366
buses, 228, 258–259
business models, self-driving vehicles, 296,
   339
business travelers, autonomous sleeping
   cars, 328

cab medallions, for taxi services, 222
California, self-driving cars in, 2
call centers, remote control of autonomous
   vehicles, 145
cameras, 133–134, 141, 143
Camry, 351
Canada, autonomous vehicles in, 208
capital, impact of innovation types on,
   351–353
car brands, become less important, 255–256
The Car Hacker's Handbook: A Guide for the
   Penetration Tester (Smith), 218
car sickness, self-driving vehicles and,
   199–200
Car2Come, 225
Car2Go, 225, 228, 300
Carlin, George, 369
Carnegie Melon, 117–118, 374
CARS (Center for Automotive Research),
   373–374
cars, love for, 29–35
Carson, Chris, 287
Center for Automotive Research (CARS),
   373–374
certainty epidemic, 366
change, 375–376, 379–380
Chanute, Octave, 9–10
ChargePoint charging box, 77–78
charging stations, 27, 40, 42, 59, 76–81,
   103, 105
children, qualifying for our future, 379
China
   autonomous trucks, 261
   competition with Toyota, 355

China (*continued*)
  controlling self-driving cars with
    thoughts, 174
  global leaders in automotive industry, 382
  innovative development, 339–340
  as leading nation in EVs, 58–61
  new developments, 2–5
  new startups, 152
  open source project of Baidu, 213
  ruthless exploitation of environment, 58
  transportation delivery system, 283
chips, 142, 220
Christensen, Clayton, 5, 25, 350, 353–356
Christopherson, Susan, 352
Chrysler, 355
cities/urban areas, 2, 304–311, 321–322
Cityscapes, 213
classification, self-driving vehicle, 118–119
clean future, batteries and accumulator,
    87–91
cobalt, in EV batteries, 72
COCO (Common Objects in Context),
    open source, 213
coffee filter effect, EV battery oxidation,
    72–74
cognitive dissonance, 50–51, 376–377
Collins, Jim, 15, 368
color of car, accident frequency and, 115
combustion engines
  efficiency of electric vehicles vs., 32,
    190–191, 247–248
  end of World War I as triumph for,
    22–23
  first automobile propulsion in 1900s
    using, 22
  as thing of the past, 40, 51, 60–62, 383
Comma.ai, 151–152, 212, 249
Common Objects in Context (COCO),
    open source, 213
communication
  with autonomous vehicles, 172–175
  with connected self-driving cars, 215–220
  interaction with machines is changing,
    171–172, 175
  vehicle-to-vehicle/vehicle-to-infrastructure,
    216
  while driving a car, 215
commuting, 32, 209, 279, 306, 316

company cars, accidents in, 114
company mission statements, 341–343
company scripts, crisis management, 345
competing against nonconsumption,
    355–356
conditional driving automation, SAE Level
    3, 118
connected cars, communication between,
    215–220
construction sites, programming for, 212
construction, transition to EVs and lost jobs
    in, 273
Consumer Electronics Show (CES), Las
    Vegas, 256
control, 196–198, 270
corporate culture, 149–150, 343–347, 362
corporations, steps of decline for large, 15–16
corruption, accidents as result of, 162
cost, of EVs
  batteries, 74
  economics of new technologies up to
    2025, 165
  price of, 95–99
  savings from less accidents, 210
  they are too expensive, 106
courteous, self-driving should be, 168
creativity, innovation and, 358–359
crisis management, 344–346
crowd-sourced machine learning, 151
Cruise Automation, 151
cryptography, cybersecurity, 332, 335
culture. *See* corporate culture
cybercrime/cybersecurity, 331–335
cyclists, self-drivers interpreting intentions
    of, 173

Dahn, Jeff, 73
Daimler, Gottfried, 23–24, 117–118
DARPA (Defense Advanced Research
    Projects Agency), Grand Challenge,
    116–117, 262
dashboard battle, 294–295
data, automotive industry, 53–56, 283–288
David and Goliath, 7–9, 25
*David and Goliath: Underdogs, Misfits, and
    the Art of Battling Giants* (Gladwell), 7
DDoS (distributed denial-of service) attacks,
    connected cars, 217–218

de Waal, Frans, 346
deadlocks, robots avoiding in frustration
    mode, 179
dealers, sales models, 296–300
decision making, traditional, 362–363
dedicated short range communication
    (DSRC) channel, 217
deep vertical integration, innovation, 47–48
Deepwater Horizon accident liability, 180
Defense Advanced Research Projects
    Agency (DARPA), Grand Challenge,
    116–117, 262
design, of self-driving vehicles
    of automatic doors, 211
    for communication, 172
    of exterior, 186–188
    FashionTech in car design, 323
    friendly, 167
    innovation and, 339–340
    of interior, 188–189
    lightweight, 191
    skeumorphism, for technological
        changes, 170
    for technology and safety regulations, 188
    ugly, 99
Detroit, 3–7, 131, 134, 256
Detroit Electric, 22, 59–60
Dickmanns, Professor Ernst, 117–118
Didi Chuxing, China, 157, 221
diesel cars. See also Volkswagen, 310, 341,
    381
DIGINET-PS project, Germany, 207
digital experience, 291
Digital Millennium Copyright Act
    (DMCA), cybersecurity, 334–335
discovery, in digital companies, 149
disengagement reports, by manufacturer,
    124–127
disposal, EV battery, 74–75
disruptive innovation
    defined, 349
    on-demand taxi service as, 222
    from ideas of nonexperts, 236
    Keeley model, 358
    path taken is difficult to foresee, 181
    players appear after acceptance of new
        technology, 13
    recognizing disruptive idea, 240–242

technological innovation vs., 25
    wanting to make world a better place,
        11–13
disruptors, creating new markets, 353
distributed denial-of service (DDoS) attacks,
    connected cars, 217–218
DMCA (Digital Millennium Copyright
    Act), cybersecurity, 334–335
doors, automatic, 211
Dowd, Mark, 307
Dragan, Anca, 144–145, 219
Drive assistance, SAE Level 1, 118
Drive.ai, 26, 125, 172, 211
DriveNow, BMW sales model, 225, 300
driver assistance system
    evolutionary approach to self-driving
        vehicles, 249–250
    installed in Tesla, and later activated,
        232–233
    preventing accidents, 150, 191–192
driving
    jobs lost in transition to EV, 268–269
    lost jobs for instructors of, 270–271
    not wanting self-driver because you love,
        195–196
    novices vs. experienced, 112–114
    programming human behaviors for, 168
    will become antiquated notion, 1
driving test
    decrease in licensed drivers, 31–32
    last person has been born who will take, 1
drunk drivers, ridesharing services reducing,
    228
Dudenhöffer, Prof. Ferdinand, 186–187, 197

economics, 350–351, 352–356
ecosystem, Silicon Valley automotive,
    256–258
ECU (electronic control unit), sensor, 142
Edmondson, Amy, 361
education, qualifying for our future, 379
efficiency innovation, 346–347, 351, 353
efficiency principle, Ockham's razor,
    356–357
electric transit, China, 61
electric vehicles (EVs)
    are future, 49–51
    arguments against, 104–108

electric vehicles (EVs) (*continued*)
  automotive technology and, 339
  batteries/accumulator. *See* batteries and
    accumulator
  changing dynamics, 63–66
  China as leading nation in, 58–61
  combustion engines vs., 32
  conclusion, 109
  dealers and clients, 296–300
  decreasing demand for oil, 328–330
  decreasing price of energy, 331
  in early 1900s, 22
  electric fleet maneuvers, 99–100
  fuel cell development, 101–104
  German manufacturers react to, 362
  gripping smart tires for heavyweight cars,
    94–95
  high-performance battery cells, 320–321
  hybrids as interim solution, 100–101
  manufacturers of, 60–62
  markets change quickly, 40
  Ockham's razor applied to innovation, 357
  price for, 95–99
  proof of success, 373
  propulsion technology, 92–94
  turning points in, 381
  why early designs tend to be ugly, 99
electricity
  arguments against EVs, 107
  cost of electric vehicles, 96
electrolyte additives, EV batteries, 73
electrolyte oxidation, EV batteries, 72–74
electronic control unit (ECU), sensor, 142
electronic identifiers, connected cars, 219
electronic security protocols, connected cars,
  217
Electronics Research Lab (ERL), 372
Elite Dangerous video game, 149
emergency solutions, good intentions
  executed badly, 82–86
emissions
  advantage of scandals for Germany/U.S.,
    242–245
  EVs and, 331
  footprints for clean future, 87–91
  ridesharing creating decrease in exhaust, 228
  self-drivers vs. combustion engine,
    190–191

emotion
  design of self-driving vehicles and,
    186–187
  evaluating drivers on autopilot, 134
  induced by cars, 264
eMotorWerks, JuiceBox, 77
encryption, cybersecurity, 335
endowment effect, overcoming cognitive
  distortions, 376
energy
  changes in petroleum industry and,
    330–331
  decreasing demand for oil, 328–330
  decreasing price of, 331
  efficiency of EVs vs. gas cars, 88–90,
    93–94, 190–192, 202
engine production jobs, 43
engineers, 156–157, 160–161
entertainment business, lost jobs in transition
  to EVs, 277
entrepreneurial thinking. *See* disruptive
  innovation
environment
  advantages of autonomous cars, 201–203
  argument against EVs, 106
Environmental Protection Agency (EPA),
  306
EPA (Environmental Protection Agency),
  306
ERL (Electronics Research Lab), 372
ethical issues
  AI develops its own behavior, not the
    engineer, 160–161
  commission in Germany for, 165–166
  decisions made by autonomous vehicles,
    169
  legislation for robots, 163–164
  machine learning decisions and,
    147–148
  prohibiting humans from controlling
    vehicles, 166
  reducing traffic accidents, 161–163
  trolley problem, 158–160
ethyl carbonate, EV batteries, 72
Europe
  Americans/Japanese moving vehicles into,
    208
  automotive technology in, 338–339

future is being planned without, 3
horse manure crisis in, 14–15
innovation gap in, 234–236
lagging in technology, 6–7
test areas for self-driving vehicles, 208
European New Car Assessment Program
    (NCAP), 213–214
European Union (EU), 184, 285
exponential growth, predicting, 38
external interference, with self-driving cars,
    178–179
extreme programming (XP), cybersecurity,
    334
eye contact, right of way via, 172

FAA (Federal Aviation Administration),
    drone regulations, 185
Facebook, 39–40, 383
failure, psychologically safe environments
    for, 361–362
Fan Hui, AlphaGo vs., 146
FashionTech, car design, 323
fear of loss, overcoming, 376
Fiat, 157, 287
Fiat-Chrysler, 103, 116, 128, 211, 372
financing, 48, 303–304
first automobile revolution, 19–22
first principles approach, 10
fleets, EV service, 100
flight, pioneers of, 9–10
FlightCar, 225
flying cars, 335–336
footprint, reducing CO2, 89–90
Ford
    acquisition of Argo.ai, 156
    normalization of deviance example, 345
    work on self-driving vehicles, 242
Ford, Henry, 358, 364, 370
foresight, 33–35
fossil fuels, lost jobs in transition to EVs,
    273–274
Fox, Mike, 40
Free University of Berlin, research, 374
Fröhlich, Klaus, 362
frustration mode, robots with, 179
fuel cell development, 101–104, 339
Full driving automation, SAE Level 5,
    118–119

future
    changing world and, 375–376
    cleaner, 87–91
    considerations for traditional manufac-
        turers, 253–262
    guessing professions in, 277–278
    overcoming cognitive distortions,
        376–377
    qualifying for, 379
    signals, trends, and foresight of, 33–35
    unconditional basic income (UBI) and,
        377–378
    willingness for change, 379–380
The Future: Six Drivers of Global Change
    (Gore), 10

gas stations, vs. number of charging sta-
    tions, 78–79
Gasoline and Automotive Services Dealers
    of America (GASDA), 40
Gates, Bill, 378
General Motors (GM)
    basing entire future on the Volt and Bolt,
        66
    buying back its own shares, 41–42
    decline of, 368
    driver assistance systems, 150
    orientation to innovative development,
        339
    self-driving vehicles, 242
    work from bottom of market to top
        (cheap to luxury), 355
German Association of the Automotive
    Industry (VDA), 266
German automobile industry
    advantage of scandals for, 242–245
    battery industry, 91
    calibration directive for taxis, 224
    challenged by Tesla, 51
    changing world and, 376
    charging point obstacles, 79
    data protection in, 284–285
    digital transformation, 292
    dissecting Tesla, 353–354
    famous entrepreneurs of, 17
    future is being planned without, 3
    impact of transition to EVs, 266
    importance of, 49

German automobile industry (*continued*)
  inadequacies of introducing EVs to,
    85–86
  innovative development and, 339–340
  invention of cars, 381
  lagging behind in technology, 6–7,
    62–65, 220
  mission statements, 342–343
  Moore's law not applicable, 370
  price-fixing scandal, 341
  record profits (2016-2018) based on
    discounts, 341
  research on self-driving cars, 374
  role in automobile revolution, 5
  safety benefits of driverless cars, 309
  sales models, 297–300
  self-driving vehicle experiments, 117–118
  Silicon Valley innovation vs., 3–4
  status of automotive technology in,
    337–339
  test areas for self-driving vehicles,
    206–208
gestures, autonomous cars interpret, 174
Gioia, Dennis, 345
girlfriend effect, reducing accidents, 162
GoMentum Station, 128, 204, 206
Google
  AlphaGo, 145–146, 161, 163
  Maps service, 189–190
Google self-driving vehicles
  Android Auto, 294–295
  battle for big data, 286, 288
  defensive driving behavior of, 160
  GPS accuracy issues, 288–290
  Koala. *See* Koala car, Google
  as leader in testing, 120, 123–125
  Open Automotive Alliance, 218
  potential consumers and trust of, 43
  role of, 189–190
  salaries for self-driving engineers, 156
  signs of change, 383
  Street View cars, 189
  testing self-driving vehicles, 115–116,
    370
  trusting more after traveling in, 175
  TV, 39–40
  X Self-Driving Program. *See* Waymo
Gore, Al, 10

GPS, 137–139, 288–290
graphics processing units (GPUs), for
  autonomous vehicles, 142
greenhouse gas emissions, gas cars vs. EVs,
  88
Greiner, Helen, 167, 187

hacker attacks, 217–218, 331–335
hand signals, autonomous cars interpreting,
  173–174
Harriman, John Emory, 335
Hastings, Reed, 364
healthcare, transition to EVs and lost jobs
  in, 275–276
heat management, EV battery, 73
heavy vehicles, deadly for rest of nation, 115
HERE, GPS accuracy issues, 288–290
Herrtwhich, Ralf, 151–152
heuristics, 377
Hewlett, William, 370
hierarchical structures, decision making in,
  362–363
High driving automation, SAE Level 4, 118
Hindenburg blimp, tragic fate of, 102
horizontal integrated systems, 149
horse-drawn coaches, 1, 19–22, 26–27
horse manure crisis, 14–15, 19–22
Hotz, George, 151, 212
How the Mighty Fall (Collins), 15, 368
human instructions, what can go wrong,
  178
Humanising Autonomy, 174
Humans Need Not Apply (Kaplan), 180–181
hybrids, 78, 100–101
hydrogen, fuel cell development and, 102
Hyundai, 53, 128, 355, 372

iCar, Apple, 283
identity theft, cybersecurity, 332
IFTF (Institute for the Future), 34–35
incentives, for electric cars and hybrids,
  82–86
income, unconditional basic income (UBI),
  377–378
incremental (gradual) innovations, 349
India, autonomous vehicles in, 208
induction charging process, 79–80
Industrie 4.0 standard, 291, 295

infrastructure
   lost jobs in traffic area, 270–271
   requirements for connected cars, 219–220
   as shortcoming of fuel cells, 103
innovation
   advantage of scandals for Germany/U.S.,
      242–245
   culture of, 339–340
   defining, 21
   in digital revolution, 39
   disruptive. *See* disruptive innovation
   economic crisis, job loss and, 350–351
   economics changing, 352–356
   efficiency innovation, 347
   expanding human consciousness, 45–46
   in first automobile revolution, 19–21
   how it happens, 236
   incremental (gradual), 349
   Keeley model, 357–359
   manufacturer outposts in Silicon Valley,
      371–373
   Ockham's razor applied to, 356–357
   psychological safe environments for, 361
   research and development budgets and,
      235–240
   research of car manufacturers and,
      234–235
   revolutionary impact of, 265–266
   second automobile revolution, 46–48
   in traditional life cycle model, 38–39
   types of, 349
The Innovator's Dilemma (Christensen), 350
instinct, in machine learning, 145–146
Institute for the Future (IFTF), 34–35
insurance, self-driving vehicle, 185, 274–
   275, 302–303
Intel, 142, 143, 156
intent, identifying, 156
intersections, accident frequency at, 311–313
intuition, in machine learning, 145–146
invention, 21
iPhone, 34–35, 37, 170
iRobot, 167
Ive, Jonathan, 325

Japan, perfection of automotive production,
   381
jobless recoveries (Christensen), 350

jobs
   after transition to autonomous vehicles.
      See professions, EVs and changes in
   German automobile industry and loss
      of, 340
   impact of innovation types on, 352–353
   unconditional basic income (UBI) and,
      377–378
   why companies lose innovative power,
      350–351
Jobs, Steve, 292, 338, 363
Johansson, Frans, 358
just-in-sequence production, 319
just-in-time production, 319
just in time production, automobiles, 47–48

Kalanick, Travis, 230, 363
Kaplan, Jerry, 180–181
Keeley, Larry, 357–358
Keeley model, 357–359
Keese, Christopher, 149
Kemble, Fanny, 46
Kentley-Klay, Tim, 209
"Kill the company" role playing, 367–368
killer questions, 367
Knoflikari (The Buttoners) movie, 209
Koala car, Google, 81, 116, 167, 173,
   186–188
Kodak, 25
Koko (sign language of), 146
Kopernikus.auto, 151
Krafcik, John, 131, 153–154
Kriesel Electric, battery heat, 73

Langley, Samuel Pierpont, 10
lasers, 130–135
last mile problem, autonomous vehicles for,
   193
latent demand, road construction and, 313
laws
   for accidents in self-driving vehicles,
      162–163, 176, 301–303
   for autonomous and self-driving vehicles,
      323–325
   cybersecurity, 334
   examples of what could go wrong, 177–181
   robot, 163–164, 181–186
   for slaves before Civil War, 180

Lean Startup (Ries), 365
Lenz, Hans-Peter, 43
Levandowski, Anthony, 116–117, 262
levels of driving automation, SAE, 118–119
licenses
  companies receiving test, 120–123
  cost for taxi, 222
  a generation that does not want driver's, 1–2, 224
LiDAR (light detection and ranging) systems
  autonomous and self-driving vehicles, 130–133
  design affected by, 188
  integrating into Google vehicles, 117
  Tesla philosophy vs. Waymo, 141
Lies, Olaf, 309
Lilienthal, Otto, 9
LinkedIn, 39–40
lithium-ion accumulator, 71–72
loans. *See* financing
locomotive (red flag traffic) act, Britain, 181
Lohner, Ludwig, 23–25
loss, overcoming fear of, 376
Luminar Technologies, LiDAR systems, 131
Lyft, 2, 25, 231, 302–303, 373

M City, Michigan, 205
machine learning
  crowd-sourced, 151
  overview of, 144–148
  system decides, not engineer, 160–161
machines
  relationships between humans and, 167, 187
  reservations about not being able to control, 196
Mahan, Steve, 309
maintenance, 98, 272
majority, innovation and, 39
maps, 137–139, 288–290
Marchetti, Cesare, 306
market
  for AI specialists, 143–144
  battery, 81
  LiDAR systems, 131–132
The Medici Effect (Johansson), 358
memory effect, EV batteries, 72–74

Mercedes, 187, 206, 339–340, 342–343
methodologies, innovation and, 339–340
Microsoft, dashboard battle, 294
military cost, ensuring energy supply, 89–90
mission statements, company, 342–343
Mobileye, 142–143, 156
mobility, 325–327, 343
Model T, 351, 358
Monaghan, Tom, 369
moonshots, Google aims for, 189
Moore, Gordon, 369
Moore's law, 369–370
Morse, Samuel, 38
motion planner, 139
motorcycles, self-driving, 262–263
Müller, Matthias, 157
Müller, Matthias, 295–296, 362, 364–365
Musk, Elon
  on approval for autonomous vehicles, 150
  attitude toward profit and loss, 84–85
  combustion vehicle cost vs. EV, 74
  education and training of, 25
  on good intentions vs. results, 83
  independence from fossil fuels, 42
  on innovation of leadership cultures, 57
  interacting with customers using tweets, 302
  no asshole rule (Sutton), 363
  on programs that change the world, 10

nanodegree, in self-driving engineering, 156–157
National Aeronautics and Space Administration (NASA), 371
National Automobile Chamber of Commerce (NACC), 30
National Highway Traffic Safety Administration. *See* NHTSA (National Highway Traffic Safety Administration)
Nauto, 122
navigation, and map solutions, 137–139
NCAP (European New Car Assessment Program), safety focus of, 213–214
NEDC (New European Driving Cycle), 83
Netflix, creating own market, 39–40
New European Driving Cycle (NEDC), 83

NHTSA (National Highway Traffic Safety Administration), 150, 152–153, 184, 213–214, 308–309
nickel, in EV batteries, 72
Nike, production of goods, 257
NIO, mission statement, 343
Nissan Leaf, 66, 74
The No Asshole rule (Sutton), 363
No driving automation, SAE Level 0, 118
Nokia, 25
normalization of deviance (Vaughan), 345–346
Norway, 58
Nuro.ai, 143
NuTonomy, 123, 192

Obama, President Barack, 181
objects, vehicle map and navigation of, 137–139
Ockham's razor, and innovation, 356–357
Open Automotive Alliance, Google, 218
Open Charge Alliance, charging stations, 79
Open Pilot, 212
open questions, vs. closed, 366
open source, 212–213, 370–371
organ donation, decrease in car accidents and, 276
Ot.to, 123, 128, 143, 156, 259–261
Otto, Nicholas, 23–24
over-the-air updates, 140–141, 292–294
overnight accommodation, transition to EVs/lost jobs in, 277

Packard, Dave, 370
Page, Larry, 25, 336
parking lots, 202, 232, 273, 304–311, 314–317
ParkWhiz, 315–316
Partial driving automation, SAE Level 2, 118
PASCAL Visual Object Class project, 213
pedestrians, 172, 175–176
Peleton Technology, 257, 261
penalty payments, Volkswagen, 41
Perceptive Automata, 174
petroleum industry, 328–331
Pindeus, Maya, 174
pioneers, 9–10, 16–17, 23–28
plugs, charging EV vehicles, 76–81
Polaroid, 25

politics, changes in petroleum industry, 330–331
polypropylene, EV batteries, 72
popcorn effect, 38
Porsche, 242, 341, 362
Porsche, Ferdinand, 23–24
power. *See also* energy
  arguments against EVs and electrical, 107
  for clean future, 87–91
  decreasing demand for oil, 328–330
  electricity price trends, 96
  environmentally friendly sources for EVs, 106
power grid, battery market and, 81–82
price-fixing scandal, German automobile industry, 341
Prius, 66
private car ownership, freedom and costs of, 231–232
private corporations, problems of ride-sharing owned by, 228–229
production, 48, 319–320, 358
professions, EVs and changes in
  guessing future, 277–278
  no-one should feel immune, 265–267
  things manufacturers should expect, 263–265
  types of lost jobs, 268–277
propulsion, 22, 92–94, 101–104
psychologically safe environments, for innovation, 361–369
Public Key Infrastructure (PKI), cybersecurity, 334–335
public streets, testing self-driving vehicles, 204
public transportation, 32, 327–328
purchasing self-driving vehicles, 185, 192–199

Quanergy, LiDAR systems, 131

R&D (research and development) budgets, 235–240
racing autonomous cars, 200–201
radar-based distance meters, 133
railroads/railway, 325–327
range anxiety, EVs and, 105
Read, Gigi, 379
recycling, 75–76, 202, 249

red flag traffic (locomotive) act, Britain, 181
regulations
　batteries and accumulator, 82–86
　for self-driving vehicles, 181–186
　taxi industry, 222–224
Reily, Carol, 211
remote control of autonomous vehicles, 145
Renault-Nissan, ridesharing, 225
rental companies, ridesharing, 225–226
resale value, and cost of EV, 97
research, 271–272, 373–374
research and development (R&D) budgets,
　235–240
resource curse, 243–244
resources
　autonomous taxis with compact design/
　　fewer, 262
　combustion engines lavishly waste, 247–248
　replacing old resources with new, 248
return on investment (ROI), efficiency
　innovation and, 352
revenues, self-driving vehicles impacting
　government, 300–301
rewards system, 12
ridesharing services
　automotive manufacturers involved in,
　　225–226
　changing our experience of transportation,
　　31
　data for improved traffic planning, 226
　effect on car sales, 224–225
　manufacturer agreements with, 55
　private ownership of vehicles vs., 227
　problems, 228–232
　regulations, 183, 222–223
　self-driving vehicles for, 132, 186
　taxi services, 221–224
　traditional automobile industry and,
　　254–256
　types of, 221
　welcome side effects, 228
Ries, Eric, 365
right of way questions, 199
Right Question Institute (Rochstein), 365
risk, 149, 361–362, 377
road construction issues, 312–313
Road to Zero, 308–309
road trains, 202–203

robotaxis
　compact design with fewer resources of,
　　262
　overview of, 192–193
　under scrutiny, 327–328
　sending cars during day to be, 210
　sustainable energy of, 202
　traditional automobile industry and,
　　254–256
Robotics and AI Department, Carnegie
　Melon, 374
robots
　abused by children as they learn empathy
　　for, 176
　demonstrating behavior inspiring trust, 167
　ethical questions about, 163–164
　laws for, 181–186
　tax on, 378
Rochstein, Dan, 365
role playing, 368
Roomba, 167
Roseking, Mark, 152–153
route planning, 139
rules
　AI learns/develops its own behavior from,
　　160–161
　autonomous vehicle reactions and, 147, 155
　machine has to know when it can break,
　　175
　Tesla's investment strategy ignores, 84
　of traffic vs. one's own behavior, 170
　transgression of, 167–169
rural areas, benefits of driverless cars, 311
Russia, autonomous vehicles in, 208–209

safety
　advantages of autonomous cars, 202
　affecting design of self-driving vehicles,
　　188
　challenges of self-driving cars, 213–214
　for connected cars, 216–218
　driver assistance systems enhancing, 150
　of electric vehicles, 86
　fuel cell development and, 102
　inside of EVs, 92–93
　machine learning decisions and, 147–148
　new metrics to evaluate, 152–153
sales, lost jobs in car, 272–273

sales models, 297–300
Sanford, Courtney, 114
Saudi Arabia, Uber in, 227–228
Schmidhuber, Jürgen, 163–164
Schwarzenegger, Arnold, 86
scripts, crisis management, 345–346
Seba, Tony, 74
second automobile revolution, 45–51, 382–383
security, for connected cars, 217–218
Sedol, Lee, AlphaGo vs., 146
Self-Driving Coalition for Safer Streets, 208
self-driving engineers, salaries for, 156
self-driving vehicles. *See* autonomous and self-driving vehicles
semantic pixel labeling, 143
semantic segmentation, vs. object recognition, 134–135
sensor data, 138–139
sensors, 134, 141–143, 172, 178, 220
sharing economy, 229, 295–296
sharing models, 370–371, 383
Shoup, Donald, 315
silent, arguments that EVs are too, 108
Silicon Germany: How We Can Manage Digital Transformation (Keese), 149
Silicon Valley
    automobile manufacturer outposts in, 371–373
    automotive ecosystem in, 4–5, 256–258
    changing world, 376
    corporate culture and, 343–344
    global leaders in automotive industry, 382
    information sharing in, 370–371
    innovation coming from, 2–4, 340, 353
    leading edge automobiles and, 337–338
    mindset in, 346–347
    as psychological safe environment, 363–364
The Silicon Valley Mindset (Herger), 343
simulations, autonomous driving, 153–155
Singapore, testing self-driving vehicles, 122
skeumorphism, 170
Slee, Tom, 230
Society of Automotive Engineers International (SAE), driving automation levels, 118–119
software problems of digital era, 27
solar power, cost of electric vehicles, 96–97

Speck, Jeff, 304
Sporkhurst, August, 23–24
SSL (Secure Sockets Layer) Protocol, connected cars, 217
standards, 76, 78, 127
Stanford Artificial Intelligence Laboratory, 373–374
Starbucks, induction charging process, 80
start-up companies, 151–152, 157, 221, 241–242
status quo, overcoming, 376–377
Stavens, David, 116
steam engines, 22
steel industry economics, 354–355
Stine, Robert L., 40
story stocks, Tesla, 42
Studebaker, 25
success, not taking for granted, 382
Superintelligence (Bostrom), 146
superuser mode, administrative control of robot, 178
supportive innovation, 351
surge pricing, Uber, 226–227
sustaining innovation, 351
Sutton, Robert, 363

talent auctions, 156
taxes, 108, 230, 300–301, 378
taxibot, self-driving motorcycle as, 262–263
taxis. *See also* robotaxis
    conflicts with car-sharing business, 295–296
    driverless fleets throughout world now, 192–193
    sharing economy in progress, 221–224
technology
    affecting design of self-driving vehicles, 188
    changing world and, 375–376
    entrepreneurs in United States, 16–18
    exponential growth of, 37–38
    innovation and, 339–340
    lagging behind in developing, 7, 337–339
    many players appear after acceptance of new, 12
    propulsion, 92–94
    signals/trends of future, 34–35
    skeumorphism in times of changing, 169
telegraph, 38

Templeton, Brad, 118, 163, 181
terrorists, autonomous vehicle use by, 199
Tesla
  autopilot continuous improvement, 149
  batteries, 74, 86
  competing against nonconsumption,
    355–356
  data network downed on, 218
  as the future, 9
  German engineers dissecting, 353–354
  German manufacturers react to, 362
  horizontal integrated systems used by, 149
  impact in Silicon Valley, 257, 338
  innovation of, 67–68
  over the air updates, 294
  R&D budget, 234–235
  sales models, 297–300
  signs of change, 380–381, 383
  skeumorphism in design of, 170
  training machine-learning systems, 140–141
  trusting more after traveling in, 175
  Waymo philosophy vs., 141
Tesla Model 3
  battery power, 71
  changing dynamics, 63–66
  dashboard battle, 294–295
  fossil fuel independence, 32, 42
  market dominance, 4, 37, 40–41, 84–85,
    251
  selling price, 98
  software updates, 293
Tesla Model S
  battery power, 74
  components installed/later activated,
    232–233
  expansion of infrastructure and, 84–85
  fatal accidents in, 150, 153
  gripping smart tires for heavyweight cars,
    94–95
  market for, 64, 223–224
test users, 39
Texas A&M, research, 374
text, communicating intentions from autonomous cars, 172
Thaler, Richard, 10–11
Thiel, Peter, 39–40
thinking
  at 180 degrees, 368–369

  in basic terms approach, 10–11
  controlling self-driving cars with
    thoughts, 174
Thomsen, Lars, 38
3-D maps, 130–131, 133
Thrun, Sebastian, 25, 116–117, 139–140,
  156–157, 373–374
timescale, when EVs are available
  2019-2045 timeline, 250–253
  heavy trucks on tour by themselves,
    259–261
  manufacturer considerations, 253–262
  manufacturers, what to expect, 263–265
  pretty pods, 262
  recycling manual cars, 249
  replacing old resources with new, 247–248
  self-driving buses, 258–259
  self-driving motorcycles, 262–263
  Silicon Valley's automotive ecosystem,
    256–258
  transition period, 249
tiredness, car accidents caused by, 114
tires, heavyweight cars using smart, 94–95
total cost of ownership (TCO), 97
Toyota, 66, 351, 355
TradeNet, 230–231
traditional automobile manufacturers
  automotive ecosystem and, 256–258
  considerations for future, 253–256
  developing autonomous trucks, 259–261
  evolutionary approach to self-driving
    vehicles, 249–250
  not wanting to risk reputations, 151–152
  partnering with startups, 157
  pretty pods, 262
  reasons for lagging behind, 240–242
  responsible for design, not parts pro-
    duction, 257
  self-driving buses, 258–259
  self-driving motorcycles, 262–263
  transition to EVs and lost jobs in pro-
    duction, 269
traffic
  accidents. See accidents
  area planning, lost jobs in, 273
  jams, 195–196
  latent demand, 313
  lights and signs, 168, 317–318

removing cars and parking lots from urban areas, 304–311
smart management for smart cities, 321–322
training, 373–374
trends, 33–35, 38
trolley problem, 158–165, 362
trucks, autonomous, 259–261
trust of self-driving cars, 175, 186–187
Turkle, Sherry, 187
Twitter, creating own market, 39–40

Uber
    acquisition of Ot.to, 143, 156
    business model, 296
    as case of non-regulation, 183–184
    competing against nonconsumption, 356
    controlling aggregated resources, 229–230
    delivery systems, 282–283
    fatal crash of, 123, 176–177
    founders, 25
    lowering costs for travelers, 229
    paying for talent, 156
    Pool rides, 225
    in Saudi Arabia, 227–228
    under scrutiny, 327–328
    surge pricing, 226–227
    taxi lobby opposition to, 295–296
    test fleet in Pittsburgh, 192
    as unique service, 373
    who pays for accidents, 302–303
    Xchange Leasing, 303–304
UCS (American Union of Concerned Scientists), 88
Udacity, 129, 139, 154, 156–157, 212, 373
unconditional basic income (UBI), 377–378
university professors, lost jobs for, 271–272
updates, over the air, 292–294
urban planning, 304–311, 321–322
Urmson, Chris, 123, 160, 288
utilities, decreased demand for oil, 328–330

vehicle-to-grid system, 81
vehicle-to-infrastructure (V2I) communication, 216–217
vehicle-to-vehicle (V2V) communication, 216, 219
Velodyne, LiDAR systems, 131

venture capitalists, 353
vertical integration systems, 149, 319–320
Vogt, Kyle, 25
Volkswagen
    loss of credibility, 18, 49, 63, 242–245
    mission statement, 342–343
    normalization of deviance and, 345–346
    penalty payments, 41
    record profits (2016-2018), 341
    vulnerable to buy out, 340
Volvo, 123, 162–163, 219, 294–295, 302
Voyage, 374
voyeurism, and car accidents, 114–115
vujà-dé, 369

Wagener, Gorden, 187
walkability of cities, 304
wave effects, 279–282
Waymo
    commercializing self-driving technology, 192–193
    communication via horn sound, 172–173
    database access from all driven vehicles, 140–141
    developed own LiDAR system, 131
    license to test on public streets, 204
    remote control, 145
    test driving 25,000 miles per week, 149
weight of car, 115, 247
Westerheide, Fabian, 143–144
wireless charging, induction, 79–80
Worldwide Harmonised Light Vehicle Test Procedure (WLTP), 83
Wright, Orville and Wilbur, 9–10, 13

X Self-Driving Program. See Waymo
Xchange Leasing, Uber, 303–304
XP (extreme programming), cybersecurity, 334

Yerkes-Dodson law, 114
Yours Is Mine (Slee), 230

zero-emission vehicles (ZEV), GM compliance, 99
Zeroth Law, 181
Zipcar, 224–225
Zuckerberg, Mark, 286

About the Author

Mario Herger has been living in Silicon Valley since 2001. He researches technology trends, writes books and consults companies on topics such as innovation, Silicon Valley mindset, foresight mindset, automotive, creativity, and intrapreneurship. He worked for SAP for fifteen years as software developer, development manager, and innovation strategist.

During his PhD in chemical engineering, he measured and calculated emissions of diesel fuel used in heating systems. And no, he never used this knowledge for evil purposes.

Now he helps companies with his keynotes and workshops to understand how they can apply the innovative and entrepreneurial spirit from Silicon Valley in their organizations, to be more innovative, discover trends earlier, and influence them. He looks at signals of emerging technology trends and how they impact society, politics, or employment. He also hosts and meets delegations from all over the globe in Silicon Valley and connects them with inspiring people and companies.